SYSTEMIC INTERVENTION

Philosophy, Methodology, and Practice

Contemporary Systems Thinking

Series Editor: Robert L. Flood
Monash University
Clayton, Victoria, Australia

DESIGNING SOCIAL SYSTEMS IN A CHANGING WORLD
Bela H. Banathy

GUIDED EVOLUTION OF SOCIETY: A SYSTEMS VIEW
Bela H. Banathy

LIBERATING SYSTEMS THEORY
Robert L. Flood

OPERATIONAL RESEARCH AND SYSTEMS
The Systems Nature of Operational Research
Paul Keys

POWER, IDEOLOGY, AND CONTROL
John C. Oliga

SELF-PRODUCING SYSTEMS
Implications and Applications of Autopoiesis
John Mingers

SOCIOPOLITCAL ECOLOGY
Human Systems and Ecological Fields
Frederick L. Bates

SYSTEMIC INTERVENTION
Philosophy, Methodology, and Practice
Gerald Midgley

SYSTEMS METHODOLOGY FOR THE MANAGEMENT SCIENCES
Michael C. Jackson

A Continuation Order Plan is available for this series. A continuation order will bring delivery of each new volume immediately upon publication. Volumes are billed only upon actual shipment. For further information please contact the publisher.

SYSTEMIC INTERVENTION

Philosophy, Methodology, and Practice

Gerald Midgley

Centre for Systems Studies
Business School
University of Hull
Hull, England

Kluwer Academic / Plenum Publishers
New York, Boston, Dordrecht, London, Moscow

Library of Congress Cataloging-in-Publication Data

The Joseph Rowntree Foundation supported the project reported upon in Chapter 14 of this book, but the material presented in that chapter represents the findings of the author, not necessarily those of the Foundation.

ISBN: 0-306-46488-8

233 Spring Street, New York, New York 10013

http://www.wkap.nl

10 9 8 7 6 5 4 3 2 1

A C.I.P. record for this book is available from the Library of Congress

Printed in the United States of America

Acknowledgments

This book is dedicated to a number of important people in my life. My partner, Wendy Gregory, has given me ten years of love, friendship, support and intellectual challenges. I am incredibly fortunate to have a relationship that can provide so much all intertwined together. My daughter, Elizabeth, amazes me every day with her love, curiosity and imagination. I hope that I can make her as proud of me as I already am of her. My friend, Greg Tuck, helped me hone my intellectual skills through many nights of philosophical and political debate as teenagers and in our early twenties, and his friendship is still an important anchor for me. Likewise, my more recent friendship with Alejandro Ochoa-Arias has had a significant influence on my research direction, and continues to do so, even though we live on different continents. Bob Flood and Mike Floyd gave me a 'leg up' onto the academic ladder by taking me on as a research student at City University, and (for me) they were the best kind of supervisors because they had the courage to allow me to follow my own path. Finally, my parents, Doreen and Peter Midgley, always gave me their support—even through eight years of struggle after graduation while I engaged in a series of low-paid and unpaid interventions. They never once told me to 'get a proper job', even though lots of other people did!

When I eventually started regular paid employment, it was on my own terms: researching systemic intervention in the Centre for Systems Studies at the University of Hull. I had long recognised Hull as one of the premier institutions for systems research, and when a position was created for me by Bob Flood and Mike Jackson, giving me complete freedom to follow my own research agenda, it was the fulfilment of a dream. My experience at Hull has taught me that it *is* possible to follow your heart, and hopefully make a meaningful contribution to the lives of others along the way. At Hull I have been privileged to work with some brilliant colleagues, post-graduate students and visiting academics from other institutions around the world, who have sparked enough ideas in me to keep me busy for more than one life-time.

I should acknowledge that Chapters 14-17 of this book present interventions conducted in partnership with others, and in some cases sections of the text contain joint-authored written material as well. I would therefore like to thank Alan Boyd, Mandy Brown, Claire Cohen, Wendy Gregory and Isaac Munlo for their contributions and their permission to use passages of joint-authored text under my own name. I would also like to acknowledge the financial support I received for various aspects of the research reported in this book: in particular, from the Economic and Social Research Council; the Royal Society; the British Council; the Joseph Rowntree Foundation; the National Natural Science Foundation of China; the Colombian Academy of Science; the King's Fund; the University of Hull; City University; Javeriana University; the European Union; Humberside Social Services; the Children's Society; NCH Action for Children; and the Manchester & Salford Methodist Mission.

Finally, I would like to extend my thanks to Ken Derham and his staff at Kluwer/Plenum who have given me first class support. I couldn't ask for more from a publishing team.

Contents

1

Systems Thinking for the 21st Century

This book has three major aims: to formulate a new approach to some vital philosophical issues that have bedevilled theoreticians in scientific and other academic communities for many years; to generate a methodology[1] for intervention (intervention being purposeful action by an agent to create change) that is consistent with this philosophy; and to provide examples from my own practice to illustrate how the methodology can be translated into action. Each one of these aims could have taken a book on its own to pursue, but I have brought them together, despite the risk of over-abbreviating each argument, because I see them as fundamentally interlinked. Together, these three aims present us with an enormous research agenda which will take more than my own lifetime to explore to the full. I therefore invite you to walk with me and, if you find this agenda meaningful, let us see where it leads.

Each of the three aims can be related to three major, interlinked developments in Western thought that have taken place during the last hundred years, all of which are still actively controversial. In my view, they have exceptional significance for where we are heading as we enter the 21st Century. Below, I provide just a couple of paragraphs on each of these developments. Then, I offer a brief review of later chapters so you can get an overview of where my argument is going. Finally, I make some general comments on what I see as the value of this book in terms of dealing with problematic issues and managing social change in the 21st Century.

[1] A 'methodology', as I use the term, is a theory about the valid and/or legitimate use of methods. See Chapter 5 for a more detailed definition.

1.1 From Mechanism to Systems Thinking

The first development in 20th Century Western thought of importance to this book was the undermining of *mechanism*. Mechanism is the view that everything can be observed and described as if it is a machine—a predictable, functional, inherently understandable object seen from a discrete distance by an independent subject. Mechanism assumes that our knowledge of the world (and the language we use to frame this knowledge) reflects reality more or less accurately (Pepper, 1942). Mechanistic *science* is therefore characterised by the use of methods for structuring reliable observations to build so-called 'objective' knowledge about the world. At the risk of over-simplifying, the idea is that the more knowledge we have, the richer will be our understanding; the better able we will be to predict what will happen in the world; and the more control we will have over our destiny. According to this view, all the things in the world (including human beings, organisations and societies) are like clockwork toys. If we can figure out how they work, then we will be able to change them according to our will, within the limits of the natural laws that they conform to.

Of course, the 20th Century saw all these mechanistic assumptions come under attack. Chaos and complexity theorists began to use new ideas in mathematics to show that much of what happens, far from being inherently predictable, is actually *un*predictable (see Gleick, 1987, Stewart, 1989, and Capra, 1996, for some useful, non-technical reviews). There is a philosophical argument amongst chaos theorists about whether unpredictability is an inherent feature of the world, or whether it stems from the inevitable limitations of human understanding (Fitzgerald, 1999). Nevertheless, whichever view is taken, mechanism is still undermined: it seems that a great deal of what we experience will remain beyond our understanding, so the dream of perfect explanations has become tarnished.

Furthermore, just about every philosopher of science who has been taken seriously in the latter half of the 20th Century has argued that we cannot know the exact relationship between human knowledge, the language we use to frame this knowledge, and reality.[2] This is because, whatever we know about reality is just that—*knowledge*, not reality itself. Also, however much evidence we accumulate about the nature of a particular phenomenon, we cannot be sure that some disconfirming

[2] Of course, this is not a new insight. It can be traced back to Kant (1787), and in the 20th Century Wittgenstein (1953) was an influential exponent of the view that there is an unquantifiable gap between knowledge and reality, meaning that we might as well give up talking about the latter.

evidence is not just around the corner (Popper, 1959). See Chapter 2 for a fuller discussion of this issue. While some (e.g., Popper, 1972, and Bhaskar, 1986) have stuck with the view that knowledge does indeed reflect reality, others (e.g., Kelly, 1955, and Maturana, 1988a,b) have given up trying to talk about reality itself and have instead built theories about the subjective and inter-subjective construction of realit*ies* (the plural indicating that there can be as many realities as human beings). Again, the realisation that the exact relationship between knowledge, language and reality is inherently unquantifiable fundamentally undermines mechanism, whether or not people stick with the view that there is indeed an external reality independent of human knowledge.[3]

Now add to these insights the transformations that have happened during the 20th Century in the various disciplines. Let me give just three examples (many more could be provided). In physics, quantum theory (e.g., Bohr, 1963; Bohm, 1980) has illuminated a relationship between the observed and the method of observation—the latter playing a fundamental role in constructing the former.[4] In biology, the theory of evolution has been revised to embrace the idea that organisms co-construct their world rather than passively adapt to it, resulting in the conclusion that organisms are inevitably a *part of* what they observe, not separate from it [see Margulis and Sagan (1987), Lovelock (1988) and Ho (1989) for some general accounts]. Finally, in psychology, scientists have argued that much of what human beings assume is factual is actually what is *socially relevant* within a particular discursive context (e.g., Middleton and Edwards, 1990; Gergen, 1991; Shotter, 1993; Harré and Gillett, 1994). These changes in 20th Century thought, which have undermined the traditional division between the human observer and what s/he observes, have shaken the mechanistic view of the Universe (which is built on this division) to the core.

---

[3] In Chapter 4 I argue that it is possible to side-step this rather futile debate about whether or not an external world actually exists. The process philosophy I will propose allows as to ask a different kind of question, and as a result all the theoretical insights that come from both a 'realist' perspective (one which says that there *is* an external world which language refers to) and an 'idealist' viewpoint (one which says that what we think of as the external world is either subjectively or socially constructed) can be regarded as valid.

[4] There are different interpretations of this phenomenon in the discipline of physics: some authors talk in terms of the method of observation constructing how objects (which have an existence in external reality) are seen, and others say that objects are *created* through the process of observation (there is no independent, external reality). Nevertheless, what seems to be beyond dispute is the fact that the method of observation does play a role in constructing the observed.

Of course, mechanism has underpinned so much of our thinking during the last three hundred years that it has become very difficult to conceptualise an alternative. Nevertheless, the undermining of it places us in a position that very few generations have the privilege of experiencing: we find ourselves at the end of one epoch, and on the threshold of entering a new one whose contours, as far as I can see, are not yet fully visible.

So, what will replace mechanism? One answer that has been offered again and again in the latter half of the 20th Century is *systems thinking*. However, we must be clear that there are many competing systems perspectives, and some of them have unwittingly inherited mechanistic assumptions (which is inevitable given three hundred years of the dominance of mechanistic thought). It is therefore one of the aims of this book to undertake a fundamental rethink of systems philosophy to deal with this problem. Of course this is no small task, and it would be arrogant of me (not to say foolish) to think that I could achieve it in just a few chapters—especially as the rethinking of systems philosophy is just one of the three aims being pursued in this book. Nevertheless, I hope that I can make a reasonable start so that we can begin to shape a credible alternative to mechanism for the 21st Century.[5]

1.2 From Observation to Intervention

The second development in 20th Century Western thought of particular relevance to this book was the realisation that, without mechanism, the traditional foundations of science were also being undermined. Science was once founded on the possibility of *independent observation*: the notion that truly objective knowledge is only possible if the observing subject is independent from the observed. If the subject is implicated in the construction of the observed, then the concept of objectivity (as we thought of it during the epoch of mechanism) begins to crumble (see, for example, Rorty, 1989). One way that the subject has indeed been implicated in the construction of the observed has been through the realisation that *value judgements* direct what the scientist sees and what s/he passes over (e.g., Churchman, 1979; Ulrich, 1983;

---

[5] Fuenmayor (1994), contrary to many of the arguments of systems thinkers (including my own), places systems thinking, along with mechanism, firmly in the older tradition of thought that is now dying. However, I suggest that this is because he has taken as his object of study a form of systems thinking that has indeed inherited many mechanistic assumptions. I hope that it will become clear in this book that another form of systems thinking can be constructed from a very different philosophical starting point.

Bhaskar, 1986; Hollway, 1989). Of course, this is a highly contentious assertion, and many people still want to cling to the possibility of objective science. However, I believe that recent controversies over the social role of science, as in the production of genetically modified organisms (ESRC Global Environmental Change Programme, 1999), have decisively shifted the debate: the discourses of objectivity and value-neutrality have been exposed as ways of talking about science that *actively prevent* scientists from seeing the value-laden nature of their actions. See Resnik (1998) for some examples of science playing a non-neutral social role.

One alternative discourse to independent observation is that of *intervention*. People involved in systems practice, action research, operational research, management science, counselling, community psychology, family therapy and community development (to name just a few applied disciplines[6]) have been talking about intervention for some time. However, it is my contention that we can develop an understanding of intervention that is meaningful across the board, from the scientific disciplines to management consultancy, from engineering to counselling for personal change. This is an understanding of intervention that is *not opposed to observation*, but recognises observation as one practice of intervention that is just as valuable and *value-full* as others. The second aim of this book is therefore to construct a *methodology of intervention*. Actually, a methodology of *systemic* intervention, as I will argue that our new systems philosophy can usefully inform intervention practice.

1.3 From Theories of Everything to Theoretical Pluralism

Finally, there was a qualitative change during the 20th Century in how people saw both theories and methods. Theories, in the mechanistic mode of thought, were either true or false—or, more humbly, falsified or as-yet-unfalsified (Popper, 1959). The fact that knowledge was seen as a more or less accurate reflection of reality meant that the search was on for 'ultimate' theories in each discipline which would supposedly explain all phenomena of relevance to that

[6] It is, of course, debatable whether these are really 'disciplines' in the traditional sense: that is, fields of knowledge demarcated by subject matter. Systems practitioners, for example, have long claimed that theirs is a *trans*disciplinary practice (see Midgley, 1996a, 1998, for a full argument); and operational research was founded on the principle of *inter*-disciplinarity (Keys, 1991). However, I have used the term as a matter of convenience: each of these applied 'disciplines' is constituted by a community of academics and practitioners with some common interests, and in this sense they are similar to the traditional scientific disciplines.

discipline. Physicists even talked (and some still do) about discovering a 'theory of everything' (see Hawking, 1988, for a non-technical account). Of course, what a 'theory of everything' means in the context of physics is a theory of the origins and laws of the Universe—not *really* everything.[7]

However, as soon as people began to realise that the gap between 'knowledge' and 'reality' is inherently unquantifiable, this brought into question whether an 'ultimate' theory could ever be found at all. Gradually, the meaning of the term 'theory' began to change: in the latter half of the 20th Century, a theory became *a way of seeing*—a means of explanation dependent on the positioning of the observer (e.g., Vickers, 1970; Francescato, 1992). All ways of seeing—all theories—are inevitably partial: they are informed by the purposes and values of the agent(s) constructing and using them. If a theory is merely a way of seeing that explains things in terms of particular purposes and values, and if a variety of purposes and values can legitimately be pursued in different contexts, then a corresponding variety of theories may also be useful. Thus, *theoretical pluralism* becomes possible. A tricky balancing act is needed between maintaining coherence and welcoming in a variety of theoretical perspectives to enrich understanding.

Similarly, *valid* method is no longer synonymous with *scientific* method. If it were possible to have absolute knowledge of reality, then all that would be needed would be valid and reliable methods of observation. Of course, for a long time there was a consensus amongst scientists that it is indeed possible to bridge the gap between reality and human knowledge, so for several centuries they successfully marginalised all methods other than those used for structuring observation.[8] Now, however, with values and subjectivity on the agenda once again, and the possibility for supporting *intervention* through the use of different methods, there is no basis for focusing on just one type of method. Methods for clarifying values, exploring subjective viewpoints, facilitating participation, visioning possible future scenarios, etc., are brought alongside methods for structuring

---

[7] This search for 'ultimate' theories has been dubbed 'isolationism' by some commentators (e.g., Reed, 1985; Jackson, 1987a; Flood, 1989a) because it produces a tendency amongst proponents of 'ultimate' theories to isolate themselves from the insights of others (unless those insights can easily be integrated into the 'master' theory). Isolationism comes about because human beings are generally not disinterested evaluators of theory. Once someone has put a substantial portion of their career into developing a theory, they have a vested interest in promoting it and ensuring that it is not undermined by competitors.

[8] In the discipline of psychology, for example, this led to the marginalisation of psychoanalysis. Only experimental psychology was accepted within the mainstream (Leary, 1980; Koch and Leary, 1985), despite the enormous influence of psychoanalysis outside the institutions of science.

observation. *Methodological* pluralism therefore becomes a partner to theoretical pluralism.

This leads me to the third aim of the book you are reading. While theoretical and methodological pluralism are both features of the methodology of systemic intervention that I propose, it is also important for me to show *how this can be practised*. My own intervention practice has been in the disciplinary area of Community Operational Research ('Community OR' for short), which involves addressing problematic issues in community contexts, so in the final section of this book I detail the concerns of Community OR and present four of my own interventions as examples of the practice of systemic intervention. No doubt my experience of engagement in Community OR has shaped some of my philosophical and methodological language, but I nevertheless believe that the fundamental concepts of systemic intervention should be just as relevant to those working in other disciplines, whether in the 'natural' or 'social' sciences—provided that they are willing to accept that their practice is inevitably value-*full*, not value-neutral. Such an acceptance means that the exploration of, and engagement with, the social context in which they operate has to be an integral part of their research—or, as I prefer to say, what scientists normally see as their research becomes just one part of their *intervention practice*.

1.4 The Structure and Contents of this Book

This book is divided into three sections, reflecting the three aims already outlined (plus many other subsidiary aims that are revealed along the way). Section One focuses on the *philosophy* of systemic intervention; Section Two on *methodology*; and Section Three on *practice*.

Section One starts with an exploration of epistemology (about the nature of knowledge and its generation). Epistemology is important because different assumptions about the nature of knowledge give rise to different methodologies, and hence very different forms of intervention practice. More details of why I regard the exploration of philosophy to be of importance to intervention practice will be provided in Chapter 2, for the benefit of practitioners who are sceptical about the value of philosophical inquiry. Chapter 3 then introduces *systems* philosophy, concentrating on the preoccupation of systems thinkers with undertaking 'holistic' or 'comprehensive' analyses. Of course, there is no such thing as a genuinely comprehensive analysis, so the defining feature of systems thinking is reflection on the *boundaries* of inclusion and exclusion. Chapter 3 also describes a variety of epistemological positions proposed by other systems theorists. We will find that each of these has problems associated with it: in one way or another, even though they all seek to challenge mechanism, most nevertheless succumb to one vital mechanistic assumption—that independent observation (assuming a dualistic separation of the observer and observed, or the subject and object) is possible.

As a result, in Chapter 4, I map out a new path for the development of systems philosophy to inform systemic intervention. A key focus of this is the replacement of subject/object dualism with a distinction between *process* and *content*. By 'process', I mean the process of *making boundary judgements* (distinctions of what exists).[9] Boundary judgements define what constitutes 'content' in any particular analysis—and we can distinguish *first-order* content (judgements about what is 'in the world') and *second-order* content (judgements about what it is that gives rise to boundary judgements in the first place). As will become apparent in Chapter 4, this means that 'subjects' are identified through exactly the same process as 'objects': both are types of *content* defined through the *process* of making boundary judgements. The supposedly 'fundamental' dualism between subject and object is thereby dissolved.

Section Two of the book then goes on to examine the methodological consequences of taking this new approach. First, in Chapter 5, I offer an argument for why those with an interest in philosophy and practice should think about methodology at all. Then, in Chapter 6, I begin to lay out my own methodological ideas, starting with the concept of *intervention*. I define intervention as purposeful action by an agent to create change, and contrast this with the concept of observation. I argue that observation, as undertaken in science, should actually be seen as a 'special case' of intervention, not as distinct from it. Next, I relate the systems philosophy outlined in Section One to the methodology of intervention, and suggest that 'systemic intervention' is purposeful action by an agent to create change *in relation to reflection on boundaries*. This leads, in Chapter 7, to the exposition of a theory of 'boundary critique' where I seek to deepen our understanding of what it means to reflect on boundaries in the context of intervention. In particular, I discuss the relationship between boundary and value judgements; the extension of the concept of boundary judgement to encompass concerns about how things *ought* to be (as well as what actually exists); the importance of wide-spread stakeholder participation in systemic intervention; and the need for agents to deal

[9] Boundary judgements using language may also distinguish what *ought* to exist, not just what already does. The distinction between 'is' and 'ought' will be discussed in Chapter 7.

with the marginalisation of particular issues and stakeholders within social contexts.

Then, in Chapter 8, I make the case for theoretical pluralism. This follows on logically from the theory of boundary critique because every theory is based on either implicit or explicit assumptions about the appropriate boundaries of analysis. Therefore, if it is possible to explore and use a variety of different boundaries during intervention, it is also possible to draw upon a variety of different theories. Of course, if *theoretical* pluralism is possible, and if different theories can be linked into different methodologies and methods, then *methodological* pluralism must be possible too. An argument for methodological pluralism is presented in Chapter 9, and the emphasis is placed on the value of this in terms of optimising flexibility and responsiveness to stakeholder concerns during intervention.

Chapter 9 also contains a review of intervention methods and methodologies which interveners may be able to draw upon in support of their systemic intervention practice.[10] Chapter 10 then gets into some detail about how interveners can mix methods from different methodological sources to meet a variety of purposes. Lastly, in Chapter 11, I address the main objections raised against methodological pluralism—one of which is that it asks too much of interveners in terms of knowledge of a wide variety of methods. I argue that this need not be the case: systemic intervention certainly requires a willingness to learn about new approaches to meet new challenges, but there should be no insistence on the development of a large amount of theoretical knowledge about methods prior to engaging in practice—learning can take place *through and around* practice.

Finally, in Section Three of this book, I focus on practice itself. I start with a short chapter (Chapter 12) aimed at philosophers and methodologists who shun practical engagement, arguing that such engagement is necessary if the consequences of philosophical and methodological ideas for intervention are to be tested. In Chapter 13, I then provide some background to my own intervention practice in Community OR (dealing with problematic issues in community contexts).

Following on from this, Chapters 14-17 contain four examples of systemic intervention, each of which is used to illustrate a different aspect of the methodology outlined in Section Two. Chapter 14 describes an intervention I undertook with a multi-agency group seeking to

[10] Inevitably this review is very limited, as there are far too many intervention methods and methodologies to review comprehensively in just one chapter of a book. However, the review will give the reader a feeling for what is available, and a starting point for further reading.

address a variety of interconnected, problematic issues affecting the provision of housing services for older people. This demonstrates how boundary critique can be practised.

Chapter 15 focuses on some work I did with nineteen different organisations who came together to plan the development of a counselling service that could be activated in the event of a disaster. This illustrates how methods from a single methodological source (rather than multiple methodological sources, which is more usual) can be chosen, taking into account stakeholders' perceptions of the problematic situation to be addressed and the characteristics of a variety of possible methods that might be considered as candidates for supporting the intervention.

Chapter 16 details an evaluation I undertook, and some planning I facilitated, with a project that sought to keep people with mental health problems and learning disabilities accused of criminal offences out of prison (so that they could get appropriate treatment rather than be placed in an institution that would exacerbate their problems). Here, the focus is on the mixing of qualitative and quantitative methods from diverse methodological sources—including some methods used in traditional 'observational' social science research.

Finally, in Chapter 17, I present an intervention I conducted with a variety of partner organisations, facilitating the identification of the needs of homeless young people (under 16 years old) living on the streets, and planning new services to meet those needs. Again, the focus of this chapter is on mixing methods—but this time a wide variety of qualitative methods designed for researching stakeholder views and feeding these into a participative planning process.

Importantly, all these interventions (except the one presented in Chapter 15) had as a principle concern how the voices of marginalised stakeholders and the issues that were important to them could be included in plans for change. The book then ends with a short concluding chapter (Chapter 18) which invites the reader to begin to practice systemic intervention and contribute to its further development.

1.5 Practical Consequences for Life in the 21st Century

So far, I have positioned this book in relation to some significant developments in Western thought that began to take shape in the 20th Century, but (as far as I can see) are not yet fully formed. I have also given a brief outline of the contents of the chapters to come. It should be evident that systemic intervention, as I conceive it, involves reflecting on the boundaries of problematic situations; sweeping in the viewpoints of a wide variety of stakeholders; ensuring that issues of marginalisation (of stakeholders and issues) are addressed; and drawing upon theories and methods to suit the purposes being pursued. I believe that such a stance is becoming increasingly necessary as we enter the 21st Century, and that the conditions are right for systemic intervention to be used much more widely in the years to come. Just three of these conditions are discussed below.

1.5.1 Interconnectedness

It is very common nowadays to observe multiple interactions between phenomena, and links between problems, making the setting of boundaries when intervening difficult and potentially contentious (see Churchman, 1970, for a well thought out analysis of interconnectedness and boundary setting). A classic example is in attempts to address global environmental issues. Ecological problems (e.g., global warming, deforestation, the reduction in species diversity) interact with social problems (e.g., the inequitable distribution of wealth between nations, and the net flow of money from poor to rich countries that results from international debt): as long as these inequalities continue, governments in the poorer countries will resist requests from those in richer countries to curb unsustainable economic growth (Shiva, 1990). In turn, ecological and social problems interact with personal ones: Gregory (1992) argues that, in liberal/capitalist societies, ethical decision making is increasingly being delegated to the level of the individual, but unless wider social and organisational change is undertaken (putting the infrastructure in place to support the ethical choices of individuals), people will find themselves more and more regularly in the stressful situation of striving and failing to reconcile competing personal, familial, social and ecological demands (also see Midgley, 1992a, for an extended argument about the connections between ecological, social and personal issues). It is only by being open to exploring the boundaries of global environmental issues, and encouraging the participation of diverse stakeholders, that a variety of possible angles can be covered, and unanticipated side-effects of intervention can be minimised. Systemic intervention therefore provides an appropriate language for framing inquiry into these kinds of issues.

However, the phenomenon of interconnectedness is not just experienced in relation to *global* issues: it is also commonly encountered in *local* situations (Ackoff, 1981; Checkland, 1981; Jackson, 1991). I came up against a good example in an intervention with a group of organisations providing housing services for older people (see Chapter 14 for details): stakeholders surfaced a great variety of problems, all of which were seen to interconnect and support each other to create a much larger *problematic situation*. It became apparent that none of the component problems would be amenable to solution unless the problematic situation was dealt with as a whole. Again, systemic intervention provides an appropriate language to both describe and address this kind of situation.

1.5.2 Scepticism about Value-Neutral Science

Another significant issue that has come to the fore as we enter the 21st Century is the increasing scepticism amongst most Western populations about the value of science and the trust-worthiness of scientists. New technologies with very significant implications are currently the subject of wide-spread research, some of the most high-profile being the technologies of genetic engineering and cloning, which promise to deliver major benefits such as improved food production, the eradication of many genetic diseases, the production of replacement organs, and increased longevity and quality of life (at least in Western countries which can afford the technology).

Of course, there are dark sides to these technologies too: for example, we simply do not know what the ecological consequences will be of introducing genetic changes into the food chain[11]; and if access to the new health technologies is determined by wealth, we could find ourselves in a situation where the 'haves' can buy extended life while the 'have nots' are allowed to suffer a 'natural' death. To an extent we are already in this situation, in that large sectors of the world's population have no access to modern health care (even adequate food, clean drinking water and basic sanitation), and the average life expectancy among citizens of Northern nations is much higher than among citizens of the South (Berger, 1974; George, 1976; Caldwell, 1977; Donaldson, 1986). However, if the same unequal allocation of resources is allowed to take place *within* the richer nations, the inequality will become much more visible to their citizens. I think that most people in the Northern countries would find this morally repugnant.[12] Then there

[11] It is fascinating to note that this cannot be determined by experimentation without bringing about the very changes the experimentation is designed to research. The problem is that genetically modified crops cannot be fully isolated from their environment, and some cross-pollination with non-modified crops is inevitable. I say that this is 'fascinating' because it is a crude example of how attempts to observe can be interventionary—whether the scientists involved wish this to be the case or not.

[12] Of course, this raises the issue of why governments of the richer countries are often willing to address health inequalities within their national boundaries, but not between nations. This is just the kind of boundary judgement that is coming into question as we enter the 21st Century, and which systemic intervention can help people explore.

are all kinds of issues surrounding the way the technology itself is developed: there is the question of whether it is ethically acceptable to create human embryos that are destined to be destroyed (just as abortion is a highly contentious issue, so is this); and there has long been controversy over experimentation on animals, which a great deal of genetic research involves.[13]

What became apparent in the 1990s is that it is no longer considered acceptable by many people in Western populations for science to be conducted in a way that is disconnected from public debates about morality (ESRC Global Environmental Change Programme, 1999). Furthermore, it has become virtually impossible, practically speaking, for scientists in these controversial areas of research to actually continue their work in this disconnected manner: campaign groups can network very effectively using the internet, and can grab the attention of the population via the media in order to raise ethical issues that rapidly become major public concerns. Again, a dramatic example of this has been the campaign against the use of genetically modified organisms (GMOs) in food production, which was taken up by the media in a big way in the UK, and rapidly spread to the rest of Europe and the USA[14]: the boycott of foods containing genetically modified materials has had a major economic impact on farmers, food manufacturers and shops, and has caused a rethink of public policy—especially in the UK, where the government has been forced into imposing a ban on production while further scientific tests are conducted.[15] While the dominant *scientific* opinion was that further experimentation was unnecessary, the dominant *popular* opinion was that the scientists had made a misjudgement. In essence there was a lack of effective communication between the two camps: there was clearly a difference of opinion on how safety should be judged, and yet there were no means other than public displays of authority and influence to resolve the problem.

This crisis in public confidence, and unwillingness to leave the discussion of ethics in the hands of scientists (who tend to marginalise

---

[13] See Singer (1990, 1991), Gray (1991a,b), Ryder (1991), Thomas and Blackman (1991) and Midgley (1993) for a discussion of the rights and wrongs of animal experimentation.

[14] In the USA, GMOs had already been in routine use for several years, yet the campaign against them suddenly took off once the American media realised what a 'live' issue it had become in Europe.

[15] Because this scientific experimentation cannot be conducted in isolation from local ecosystems (see footnote 11), direct action groups have begun to destroy the experimental crops. There is therefore no certainty that the research will ever be completed, and in this situation the UK Government will either have to use the Police to protect the crops or they will need some other legitimating mechanism to enable a final decision to be taken—possibly some kind of public inquiry? We will have to see.

ethics into a separate discourse[16]), provides a fertile environment for the introduction of systemic intervention. After all, if scientific activity (such as the controversial research discussed above) came to be seen as just one aspect of systemic intervention, then scientists and interested social groups would have a language with which to deal with ethical issues in a more participatory manner. This effectively means breaking down two barriers: the one set up by scientists who wish to maintain a strong distinction between observation and intervention (see Chapter 6 for an argument in favour of seeing observation as a 'special case' of intervention); and the one between the 'natural' and 'social' sciences. Systems thinkers have argued for many years that this is a false distinction (e.g., von Bertalanffy, 1968; Miller, 1978), but their reasoning tends to be that the same theories are relevant in both domains. My own reasoning is quite different: as I see it, the distinction needs to be challenged because all natural scientific research has a social context, and a value-basis upon which it is conducted. If the social roles of natural science are to be acknowledged, and the problematic issues raised by many research agendas are to be the subject of participatory debate, then natural scientists are going to need to master some of the methods and techniques for facilitating debate that social scientists have developed in the latter half of the 20th Century (see Chapter 9 for details). At the very least, they will need to form inter-disciplinary alliances so that they can work together with others more used to facilitating debate as part of their research activities. As I see it, natural scientists should not view ethical debate as *restricting* their research: on the contrary, if scientific experimentation is genuinely viewed as an aspect of systemic intervention, then there are enormous opportunities for public participation in forming *new* research agendas.[17]

1.5.3 The Politics of Social Exclusion

The third condition that we find at the beginning of the 21st Century that provides fertile ground for the growth of systemic intervention is the emphasis amongst increasing numbers of governments

---

[16] Popper (1959), whose philosophy of science was so influential in the mid-20th Century, actually argues that pursuit of the ideal of truth should be the primary interest of science, so ethics comes to be seen as a separate concern.

[17] I have not yet had the chance in my own systemic intervention practice to work on 'natural' scientific agendas in this way, so I cannot give a substantive example of how debate and experimentation might interact. However, there is a great opportunity for research here, which I hope that people will take up.

around the world (and not just in Western countries[18]) on countering 'social exclusion'. This is a term that is used in many different contexts, but it always refers to the exclusion of a group or class of people from the 'mainstream' of society: e.g., people in poverty are excluded from the use of many banking services, and they generally cannot afford computers so are not benefiting from the internet revolution; many long-term unemployed people, people over 55, disabled people, people with mental health problems, and people lacking basic literacy skills are excluded from participation in the labour market; homeless people are excluded by many organisations from accessing services because, without a place of abode, they are not easily contactable; the children of travellers are excluded from educational opportunities because schools stay in one place so the children must move from class to class; the majority of people in Southern nations are excluded from the health care enjoyed by many people in the North—the list of situations of social exclusion is almost endless. Some have global implications and others are more localised.[19]

There is a normative assumption in some thinking about social exclusion that the excluded *should* participate in the mainstream institutions of society at all costs. Of course, issues of cultural diversity and choice are raised here: there are some individuals and groups who *choose* a position in the margins, and it is becoming increasingly apparent to many people with responsibility for implementing social inclusion policies that the legitimacy (or otherwise) of such choices needs to be the subject of debate between the interested parties. To complicate matters even further, some of the choices made by excluded individuals and groups are not taken freely: for example, each year in the UK many thousands of children choose to live on the streets (Stein

---

[18] Ochoa-Arias (1999) makes some interesting comments about the Venezuelan government's rhetoric about finding a 'third way' between capitalism and socialism. This is precisely the same language used in the UK by many political commentators. Although the term 'third way' has been around for some time, it seems to have been popularised in recent years by Giddens (1998). One important focus in Giddens's book is how governments and organisations can work to counter social exclusion within a market economy.

[19] My own preference is to talk about *marginalisation* rather than exclusion, because the term 'exclusion' suggests that some groups are fully outside the boundaries of normal society. In my view, they are neither fully included nor excluded, and their marginal positions are a function of conflicts between wider social discourses that have become institutionalised in cultural, legal and organisational systems. I won't go into detail here, as the theory behind this observation is presented in Chapter 7. Nevertheless, the point should be made that marginalisation (or social exclusion) is not a superficial problem (it seems to me that some people treat it this way): many forms of social exclusion lie at the heart of the organisation of society, and therefore require some quite fundamental (but not always economic) changes if they are to be seriously addressed. However, I hope to make this the focus of a future book—it will not be addressed in this one.

et al, 1994, 1999). In the majority of cases this is because they are escaping abuse at home, and living on the streets is the lesser of two evils. Therefore, the fact that these children have chosen to live on the streets does not indicate that this is where they really want to be: it is a reflection of the lack of alternatives available to them.[20] See Chapter 17 of this book for details of an intervention specifically designed to address this problem.

Clearly, if issues of equality and access are on the agenda in the form of the discourse of social exclusion, then a language of systemic intervention that focuses attention on exploring issues of marginalisation, and choosing or designing methods with the inclusion of marginalised stakeholders in mind, should be well received. In my view, one of the most important contributions of systemic intervention in the 21st Century needs to be the reconceptualisation of social exclusion to take account of some of the dilemmas it raises, and the design of methods to address it at all levels of society.

1.6 Conclusion

It should be clear from the above that the subject matter of this book connects with major developments in Western thought that are still in the process of unfolding. We cannot know what the final results of this unfolding will be, but we can nevertheless contribute to it—putting forward arguments for change and seeing where the ensuing debate takes us. This book is intended as just such a contribution.

However, there is also a very practical side to this book: the methodology of systemic intervention that I propose is intended to make a positive difference by providing a useful language that I hope will make a start in enabling change agents to do a number of things: bring together science and ethical reflection in one practice; conceptualise complex situations characterised by interacting issues and multiple, conflicting viewpoints; reflect on values and boundaries of inclusion, exclusion and marginalisation (of stakeholders and issues); sweep into intervention the viewpoints of a wide variety of stakeholders, including those who find themselves marginalised; and choose and/or design methods that provide the means to engage with others in a flexible and responsive manner, thereby facilitating the development of new social agendas and plans for change that can command wide-spread support from those affected by them.

---

[20] For anyone who doubts that this is a forced choice, I recommend Boyd *et al* (1999a) who have researched the issues from the perspectives of the children themselves.

Admittedly this is an ambitious project, but if you find what I have to say meaningful, I hope you will join me in its further development.

I

Philosophy

2

Why Philosophy?

The book you are reading is structured so that the philosophy of systemic intervention is discussed first, methodology second, and practice last. In my view, this ordering does not reflect the importance of these subjects: I see philosophy, methodology and practice as mutually supportive areas of study, where a problem in any one might signal the need for revision in either or both of the other two. This contrasts with the approach of some writers (e.g., Fuenmayor, 1991a-c) who believe that there is a strict hierarchy, with philosophy being foundational, methodology following from this, methods being based on the methodology, and practice being the implementation of those methods.[21] For these writers, getting the philosophy right is an absolute priority, as everything else depends upon it. The idea that encountering a problem in practice may signal a philosophical inadequacy is not conceivable from the point of view of those who believe in this hierarchical relationship.[22] However, although I see philosophy, methodology and practice as mutually supportive, it is very difficult to discuss them simultaneously without jumping from argument to argument, causing confusion for the reader—hence they are discussed in separate sections of this book, although regular references will be made from one section to another.[23]

---

[21] Fuenmayor does not explicitly discuss his belief in a hierarchy in his 1991a-c works, but the focus of his writing is very much on philosophy, with everything else flowing from this. I have broached the subject of hierarchy with him in conversation, and he argues strongly in favour of it.

[22] It is clear that people often do challenge philosophy on methodological and/or practical grounds, despite this conventional hierarchy. A good example is in discourses of environmental sustainability, where the desire to talk about the need for human beings to respect ecosystem limits has created a backlash against non-realist philosophies of science (those which do not accept that our knowledge reflects a real world) (Soper, 1995), and in some cases it has even been argued that there is intrinsic *value* (rather than truth) in the external world which human beings can have knowledge of (e.g., Rolston, 1983; Goodwin, 1992).

[23] The belief in a hierarchy that makes philosophy foundational will not be explicitly

My assumption that philosophy, methodology and practice are mutually supportive also contrasts with the approach of other writers who take the very opposite view from the hierarchical one, and argue against the value of philosophy (e.g., Hutchinson, 1996; and to a lesser extent Ormerod, 1996[24]). The current chapter of this book is aimed at readers who are sympathetic to this anti-philosophical stance: those who have an interest in intervention and question the value of thinking in philosophical terms, but are still open to debate on the matter.

2.1 Two Grounds for Exploring Philosophy

One reason for thinking about philosophy is that philosophical assumptions are often used to justify what can be considered valid or legitimate practice. Below, I will give a couple of examples to illustrate. The works of Popper (1959, 1972), Kelly (1955, 1970) and Habermas (1984a,b) will be compared, as these writers present distinct philosophies with explicitly different implications for methodology and practice (these writers have also been chosen because of the influence of their work on the development of systemic intervention over recent years). Of course, it is always possible to argue that valid or legitimate practice is *restricted* (rather than defined) by philosophy, and that if we were to dismiss philosophy as irrelevant we could just get on with whatever practice we prefer. After comparing the various philosophies and their implications, I will argue against this anti-philosophical position on two grounds:

The first of these is *strategic*. I suggest that, as long as there are dominant practices that are justified with reference to philosophy, people seeking to advocate an alternative practice risk defeat unless they engage in philosophical discourse. This is a primarily practical point concerning the development of an appropriate strategy for communicating with those who work in institutions that currently define and perpetuate ideas running counter to the ones proposed in this book. In my view, far from liberating intervention, champions of anti-philosophy weaken the cause of those who argue against orthodoxies in favour of systemic practice. Essentially, these anti-philosophers do not account for *why* those who support orthodox positions favour

countered in this book. My discussion of philosophy, methodology and practice in that order will be acceptable to writers who hold this belief, making it likely that they will engage with the arguments regardless of our disagreement over hierarchy.

[24] It is perhaps more accurate to say that Ormerod (1996) is *sceptical* about the value of philosophy. He is not completely antagonistic to it.

philosophical discourse—in my view, the latter have good reasons for doing so, and I say this even though I disagree with many of the particular philosophical positions they base their work upon.

The second ground for challenging the anti-philosophical position is that philosophical arguments *are* meaningful to debates on methodology, method and systemic practice. I will give an example from Spash (1997) of one particular philosophical debate that has significant implications for the construction of methods for decision support. In my view, Spash clearly shows why philosophy should not be ignored.

2.2 *Terminology*

Now, before starting the comparison between Popper, Kelly and Habermas (which should help the reader to see the connection between philosophy and practice), I first need to clarify two pieces of terminology. Throughout this first section of the book I refer to *ontology* and *epistemology*. Ontology is a branch of philosophy concerned with the nature of reality. An example of an ontological statement is, "reality consists of objects in relationship with one another". Epistemology, in contrast, is concerned with the nature of knowledge. An example of an epistemological statement is, "we can only know our personal constructs, not external reality". The two terms begin to get confused when we acknowledge the tight connection that we find in most modern philosophical positions between knowledge and reality. For example, saying "we only have access to knowledge constructs, not external reality" is a statement about both knowledge (epistemology) and the *reality* that we only have access to knowledge constructs (ontology). Hence, some authors do not distinguish between the two terms, or use a hybrid term like "onto-epistemology" (Fuenmayor, 1991a-c). In this chapter, when reviewing the works of other authors, I use their own terminology.

2.3 *Popper's Critical Fallibalism*

We may begin the review of philosophical positions with the work of Popper, who is widely regarded as a key figure in the 20th Century philosophy of science. The essence of his position is as follows.

Popper starts from the premise that knowledge, and the language that frames this knowledge, reflects the real world. For example, we use the word 'table' because there is something in the world that this refers to. The idea that knowledge and language refer to a real world, and there is at least a rough correspondence between knowledge and reality, is what defines a *realist* ontological position. Realism might be regarded as a 'common sense' ontology, in that most people take for granted that the words they use refer to things that other people see and refer to in more or less the same way.

Now, for most of the 20th Century, it has been accepted that there is a key problem with the realist position that must be dealt with. This is the problem that one can never be *absolutely* sure about the extent to which knowledge is accurately reflecting reality. The classic illustration of the problem of uncertainty about reality, adapted from Popper (1959), involves a man sitting by a lake watching swans go by. Altogether, he sees one hundred white swans. Does this mean he can conclude that all swans are white? If (hypothetically) he has never seen a swan before, and knows nothing about them other than the fact that swans exist, he might happily turn his back on the lake, quite satisfied that all swans are indeed white. Then he will never see the black swan that floats by just seconds later. In this case, the statement 'all swans are white' is a product of the way that the man has observed the swans (being satisfied that one hundred swans is a representative sample). It is not an accurate reflection of reality.

This problem affects all attempts to describe the world. However exhaustively we test our assumptions, we can never be absolutely certain that the disconfirming evidence is not appearing just as we are turning our backs. Hence Popper (1959) argues that we can never *prove* a theory—all we can do is falsify it. Seeing just one black swan falsifies the statement that 'all swans are white'.

However, in later writings, Popper (1972) also questions the status of falsification: it is always possible to theorise that what at first *appears* to be a black swan is not really a swan at all, but another kind of bird. Thus, the statement 'all swans are white' is preserved despite a supposedly falsifying observation. Popper's conclusion is that certainty is impossible, and we must admit that our knowledge is fallible. We also have to be critical in our attitude: the continual questioning of assumptions is the only means we have to refine knowledge. Popper therefore ends up with a methodological position that he calls *critical fallibalism*: knowledge should always be open to questioning, and the questions should be guided by the ideal of truth (truth being an ideal, rather than actually attainable, because of the inevitable fallibility of knowledge).

There are clear implications for methodology and method that flow from this position. The method that Popper (1959, 1972) recommends can be summarised as follows:

(i) Propose a hypothesis about what might be giving rise to an observable phenomenon. A hypothesis may be based upon one theory in opposition to another, or it may go against all established theory.

(ii) Ask, what other observations should be made if the hypothesis is to be supported, and what observations would falsify the hypothesis (and possibly support an opposing theory)?

(iii) Set up a situation (an experiment) where reliable observations can be made that either support or falsify the hypothesis.

(iv) If the hypothesis is supported, this may be used to discriminate between theories and/or form the basis of establishing a new theory.

Consider for a moment what Popper's methodology excludes from valid practice. It excludes, for example, any analyses of individual decision making based on the idea that the subjective perspective of the individual is the primary focus for exploration. Interventions (rather than observational analyses) to *facilitate and support* individual or group decision making are also ruled out (Popper's methodology is exclusively concerned with observation). Furthermore, this methodology excludes the support of groups who wish to explore the moral legitimacies of alternative planning scenarios, where the primary focus is the right thing to do rather than what is or is not true. Explorations of morality and/or subjective perspectives cannot be considered valid practices from a critical fallibalist position because the guiding assumption of this approach is that the generation of knowledge about the 'outside' (real) world should be the primary focus.

Popper's main concern is the development of a philosophy of *science*—and in his view science should be separated from moral and subjective concerns. Indeed, he argues that this separation is vital to the maintenance of an 'open' society, where the pursuit of truth can be divorced from political interests (Popper, 1966). This is in marked contrast to other writers who argue that such a separation is not only unattainable, but also pernicious because the illusion of its attainment blinds us to the ways in which 'truths' are normatively constructed by forces of power, and prevents us from appreciating alternative 'truths' that may only be surfaced through moral and/or subjective explorations (see, for example, Marcuse, 1964; Habermas, 1972, 1976, Foucault, 1980; Ulrich, 1983, 1996a; Bhaskar, 1986; Rorty, 1989).

Having discussed Popper, and the way in which he justifies his methodology with reference to philosophy, let us move on to consider the work of Kelly (1955), which stands in stark contrast to Popper's thinking.

2.4 Kelly's Personal Construct Theory

Kelly (1955, 1970) does not accept that knowledge reflects the real world, however imperfectly. Kelly questions the whole distinction between 'knowledge' and 'world', arguing that all we can possibly have access to is knowledge. Therefore, he suggests that there are as many worlds, or realities, as there are human beings. This is a radically subjectivist ontological stance that not only contrasts with realism, but also runs counter to other anti-realist thinking being developed in the 1950s such as that proposed by Wittgenstein (1953). While Wittgenstein concentrates on language as the active force that brings forth realities (in his view everything we know about ourselves and the world is given in the words we use), Kelly prefers to focus on the individual alone. This is because he perceives individuals as unique: while they may share some common ground, Kelly finds the differences between them much more striking than their similarities. While he does not rule out a social or linguistic dimension to the construction of realities, he regards this as relatively unimportant compared with the bringing forth of markedly different realities at the level of the individual.

The other important aspect of Kelly's (1955) philosophy is his claim that an individual works to construct his or her reality primarily in terms of *activities*. Thus, what is of primary importance in the construction of a reality is the alternative paths for action that the individual has taken, or could take in the future. It is the things that are perceived as impacting on decision making for action that become part of a reality.

For Kelly, these philosophical assumptions have an important implication for understanding what constitutes valid practice. Methodologically, it is not valid to focus (like Popper) on the generation and testing of hypotheses, because for Kelly there is no external reality for us to hypothesise about. Rather, it is only valid to explore individual subjectivities—and then only in a way that is action-focused. Given two or more possible paths for action, we may ask how the individual evaluates the situation in which the decision between them must be taken. This is done by listing the 'personal constructs' (subjectively perceived variables) the individual may take into account in making a decision, and then working out mathematically which action is preferable given the personal constructs being considered.

It should be readily apparent that Kelly's methodology not only excludes from valid practice any explorations of the 'truth', but also outlaws methods supporting the collective critique of norms. This, according to Holland (1970), is an inevitable consequence of Kelly's philosophical assumption that the individual is the primary generator of meaning.

2.5 *Habermas's Three Worlds*

Our final example of the use of philosophy to define valid practice is taken from Habermas (1976, 1984a,b). Habermas argues that there are four implicit validity claims inherent in any sentence intended for communication. Giddens (1985) neatly summarises his position:

> "When I say something to someone else, I implicitly make the following claims: that what I say is intelligible; that its propositional content is true; that I am justified in saying it; and that I speak sincerely, without intent to deceive" (p.99).

The first of these implicit claims, that what I say is intelligible, is simply a precondition for effective communication. However, the other three, when made explicit, can all be questioned and justified through rational argumentation.

According to Habermas, it is these three claims that refer directly to three 'worlds': the claim that my statement's propositional content is true relates to *the external natural world*; the claim that I am justified in making the statement relates to *our social world*; and the claim that I speak sincerely relates to *my internal world*. These 'worlds' are, of course, tightly interconnected. Importantly, it is the nature of language that allows us to differentiate the 'natural', 'social' and 'internal' when we enter debate.

Because, in Habermas's ontology, there are three worlds (natural, social and internal), and language is structured so that we can differentiate them, a theory of rationality ensues. A position is rational when it is possible to distinguish the natural from the social, the social from the internal, and the natural from the internal. According to Habermas, an example of an irrational position is a myth where, say, a view of nature is taken that implies a view of social rightness. One instance is the idea that competition is essential for evolution so it is necessarily right to act in a competitive manner when engaging in social relationships. In Habermas's terms, this idea should be seen as both mythical and irrational: competition *may* be essential to evolution, but that does not mean that it is necessarily right for us to act competitively all of the time. The 'rightness' (morality) of any proposed competitive act needs to be assessed separately from the truth of the claim that competition is a natural part of evolution: it is possible to argue that social stability (requiring co-operation rather than competition) is required in a given social situation rather than biological evolution, but this argument can only be surfaced if the truth and rightness claims are evaluated independently.

From this theory of rationality Habermas derives a methodology (although he does not explicitly use the term). He argues that people should engage in *communicative action*, where they set up 'ideal speech situations' (situations free of power relations, allowing open debate) in which rational argumentation can take place that involves statements about, and challenges to, information relating to all three of the natural, social and internal worlds.

Habermas's position is certainly broader than that proposed by Popper (who, in Habermas's terms, places a primary emphasis on natural world exploration), and it is also broader than Kelly's (Kelly prioritises the internal world of the individual). Nevertheless, the three worlds of exploration are framed within a theory of the need for debate. Therefore, scientific activity (in Popper's terms) or subjective exploration (in Kelly's terms), when divorced from wider debates, are not strictly legitimate from a Habermasian point of view.

2.6 The Need for Philosophy

The philosophical thinking of these three writers clearly has implications for their construction of methodology, and hence their idea of what constitutes valid or legitimate practice. However, it could be argued that, if we wish to avoid limiting practice, all we need to do is dismiss philosophy and then practice as we wish. Indeed, this is just what some authors (e.g., Hutchinson, 1996) advocate. In answer to this, I first need to say that avoiding limitations on practice is not, as far as I can see, the goal we should be aspiring to. The goal we should aim for is to construct practice in such a manner that it has outcomes that we regard as positive, and which can be defended as such against critics. This does not mean avoiding all forms of limitation, but *selecting* our limitations (or choosing guidelines) in a self-conscious and defensible manner so as to avoid outcomes that we may view as negative. There is a need for methodological guidelines (that may in part be derived from philosophical deliberation) to ensure valuable and defensible practice.

Why, then, should we accept guidelines with their origins in philosophy? The argument against philosophy assumes that, if we divest ourselves of it, practice will be broader and more flexible. However, the opposite may actually be the case. Over the coming pages I will present two grounds for engaging with philosophy.

2.7 *The Strategic Reason for Engaging with Philosophy*

In a society where most scientific practice is based on a methodology of experiment and observation, it would be all too easy (without a compelling counter-argument) to be dragooned into conforming to the norm. An example from my own experience will illustrate.

As a newly graduated psychologist in the early 1980s, I had a very limited understanding of the philosophy of science (focusing almost exclusively on the work of Popper, 1959). I had been well trained in the techniques of experiment and observation, and had a battery of statistical tests at my disposal. I started out by undertaking research on a self-employed basis, and one of my first projects was to design evaluations of four mental health facilities serving an inner city area of London. I had three months to produce a proposal, and hoped that I would then be funded for three years to undertake the work.

I had an intuitive awareness that the quantitative methods that I had been taught were not going to be adequate to the task, and that the subjective views of service users were going to be important in judging the success or failure of services. I therefore proposed a mix of qualitative (semi-structured interview) and quantitative methods, but no controlled studies. In the experimental tradition, the validity of research findings is said to depend on having made an effective comparison between a 'treatment group' (receiving a service) and a 'control group' (not receiving anything). This is because it is assumed that it *is* possible to have knowledge of the real world, however imperfect, and the task of the scientist is to make valid observations. Validity is ensured by measuring the effect of a change on people (e.g., the provision of a new mental health service). One way to measure the effect of the change is to compare the group receiving the service with an (as near as possible identical) group who is not. I did not plan any controlled studies because conversations with staff in the four facilities revealed two significant problems with this approach. First, I was told that the facilities were required by the Health Authority to work with a minimum number of clients. Failure to do so would result in closure, and refusing a service to 50% of clients would make closure inevitable. Second, there was sufficient anecdotal evidence to suggest that the services being offered by the four facilities were extremely valuable, and both the staff and I had ethical objections to refusing people access given this situation.

To everyone's surprise (all the facilities had been given the impression by the Health Authority that approval of the research was a formality), my proposal was rejected by the Authority's research committee on the grounds that the use of controls was essential if the results were to be regarded as scientific. I visited the Chair of the committee to try to change his mind, but my ethical objection to the use of a control was dismissed. The logic of his argument was that there was absolutely no evidence of the efficacy of the services (the anecdotal evidence did not count because it was not scientific), so I would not be refusing clients anything of value. The services would only *become* valuable if my research showed that they had value through controlled study. The staff's issue of needing to meet admission targets was also dismissed as 'an organisational matter', and therefore irrelevant to scientific inquiry.

I found it very hard to argue against his points, as everything he was saying reflected what I had been taught in my own research training. My inclusion of qualitative methods arose from an *intuition* that the views of clients were important (it did not emanate from a well worked out theoretical position). I therefore found it impossible to defend my wish to use qualitative methods against the argument that subjectivity should be excluded if research is to be valid. Similarly, talking with the Chair of the research committee made me feel that my ethical stance against control groups was irrational. I left the meeting feeling rather confused.

Had I had command of some alternative philosophical arguments I would not necessarily have convinced this particular man to change his mind. He appeared quite entrenched in his beliefs. However, fluency in the language of philosophy would at least have given me the opportunity to present a convincing case to others who might have been more open to persuasion, thus enabling the use of a different approach in future.

2.8 *The Importance of Philosophical Argument*

However, while the strategy of engaging with philosophy to support ones own ideas against those of opponents is necessary, in my view it is not the most important reason for doing so. The most important reason is that philosophical arguments can help us to see practice in quite a different light than we might otherwise, and the new perspectives we can gain from this can be very valuable. I want to illustrate this with an example from the work of Spash (1997). Spash criticises authors in the environmental management literature who use methods of cost-benefit analysis to support the collaborative efforts of business planners and environmentalists in making business decisions that take account of environmental issues, and his grounds for doing so are essentially philosophical.

Spash (1997) argues that the idea of making trade-offs between competing objectives is based on 'utilitarian' philosophy. Utilitarianism asserts that, when dealing with conflicting interests, the fairest possible result will be obtained if the greatest good for the greatest number of people is pursued. If utilitarianism informs analysis, then it is perfectly acceptable to set up trade-offs where environmental damage anticipated by a minority is accepted because the majority see some other benefit accruing to them. This is the kind of logic that fits with the point of view of a business operating within a capitalist economy: it is to the short- and medium-term advantage of a business organisation to satisfy the majority of its stakeholders. If there have to be 'losers' in a trade-off, then less commercial damage will accrue in the short-to-medium-term if only a minority become disaffected.

However, according to Spash, most environmentalists use a 'rights-based' (or 'deontological') approach that is incompatible with this utilitarian trade-off rationality: they assume that it is necessary to maintain a minimum level of environmental integrity in any given situation, and this minimum level should never be compromised or traded for gains elsewhere. Spash gives the establishment of National Parks in the USA as an example of 'deontological' environmentalist intervention: the land was seen as having absolute value (or a 'right' to be left untouched) that should not be traded off for human advantage. In contrast, according to Spash, the US Forest Service, which now manages the National Parks, often operates with a utilitarian rationality which emphasises multiple land uses where decisions have to be made on which combination of uses provides the greatest good. Of course, if the majority of environmentalists think in deontological terms, cost-benefit analysis is asking them to do something that is alien to their way of thinking. The result will either be a lack of participation and agreement (failure of the methods), or domination of the environmentalist rationality by the commercial one (success of the methods at the expense of respect for the views of all stakeholders).

Here we have a clear example of how philosophical analysis can inform our understanding of methodology. Without this kind of analysis, methods like cost-benefit analysis appear to provide a 'neutral' means to realise non-neutral but consensually accepted ends. However, once we realise that such methods already embody a particular rationality that favours some stakeholders over others, we can see that they are not neutral, and we are therefore in a better position to consider appropriate alternatives (if we wish to do so).

2.9 Conclusion

Although there are some who suggest that paying too much attention to philosophy restricts practice, I argue the reverse: that philosophy may throw new light on issues of methodology, method and practice. Indeed, methods that appear superficially useful may harbour hidden, potentially dangerous assumptions that can only be revealed through philosophical analysis (as in Spash's, 1997, critique of cost-benefit analysis). Also, ignorance of philosophy can lead to confusion and defeat in the face of a clearly presented argument against one's own position. Hence, it would be all too easy, in a society in which experimental approaches still dominate discourses about research[25], to conform to the norm despite the discomfort this induces. Philosophical discourses provide one (but not the only) arena in which to judge the quality of methodological ideas. If we find contradictions between philosophical and methodological perspectives, then this is an indication that the adequacy of one or both of them might need to be reconsidered.[26] Indeed, as long as the tradition of observational science remains dominant, and is supported with reference to philosophy, we have little option but to fight the battle for alternatives on the same ground. This is not giving in to orthodoxy. On the contrary, I suggest that there are significant weaknesses in the orthodox philosophical (as well as methodological and practical) positions that, when we expose them, make the case for other positions all the more compelling.

---

[25] Although they are thankfully less dominant now than fifteen years ago.

[26] In my own case, the experience related above led to further philosophical and methodological study, and has ultimately resulted in the production of this book. Many others have had similar experiences (see, for example, Hollway, 1989), and our combined effort is actually shifting the debate. Today, students are better prepared to deal with issues relating to the philosophy of science, and now I believe that a shift in the research culture towards one in which methodological pluralism is widely accepted is achievable.

3

The Systems Idea

Having looked at the issue of why philosophy might be important, let us now begin to ask, what is *systems* philosophy? First, I will give a short answer that just scratches the surface of the meaning of 'system', focusing in particular on the motivation of systems thinkers to be as comprehensive as possible in their analyses. As it is impossible for any analysis to be totally comprehensive, this leads on to a consideration of boundary judgements: judgements about what is to be included or excluded from analyses. I will argue that the boundary concept is fundamental: it is the core idea of systems thinking.

Having surfaced this core concept, I will then be in a position to tackle another central issue for systems thinkers: the opposition of systems thinking to mechanism (discussed in Chapter 1) and reductionism (looking for simple causal relationships between variables rather than trying to understand a wide range of interactions that can only be satisfactorily explained in terms of the functioning of whole systems). We will find that both mechanism and reductionism assume 'subject/object dualism'—the radical separation of the observer from the observed, or the subject from the object, which produces the illusion of perfect objectivity. It is subject/object dualism that many writers say lies at the heart of a scientific practice, developed over centuries, that has marginalised the consideration of values and subjectivity in the production of knowledge. The focus of the bulk of this chapter will therefore be on dualism rather than mechanism or reductionism.

I will first describe subject/object dualism as it has been represented in the systems literature. I will then go on to detail four significant attempts by other systems thinkers to produce theories of ontology (about the nature of reality) and epistemology (about the nature of knowledge) which try to escape it. However, I will argue that each of these systems theories is problematic: in the end, despite valiant efforts, they do not break free of the dualism of subject and object. As a result, I argue that a new systems perspective is required. In Chapter 4, I return to the boundary idea which I use to generate a theory that, in my view, does indeed escape from subject/object dualism: it does so by setting up a different opposition—between the 'content' and 'process' of analysis. The implication of the 'content/process' oppositional pair is that both subjects and objects have the same status: they are both general types of 'content'. There is inevitably a kind of dualism still present in this idea, but not the subject/object dualism that has been so pernicious in the history of science. The precise meaning of all of this should become clearer in Chapter 4. So let us begin with an initial (albeit sketchy) answer to the question, what is *systems* philosophy?

3.1 The Meaning of 'Systems Philosophy'

One answer is that, if something can be described as 'systemic', it is (as far as possible) comprehensively understood. I say that this is 'one answer' because it is peculiar to a single strand of systems thinking—general systems theory (e.g., von Bertalanffy, 1956), which was popular in the mid-20th Century and which still has strong proponents, especially in the USA (e.g., Miller, 1978). General systems theory is premised on the idea that it is possible to offer a common language for all the various scientific communities. Through this language, general systems theorists suggest that scientists can transcend the limitations of their fragmentary disciplines while still preserving and enhancing their specialised knowledge. This common language is the language of 'isomorphies', which assumes that the purpose of science is to pursue the truth about reality. In von Bertalanffy's words:

> "A unitary conception of the world may be based, not upon the possibly futile and certainly far-fetched hope finally to reduce all levels of reality to the level of physics, but rather on the isomorphy of laws in different fields.... (T)his means.... that the world, that is, the total of observable phenomena, shows structural uniformities, manifesting themselves by isomorphic traces of order in its different levels of realism" (von Bertalanffy, 1956, p.8).

In the view of general systems theorists, the purpose of transcending disciplinary boundaries is to gain a more comprehensive understanding of the phenomenon under study. Disciplinary boundaries restrict the scope of inquiry by allowing specialist languages to develop that do not 'fit' with the specialist languages of other disciplines. Lovelock (1988) gives a good example when he cites the separation between biology and geology which for a long time obstructed the emergence of knowledge about the *co-evolution* of biological and geological forms. As a result, both geological and biological theories of evolution have been impoverished. For general systems theorists, the language of isomorphies is seen as the antidote to this kind of problem. The world can be described as a hierarchy of systems (e.g., cell, organ, organism, family, community, ecosystem, planet, solar system, galaxy), each of which contains, and is contained by, other systems (see, for example, Miller, 1978). A hierarchy of nested systems is therefore somewhat like a Russian doll—although, unlike a Russian doll, the larger systems are just as dependent on the existence and healthy functioning of the smaller ones (their components) as the smaller ones are on the larger ones that they are a part of (their environment). Isomorphies can be found across these various systems. In other words, the emphasis is on their similarities rather than their differences (Flood and Robinson, 1990). In general systems theory, a systems view is said to be the most comprehensive view that it is possible to attain.

Elsewhere in the world, and particularly in Europe, general systems theory has had less influence on the development of systems thinking. It was another strand of systems thinking that was influential in Europe, and this embodied quite a different understanding of 'system'. The key author in this alternative tradition was C. West Churchman (1968a,b, 1971, 1979), whose fundamental ideas have been taken in a variety of different directions (e.g., by Mason and Mitroff, 1981; Checkland, 1981[27]; Ackoff, 1981; Ulrich, 1983; and Midgley, 1992b).

Prior to the work of Churchman, many people (most notably general systems theorists) assumed that the boundaries of a system are 'given' by the structure of reality. In contrast, Churchman made it clear that boundaries are social or personal constructs that define the limits of the knowledge that is to be taken as pertinent in an analysis. There is also another important element of Churchman's understanding of 'system'. When it comes to *social* systems, pushing out the boundaries of analysis may also involve pushing out the boundaries of who may legitimately be considered a decision maker (Churchman, 1970). Thus, setting boundaries defines both the knowledge to be considered pertinent *and* the people who generate that knowledge (and who also have a stake in the results of any attempts to improve the system). This means that there are no 'experts' in Churchman's systems approach, at least in the traditional sense of expertise where all relevant knowledge is seen as emanating from just one group or class of people: wide-spread stakeholder involvement is required, sweeping in a variety of relevant perspectives. Far from signalling 'comprehensive understanding', Churchman's view is that the systems idea highlights the bounded

[27] Checkland (1981) actually draws upon general systems theory too, but in my view this aspect of his work could be removed without any damage to the whole.

nature of all understandings, and hence the inevitable *lack* of comprehensiveness. The key to systems thinking is to sweep in sufficient information (from various different viewpoints) to paint a rich and complex picture, but without compromising the meaning of an analysis by over-inclusion, thereby paralysing action (Ulrich, 1983).

The above two examples appear to be opposites: general systems theory assumes that systems exist in the real world and that our models of them represent (as near as possible) comprehensive knowledge, while Churchman's idea is that a system is a personal or social construct and its boundaries highlight the inevitable *lack* of comprehensiveness. What they both have in common, however, is the focus on comprehensiveness as an *ideal*. Even though Churchman focuses on the *lack* of comprehensiveness apparent when a system is defined, he does so to highlight the continual need to generate further insights by "sweeping in" more information into our understanding of a situation. While *full* understanding is unattainable, *greater* understanding than we currently have at any particular moment is always possible—although, as Ulrich (1983) points out, the purposes of any analysis will eventually impose limits on the sweep-in process.

3.2 The Boundary Concept

In the tradition of Churchman (1970) I suggest that, once we acknowledge that no view of the world can ever be comprehensive, the boundary concept becomes crucial. Where exactly boundaries are constructed, and what the values are that guide the construction, will determine how issues are seen and what actions will be taken. If the boundary concept is so important, let us take some time to explore its fundamentals.

I will start with the basic understanding of boundaries proposed by Spencer Brown (1972). Spencer Brown presents a simple rectangle containing a circle (Figure 3.1). The line around the circle is its boundary: it divides the circle from that which is outside it. Another example (Figure 3.2) is a rectangle bisected by a line. The line represents a boundary demarcating where one shape (*A*) ends and another (*B*) begins. This no doubt seems obvious, but the consequences of this simple understanding of boundary are quite profound, as we shall see later in this book (Chapters 4 and 7).

It is important to be aware, when considering the concept of boundary, that a boundary does not simply mark what is included within it. It also marks what is excluded. However, for there to be any

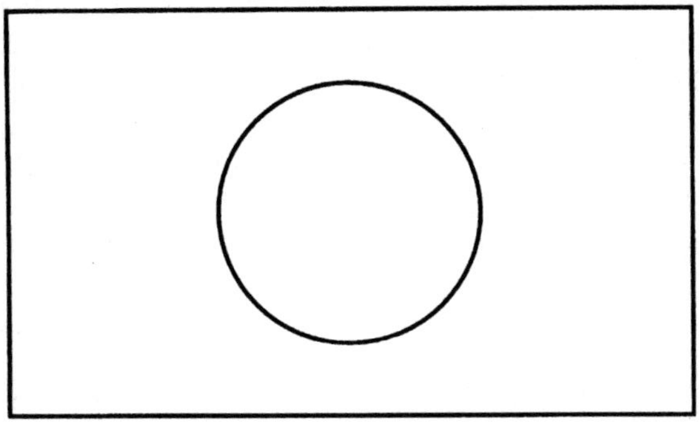

FIGURE 3.1: Representation of a bounded object (after Spencer Brown, 1972)

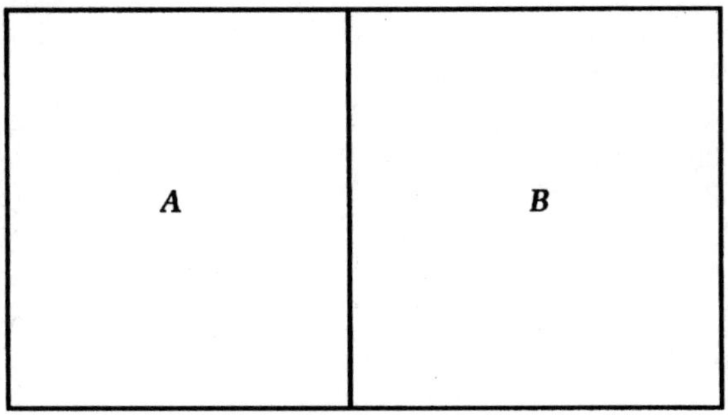

FIGURE 3.2: Representation of a demarcating boundary (after Spencer Brown, 1972)

awareness of what is excluded, a second boundary must be apparent. For example, in Figure 3.1, we only know that the rest of the rectangle is excluded from the circle because the rectangle is also bounded. And, in Figure 3.2, we only know that there is another shape (*B*) that fits with *A* to form a rectangle because both *B* and the rectangle are bounded. Beyond the rectangle, we may also become aware of other boundaries (the edge of the paper, for example, or the boundary surrounding the book, or the reader and book together). Everything is distinguished from that which it is not, and that which it is not comes to be distinguished in turn with reference to another boundary (see Midgley, 1992b, for more details of this argument).

Of course, a great deal more complexity can be constructed around the boundary concept. For example, a difference can be identified between the boundary of a human body as used in biology (which is observable as skin, and its existence is generally the subject of consensus), and the boundary placed around a social class (say the 'working class') in a sociological analysis (the positioning and meaning of which is fuzzy and subject to heated debate). Also, later in this book (Chapter 7), I will explore the relationship between boundaries and values. For now, however, we will stick with the simple picture of a boundary (painted above) that marks a distinction between an object and that which it is not. I suggest that, because of the almost inevitable uncertainty in analyses surrounding where boundaries should be placed to optimise comprehensiveness, the boundary concept must lie at the very heart of systems thinking.

However, the boundary concept is not only central to systems thinking, it also underlies mathematics. It is worth taking a one paragraph detour in the argument to clarify this point, because there has been a tendency in writings about systems thinking and operational research to talk about mathematics as 'hard' (expressing the laws of nature) and boundary judgements as 'soft' (the result of subjective and/or inter-subjective judgements that may be open to debate) (e.g., Checkland, 1981). The realisation that boundaries underlie mathematics as well as subjective and inter-subjective judgement breaks down this dichotomy. The traditional view is that mathematics is the most 'fundamental' science because equations represent generalisable relationships that are observable in the real world. For writers taking this line, mathematics provides a language that represents reality in the purest possible form. However, Spencer Brown (1972) argues that numbers can only exist *because of the prior existence of boundaries.* With reference to Figure 3.2, the numbers 1 and 2 only have meaning in relation to the two shapes (*A* and *B*) distinguished by the boundary. In other words, things can only be counted because they exist prior to numbers, and their nature as things depends on them being distinguished by a boundary from that which they are not.

At this point in the argument we have identified a theme that is common to all forms of systems thinking: the aspiration to comprehensive understanding. We have also clarified a core systems concept, boundary, that becomes important precisely because of the impossibility of comprehensive understanding. It is necessary to explore different possible boundary judgements in order to optimise the inclusion of information in analyses.

Let us now move on to address a second philosophical issue that has been played out again and again by systems thinkers throughout the 20th Century: the opposition of systems thinking to mechanism and reductionism—and by implication, subject/object dualism.

3.3 *The 'Enemies' of Systems Thinking: Mechanism, Reductionism and Subject/Object Dualism*

As discussed in Chapter 1, throughout the 20th Century systems thinkers have consistently challenged mechanism—the belief that phenomena are analogous to clockwork toys: predictable, functional, inherently understandable objects seen from a discrete distance by an independent subject. One aspect of the mechanistic worldview that they have paid particular attention to is *reductionism* (narrowing attention to linear, causal relationships between variables, thereby failing to see that these relationships can only be adequately understood as aspects of the operation of wider systems). Reductionism follows on logically from mechanism in that, if someone believes that systems are no more than the sum of their parts, it makes sense to decompose them into those parts to increase understanding. For example, a reductionist approach to physiology may view a human being as a simple collection of organs; a reductionist approach to the function of a particular organ (e.g., a kidney) may decompose it into a set of bio-chemical reactions; and a reductionist approach to bio-chemical reactions may involve an examination of the physics underlying chemical properties. Von Bertalanffy (1956) and Fodor (1974) both identify the desire of some scientists to reduce all disciplines to physics as the ultimate form of reductionism.

Reductionism has long been seen as the traditional 'enemy' of systems thinking. The battle between systemic and reductionist positions has been played out again and again in the literature (e.g., von Bertalanffy, 1956, 1968; Ackoff, 1974; Churchman, 1979; Checkland, 1981; Fuenmayor, 1991a). Sometimes the label 'reductionist' is employed in a loose manner to denote any approach that is not comprehensive. However, this way of using the term can easily be dismissed with the observation that *no* position is comprehensive (see earlier). In my view, it is necessary to be a little more specific in defining reductionism.

Reductionism, as I see it, is the reduction of phenomena to simple, objective, causal relationships. The traditional scientific method is reductionist in the sense that it requires one to look for uni-directional, cause and effect relationships (or, when this cannot be achieved, statistical associations between phenomena), screening out any 'unnecessary' complexity. Thus, it is possible to take a task (say, driving a car) and ask if performance is impaired when alcohol is consumed. By taking a random sample of drivers and giving some people a measure of alcohol and others a measure of water, it becomes possible to identify the effects of the alcohol through a comparative study. We can therefore ask if alcohol *causes* bad driving: i.e., we can ask if there is a uni-directional relationship between consumption of alcohol and poorer performance.

Of course, this kind of scientific experimentation can be very useful (for instance, when making a policy decision about whether to make drink-driving illegal), but if it is regarded as the *only* valid way of looking at phenomena, our understanding will be greatly restricted. For instance, traditional scientific experimentation is limited in what it can contribute to describing the transport system as a whole, including the various roles of human beings within it—and all the activities associated with drink-driving, including the social behaviour of drinking, the subjective rationalisations of drinkers, and the punitive and preventative measures of the Police, can be seen as part of this system. As all these aspects interact, we can observe 'emergent properties' of the system.

An emergent property is one that results from the interaction of a system as a whole rather than from one or two of its parts in isolation. This idea is alien to reductionist analysis, and yet it is essential to our understanding of so many phenomena. Again the drink-driving issue will illustrate. Emergent properties include deaths and injuries on the road; confrontations between motorists and the Police which arise when drunk drivers feel that they are being 'unfairly' targeted; and public outrage when the most callous instances of manslaughter through drink-driving gain national publicity.

Moving away from drink-driving, we also see that the road transport system has its own more general emergent properties too, such as the movement of people and goods from one place to another (enabling all kinds of activities that would otherwise be impossible); pollution; and a reduction in the measured intelligence of children living with high levels of lead emissions. Each of these properties of the road transport system comes about precisely because it functions as a system—as an organised whole. While there are also many elements acting against the smooth functioning of the system, these do not ultimately prevent it from 'working', in terms of giving rise to its emergent properties. However, the question always remains, is the balance between the desirable and undesirable emergent properties acceptable, and who should make this judgement?

To eliminate an undesirable emergent property, there is often a need to look at the issues holistically if unanticipated side-effects are not to occur. For example, in the UK, reductionist analyses of the role of seatbelts in preventing accidents led to the introduction of a law making the wearing of seatbelts compulsory. Surprisingly, this did not have the anticipated effect of reducing the number of deaths and injuries on the roads because drivers felt safer and therefore took greater risks! Had a more systemic analysis been undertaken, this attitude to risk might have been accounted for in policy formulation.

Even when it appears that there is a key part of a system implicated in creating an undesirable emergent property, and that removal of this part will eliminate the problem, it is often necessary to undertake a systemic analysis. An example is the elimination of lead from petrol: although lead was identified as the culprit in affecting the measured intelligence of children, in order to introduce unleaded fuel, changes had to occur in the wider system. Campaign groups needed to work with the media to raise the issue in the public consciousness; the public had to put pressure on government; governments made it clear to the motor industry that change was afoot; and the oil companies needed to act in partnership with car manufacturers to enable co-ordinated action to be taken so that all new cars were able to use the unleaded fuel.

The concept of 'emergent property' is essential to systems thinking, so let me give a couple more specific examples to make sure its meaning is clear. Life, for instance, can be seen as an emergent property of organisms as whole systems and cannot be explained by the independent functioning of their organs (von Bertalanffy, 1968). Or, to return to the theme of transport, a speed of 100 k/h can be viewed as an emergent property of a car with a person behind the wheel (which can be described as a system): the person could not attain that speed without the car, and the car could not do it without the driver. Only the car and driver as a system could produce the emergent property of that speed. Of course, it could be argued that the road, fuel, garages, etc., should also be included because without those elements the system would not have the emergent property. This takes us back to the earlier issue of boundaries, but it merely reinforces the point that reductionist analyses of uni-directional cause and effect cannot explain emergent properties: only allowing for complex, multi-directional interactions, and seeing things as systems, can do that (acknowledging, of course, that what constitutes a 'whole' system is dependent on a boundary judgement—there is no such thing as a complete whole).[28]

[28] Another 'take' on this issue is offered by Fodor (1974) who points to the patent absurdity of describing all phenomena in terms of their smallest possible parts. For example, can the success or failure of a business organisation really be described meaningfully in terms of the laws of physics? Fodor looks at this as a linguistic problem: a particular language has evolved to discuss physics, and another language to discuss organisational dynamics. To try to explain organisational dynamics using the language of physics would present the speaker with an impossibly complex task. Fodor argues that different languages necessarily relate to different hierarchical levels of analysis, making reductionism inherently problematic.

So it would appear, at first sight, that reductionism—or at least the use of reductionist analyses to the exclusion of all else—is indeed the enemy of systems thinking. However, I wish to argue that reductionism is actually a relatively superficial manifestation of a deeper problem: subject/object dualism.

The term 'subject/object dualism' refers to the separation of the observer (subject) and observed (object). In a dualist perspective the observer is somehow independent of the observed, standing outside it, so s/he does not influence it in any way. It is only if the observer can be said to be independent from the observed that it is possible to claim pure objectivity. Once the observer influences the observed, then observation comes to be a property of both, and any pretence to absolute objectivity disappears.

The birth of subject/object dualism is commonly attributed to Descartes (1642, 1644), but my own reading of Descartes does not support the very simplistic separation of mind and matter that some people claim he proposed (see also Rothschuh, 1973): Descartes suggests that the 'soul' and the 'material world' have a separate existence, but the soul impacts upon the material world through the human will. However, rather than side-track into the subtleties of interpreting Cartesian philosophy, it is sufficient for my purposes to note that a great deal of scientific activity, from the time of Descartes onwards, has been characterised by a naive objectivism which does indeed assume the independence of the observer and observed. It has been widely accepted that, as long as proper controls are exercised to ensure that observations are not 'contaminated' by the activities of the observer, then these observations can be regarded as objective. In this chapter, we will concern ourselves with what I will call 'naive dualism' (the radical separation of observer and observed) rather than more refined versions[29], as it is the former that has really been seen as problematic by systems (and other) thinkers throughout the 20th Century.

So let us examine how reductionism assumes subject/object dualism. If one takes an anti-reductionist perspective, saying that everything can be seen as interacting with everything else (and boundaries are constructs allowing the inclusion and exclusion of elements in analyses rather than being real markers of systemic closure), then truly independent observation is simply impossible. The observer will always be connected with the observed in some way, however indirectly. Only if one accepts the radical separation of observer and

[29] The one proposed by Descartes (1642, 1644) may be seen as a more refined version of dualism, if we avoid caricaturing it.

observed is it possible to say that there is such a thing as pure objectivity. Because reductionist analyses are looking for *objective*, uni-directional, causal relationships (see my earlier definition), they must indeed assume a naive dualist philosophy.

Therefore, rather than tackle reductionism in this chapter, let us look at how systems thinkers have sought to overcome naive subject/object dualism. If we can find a coherent, alternative position that is worth committing ourselves to, then this should enable us to reframe the debate on reductionism.

3.4 The Struggle against Subject/Object Dualism

Of course naive subject/object dualism has been widely challenged—and not just by those who call themselves systems thinkers. One of the most famous challenges comes from the discipline of physics. Einstein (1934) claims that:

> "The belief in an external world independent of the perceiving subject is the basis of all natural science. Since, however, sense perception only gives information of this external world or of 'physical reality' indirectly, we can only grasp the latter by speculative means. It follows from this that our notions of physical reality can never be final. We must always be ready to change these notions—that is to say, the axiomatic sub-structure of physics—in order to do justice to perceived facts in the most logically perfect way".

For Einstein, then, our inability to know the world 'as it really is' means that human "speculation" has to be an integral part of physics. Once proposed, these ideas took root in physics through the development of quantum theory, which challenges the conventional separation of the observer from the observed by demonstrating that the former cannot help but influence the latter (Bohr, 1963; Bohm, 1980). Indeed, quantum theory proposes the existence of sub-atomic particles that are not directly observable at all, so these propositions must be based on something in addition to empirical evidence—metaphysics (the non-empirical realm of ideas). Thus, the scientific orthodoxy identified by Einstein (1934), that "the belief in an external world independent of the perceiving subject is the basis of all natural science", came to be overthrown. The worlds of physical and metaphysical reality were seen not only to meet, but to be inseparable (Prigogine, 1989).

However, the challenge to naive subject/object dualism not only has a scientific basis but also a moral one. Once it is accepted that the observer is implicated in constructing whatever is observed, then it can be asked, what motivates the observer to be looking at one thing rather than another? In the realm of the natural sciences, the moral dimension to observation is not always so immediately discernible, but in the social sciences the question of what is being observed (and particularly what is being included in, or excluded from, the boundaries of analysis) is quite often explicitly moral. An example is an analysis of the economics of logging a stretch of rain forest, which may generate income for the logging company, provide salaries for its employees, and cheap wood for the manufacture of products to be exported to the other side of the world. However, such an analysis, bounded solely in terms of economic considerations, may be looked upon as immoral by tribal people who are displaced from their ancestral lands, and by conservationists concerned with the preservation of species diversity.

The problem is that subject/object dualism is so ingrained in Western thought that it is very difficult to even identify in some instances, let alone challenge. However, the prize for doing so is great: rooting out naive subject/object dualism will strengthen the critique of so-called value-free science (this critique has been gathering momentum for over one hundred years) so that the values flowing into observations can be made more visible. Ultimately, I believe that full acceptance of value-*full* science will take us beyond mere observation to an understanding that science, and indeed all activities which shape knowledge, is primarily concerned with *intervention*, not observation (but see Chapter 6 for an extended argument).

So let us see what systems thinkers of various persuasions have had to say about subject/object dualism. In all, I will review the work of four authors:

- Ludwig von Bertalanffy on general systems theory;
- Gregory Bateson on the theory of mind;
- Humberto Maturana on the theory of autopoiesis; and
- Ramsés Fuenmayor on interpretive systemology.

These are by no means the only systems thinkers who have addressed the problem of naive subject/object dualism, but I regard them as key authors, in the sense that they have all written extensively on the subject and have developed subtle and influential positions. After the work of each author has been presented, I will highlight problems and/or issues that they have not addressed. Then, in Chapter 4, I will propose an alternative systems philosophy that, in my view, moves beyond subject/object dualism and also deals with the issues passed over by the authors reviewed in this chapter.

3.5 General Systems Theory

Ludwig von Bertalanffy is widely credited with founding general systems theory in the mid-20th Century, despite the previous writings of Bogdanov (1913-1917)[30] arguing for a broadly similar position. However, rather than get into the argument about who *really* originated general systems theory (see Gorelik, 1987, and Dudley, 1996, for this), it should suffice to acknowledge that von Bertalanffy *popularised* the ideas in the English-speaking world, and his book, *General Systems Theory* (published in 1968), is widely regarded as a classic systems text (even by those who disagree with it and operate from different paradigms).

Von Bertalanffy (1968) makes several references to subject/object dualism: in particular, his interpretation of the Cartesian view that the material world, including living beings, operates as if it were a giant machine (with the 'soul' excluded, observing from afar). An organism, then, is analogous to a clockwork toy. Von Bertalanffy (1968) has serious reservations about this mechanistic view and asks what, if this is the case, "is the difference between a normal, a sick and a dead organism?" (p.146) His answer is that

> "....the difference is not definable on the basis of so-called mechanistic theory.... the laws of physics do not tell a difference, they are not interested in whether dogs are alive or dead. This remains the same even if we take into consideration the latest results of molecular biology. One DNA molecule, protein, enzyme or hormonal process is as good as another; each is determined by physical and chemical laws, none is better, healthier or more normal than the other.... Nevertheless, there is a fundamental difference between a live and a dead organism.... In a living being innumerable chemical and physical processes are so 'ordered' as to allow the living system to persist, to grow, to develop, to reproduce, etc." (von Bertalanffy, 1968, pp.146-147).

The most important aspect of von Bertalanffy's answer to mechanism (and the naive subject/object dualism it assumes) is contained in the final sentence in the above quotation: it is the 'ordering' of living systems that is important. A large chunk of von Bertalanffy's book, *General Systems Theory,* is dedicated to explaining what this 'ordering' is. The key to it is the idea of an *open system* (a concept also explored by a number of his contemporaries, such as

---

[30] These dates are taken from the 1996 edition of Bogdanov's *Tektology,* edited by Peter Dudley and published by the Centre for Systems Studies Press. However, in another work, Dudley (1996) acknowledges that there is some controversy surrounding when Bogdanov's writings first became public, and he suggests that the *Tektology* was first published in Russian in 1912 and then in German in 1913.

Koehler, 1938, and Kremyanskiy, 1958). Simply put, an open system is one which assimilates inputs from its environment, and excretes waste products, in order to maintain its identity. According to von Bertalanffy, all living organisms (as well as a variety of non-living entities) can be described as open systems. Whereas, in most of the Universe, the second law of thermodynamics applies (the tendency for energy to distribute itself evenly, producing increasing levels of disorder), in open systems both energy and order are maintained (see also Prigogine, 1947; Prigogine and Stengers, 1984). Furthermore, open systems are teleological: they are purposive, adaptive and/or goal-directed (von Bertalanffy, 1968; Sommerhoff, 1969).[31]

It is in this final characteristic of open systems, teleology, that the difference between von Bertalanffy's ontology and Cartesian dualism really becomes apparent. For von Bertalanffy, intention (or 'will', as Descartes would describe it) does not originate from outside the material world. It is an intrinsic part of it. Open systems (including living organisms like human beings) are *self*-organising, and are therefore non-mechanistic.

3.5.1 Critique of General Systems Theory

The idea of self-organisation now has widespread support amongst scientists (even if this is not the case for general systems theory as a whole), but we have to ask ourselves just how far von Bertalanffy took this analysis. In my view, he did not take it far enough to complete the challenge to subject/object dualism. This becomes evident when we look at von Bertalanffy's thoughts on the nature of human knowledge about the world. First of all, he makes the following claim:

> "Thus 'I' and 'the world', 'mind' and 'matter', or Descartes' '*res cogitans*' and '*res extensa*' are not a simple datum and primordial antithesis. They are the final outcome of a long process in biological evolution, mental development of the child, and cultural and linguistic history, wherein the perceiver is not simply a receptor of stimuli but in a very real sense *creates* his world..." (von Bertalanffy, 1968, p.224, emphases in the original).

While this suggests that the perceiver is actively engaged in creating reality, is it really a shift from the idea of objective knowledge (independent from the observer)? As I see it, the answer is 'no', as revealed in the following:

[31] It would be possible to give a great deal of detail about the characteristics of open systems, but for the purposes of this discussion of systems philosophy I believe that what I have provided is sufficient.

> "Our perception is essentially determined by our specifically human, psychophysical organization.... Linguistic, and cultural categories in general, will not change the potentialities of sensory experience. They will, however, change apperception, i.e., which features of experienced reality are focused and emphasised, and which are underplayed.... Suppose a histological preparation is studied under the microscope. Any observer, if he is not colour-blind, will perceive the same picture, various shapes and colours, etc., as given by the application of histological stains. However, what he actually sees, i.e., what is his apperception (and what he is able to communicate), depends widely on whether he is an untrained or trained observer" (von Bertalanffy, 1968, pp.248-249).

This is essentially the same position adopted by Popper (1959): while we cannot have absolute knowledge of reality, our observations will, to all intents and purposes, be identical—provided that human 'interference' can be controlled. If we take a leap of faith and assume that our observations reflect a real world, more or less imperfectly depending on our success at controlling human interference, then we have a version of realist philosophy that is quite close to the Cartesian thinking that von Bertalanffy criticises. For sure, it is not an absolutely *naive* realism because it does accept that human beings play a part in constructing different views of reality. Nevertheless, this part can be controlled through the use of appropriate methods. Thereby, independent observation is achieved, separating the observer and observed in an *almost* Cartesian manner. We see that a form of subject/object dualism creeps back in, albeit a form that positions the observer within the world rather than outside it.

We can also ask what else is missing from von Bertalanffy's attack on subject/object dualism. One obvious omission is a theory of language. Of course, general systems theory is a child of its time: when von Bertalanffy was writing, the view that a theory of language may help us escape subject/object dualism had not yet seeped out from the discipline of philosophy into the sciences (except, perhaps, into psychology). Wittgenstein's (1953) views, in particular, were highly influential in the latter half of the 20th Century—mostly after von Bertalanffy had stopped working. In essence (and at the risk of over-simplification), Wittgenstein argues that language provides a set of socially shared categories with which to think and communicate, and all theories of both mind and matter are inevitably expressed in language. Indeed, it is not even possible to have the concepts of 'mind' and 'matter' without language. This effectively dissolves Cartesian dualism: the content of *both* mind and matter are constructed through language. Some authors also claim that language is self-referential—words refer to other words rather than an external 'reality' or even 'consciousness'—so it is impossible to grant ontological status to anything except language (for various arguments surrounding this view, see von Foerster, 1984; Gergen, 1991; Shotter, 1993; Simons and Billig, 1994; and Harré and Gillett, 1994).[32]

3.6 *The Theory of Mind*

Let us now move on to look at the work of Gregory Bateson (1972, 1979), who is widely regarded as a key figure in cybernetics. Although Norbert Wiener is usually cited as the 'founder' of cybernetics, Bateson was part of a group of thinkers (also including Wiener, 1948, and Ashby, 1956) who were highly influential in developing cybernetic ideas and bringing them into a range of pure and applied disciplines during the mid-to-late 20th Century. The key idea in cybernetics is *feedback*: it is possible to identify causal 'loops' where a system makes a change in its behaviour and receives information back from its environment about the effects of this behaviour, which is then used to determine future actions.

A simple example is a thermostat which controls a radiator. If the room is cooler than the setting on the thermostat, the thermostat triggers the radiator into action. The room then warms up, and all the while information about the temperature is fed back to the thermostat. When the room heats up to the temperature at which the thermostat is set, the thermostat triggers the radiator to switch off. This then has a further effect on the temperature of the room, and information about the lowering temperature is fed back to the thermostat, at some point causing it to switch the radiator on again. Theoretically, this can go on indefinitely. While this is a simple mechanical feedback loop, the same basic principle is said to operate in all kinds of natural and social systems.

Gregory Bateson saw the cybernetic idea as providing the basis for a fundamental challenge to subject/object dualism, and set out to advance a novel *theory of mind* in which a 'mind' is said to extend beyond the boundaries of the brain and body to take in multiple feedback loops linking organisms with the rest of the natural world. Now, as with open systems which were discussed under the heading 'general systems theory' (above), I will not go into great detail about the theory of mind, but will instead present its basic principles. This will enable us to see how Bateson sought to transcend naive subject/object dualism and establish a more holistic philosophy.

[32] However, despite the importance of Wittgenstein's insight, I argue in Chapter 4 that his linguistic philosophy just complicates the picture: it does not really help us escape subject/object dualism.

Although Bateson wrote a great deal about his theory of mind over the years, throughout this section I will draw upon quotations from just one source: the text of a talk delivered in 1970. I have focused upon this text because it provides such a clear exposition of the theory of mind, profusely illustrated with anecdotal examples.

A key concept in Bateson's theory is "difference": a difference is a demarcation of one thing from another. This therefore has a similar (but not identical) meaning to Spencer Brown's (1972) idea of boundary (reviewed earlier in this chapter). However, while Spencer Brown chose not to specify the origins of boundaries (i.e., whether they are the product of perception; embodied in language; occur in the natural world; etc.), Bateson has some very specific things to say about where differences come from:

> "A difference is a very peculiar and obscure concept. It is certainly not a thing or event. This piece of paper is different from the wood of this lectern. There are many differences between them—of color, texture, shape, etc. But if we start to ask about the localization of these differences, we get into trouble. Obviously the difference between the paper and the wood is not in the paper; it is obviously not in the wood; it is obviously not in the space between them, and it is obviously not in the time between them. (Difference which occurs across time is what we call "change.") A difference, then, is an abstract matter.... Difference travels from the wood and paper into my retina. It then gets picked up and worked on by this fancy piece of computing machinery in my head" (Bateson, 1970, pp.457-458).

A further quotation gives more detail:

> "I suggest to you that the word "idea", in its most elementary sense, is synonymous with "difference." Kant.... argues that in a piece of chalk there are an infinite number of potential facts. The.... piece of chalk can never enter into communication or mental process because of this infinitude. The sensory receptors cannot accept it; they filter it out. What they do is select certain *facts* out of the piece of chalk, which then become, in modern terminology, information. I suggest that Kant's statement can be modified to say that there is an infinite number of *differences* around and within the piece of chalk. There are differences between the chalk and the rest of the Universe, between the chalk and the sun or the moon. And within the piece of chalk, there is for every molecule an infinite number of differences between its location and the locations in which it *might* have been. Of this infinitude, we select a very limited number, which become information. In fact, what we mean by information—the elementary unit of information—is a *difference which makes a difference*" (Bateson, 1970, p.459, emphases in the original).

Both of the above quotations raise a significant issue of relevance to this discussion of systems philosophy in that they *appear* to conflate what is in the mind of the observer with what is in the world:

difference is abstract, yet travels in the real world; it is synonymous with "idea", but potentially infinite differences exist in the piece of chalk itself. However, this is sorted out, and Bateson's challenge to subject/object dualism made clear, with the following statement:

> "There is.... an important contrast between most of the pathways of information inside the body and most of the pathways outside it. The differences between the paper and the wood are first transformed into differences in the propagation of light or sound, and travel in this form to my sensory end organs. The first part of their journey is energized in the ordinary hard-science way from "behind." But when the differences enter my body by triggering an end organ, this type of travel is replaced by travel which is energized at every step by the metabolic energy latent in the protoplasm which *receives* the difference, recreates or transforms it, and passes it on.... Be that as it may, this contrast between internal and external pathways is not absolute. Exceptions occur on both sides of the line.... [Nevertheless], in spite of these exceptions, it is still broadly true that the coding and transmission of differences outside the body is very different from the coding and transmission inside, and this difference must be mentioned because it can lead us into error. We commonly think of the external "physical world" as somehow separate from an internal "mental world." I believe that this division is based on the contrast in coding and transmission inside and outside the body. The mental world—the mind—the world of information processing—is not limited by the skin" (Bateson, 1970, pp.459-460, emphases in the original).

So, Bateson's novel challenge to the dualism of mind and matter is that 'mind' extends throughout matter. It is not localised within organisms alone (which is the conventional place for mind). Therefore, when Bateson talks about difference being "abstract" and synonymous with the concept of "idea" (see earlier), these terms are not being used in their usual sense, referring to mind as *opposed* to matter, but refer to mind that is *immanent in* matter. To understand something more about the nature of the distribution of mind throughout matter, we can return once again to Bateson's original writings:

> "Consider a tree and a man and an axe. We observe that the axe flies through the air and makes certain sorts of gashes in a pre-existing cut in the side of the tree. If now we want to explain this set of phenomena, we shall be concerned with differences in the cut face of the tree, differences in the retina of the man, differences in his central nervous system, differences in his efferent neural messages, differences in the behavior of his muscles, differences in how the axe flies, to the differences which the axe then makes on the face of the tree. Our explanation (for certain purposes) will go round and round that circuit. In principle, if you want to explain or understand anything in human behavior, you are always dealing with total circuits, completed circuits. This is the elementary cybernetic thought. The elementary cybernetic system with its messages in circuit is, in fact, the simplest unit of mind... We get a picture, then, of

> mind as synonymous with cybernetic system—the relevant total information-processing trial-and-error completing unit. And we know that within Mind in the widest sense there will be a hierarchy of sub-systems, any one of which we can call an individual mind" (Bateson, 1970, p.465-466).

3.6.1 Critique of the Theory of Mind

In my view, Bateson is rightly remembered as one of the most creative and challenging thinkers of the 20th Century. His theory of mind is counter-intuitive, as the most influential theories usually are when they are first proposed, but is well supported with a wealth of detail. However, we need to ask how successful Bateson has actually been in challenging subject/object dualism. To begin to answer this question, we can first of all return to the last quotation in which an example of a cybernetic system was provided: a man chopping down a tree with an axe. Once the circular information pathway has been described, Bateson says that "Our explanation (for certain purposes) will go round and round that circuit". The words "for certain purposes" are highly significant here. The role of purpose in cybernetic systems (or minds) is clarified further in the following quotation:

> "Suppose I am a blind man, and I use a stick. I go tap, tap, tap. Where do *I* start? Is my mental system bounded at the handle of the stick? Is it bounded by my skin? Does it start halfway up the stick? Does it start at the tip of the stick? But these are nonsense questions. The stick is a pathway along which transforms of difference are being transmitted. The way to delineate the system is to draw the limiting line in such a way that you do not cut any of these pathways in ways which leave things inexplicable. If what you are trying to explain is a given piece of behavior, such as the locomotion of the blind man, then, for this purpose, you will need the street, the stick, the man; the street, the stick, and so on, round and round. But when the blind man sits down to eat his lunch, his stick and its messages will no longer be relevant—if it is his eating you want to understand" (Bateson, 1970, p.465).

It appears that cybernetic pathways only have meaning *in relation to purposes*. But where are purposes located? This question is not explicitly answered in his 1970 work. However, if we give Bateson the benefit of the doubt, we would have to say that they are located "in the mind"—in the wider, Batesonian sense of the term. But a critic who refuses to give Bateson the benefit of the doubt might point out that the independent observer has crept back in here (in the form of the narrower, more usual conception of mind), allowing subject/object dualism to resurface. My own view is that Bateson is too sophisticated a thinker to have fallen so easily at the first hurdle, and I suspect that, in his 1970 work, he just failed to re-assert his view that purposes have a systemic reality (being explicable in terms of cybernetic feedback loops) rather than being located within a mind in the narrow sense of the term.

Certainly, the feeling I get when reading Bateson's *Steps to an Ecology of Mind* (a major collection of essays), is of a man struggling with the enormous complexity of his subject—a complexity which would slip the grasp of most of us. For example, in his 1970 work, he completely rules out the possibility of taking a realist position (in which knowledge reflects an external reality, albeit imperfectly), yet a year later (in 1971) he talks about the practice of science in a manner that is indeed implicitly realist. Compare the following two quotations:

> "We say the map is different from the territory. But what is the territory? Operationally, someone went out with a retina or a measuring stick and made representations which were then put upon paper. What is on the paper map is a representation of what was in the retinal representation of the man who made the map; and as you push the question back, what you find is an infinite regress, an infinite series of maps. The territory never gets in at all.... Always the process of representation will filter it out so that the mental world is only maps of maps of maps, ad infinitum. All 'phenomena' are literally 'appearances'." (Bateson, 1970, pp.460-461).

> "....'data' are not events or objects but always records or descriptions or memories *of events or objects*. Always there is a transformation or recoding of the raw event which intervenes between the scientist *and his object*.... In a strict sense, therefore, no data are truly 'raw,' and every record has been somehow subjected to editing and transformation either by man or by his instruments" (Bateson, 1971, pp.ix-xx, my emphases).

If you remember, it was von Bertalanffy's (1968) realism that let subject/object dualism slip back in (albeit a different version of dualism than the one he criticised, placing the mind within the material world rather than outside it): von Bertalanffy accepted the possibility of independent observation, thereby allowing a separation of observer and observed. In the first quotation above, Bateson rejects realism, and thereby von Bertalanffy's form of dualism: "The territory never gets in at all". In the second quotation, however, he makes exactly the same assumption as von Bertalanffy: that data are indeed records of actual events or objects, albeit not perfect reflections of them. So, we are left with a rather equivocal picture. Bateson can either be aligned with von Bertalanffy in accepting a biologically-situated, subject/object dualism, or he can be seen as taking an entirely new line.

I have to say that, when reading through Bateson's work, I find more evidence of subject/object dualism than otherwise. Arguably the most explicit evidence comes in the form of his view of the relationship of consciousness to the physical world. Witness the following two quotations:

> "If, as we must believe, the total mind is an integrated network...., and if the content of consciousness is only a sampling of different parts and localities in this network; then, inevitably, the conscious view of the network as a whole is a monstrous denial of the *integration* of that whole.... What the unaided consciousness can never appreciate is the systemic nature of mind" (Bateson, 1967, p.145, emphasis in the original).

> "....if consciousness deals only with a skewed sample of the events of the total mind, then there must exist a *systematic* (*i.e.*, nonrandom) difference between the conscious views of self and the world, and the true nature of self and the world" (Bateson, 1968, p.450, emphases in the original).

Bateson's dualism is not between mind and matter, but between *consciousness* and matter. Mind comes to be distributed throughout matter, but consciousness is variably aware of this reality depending on the aids it has available to it. Here we also see the same realism embraced by von Bertalanffy (the possibility of at least an approximation to independent observation): according to Bateson (1967), art can be an aid to systemic awareness—as, presumably, can the language of cybernetics and systems—bringing us closer to appreciating external reality than we might otherwise be. Finally, despite some equivocation and the possibility of alternative interpretations, I have to conclude that Bateson, like von Bertalanffy, does indeed admit a biologically-situated, subject/object dualism.

One final point should be made before concluding this discussion of the work of Bateson. Unlike von Bertalanffy, Bateson does actually consider the issue of language, and makes explicit reference to Wittgenstein (1953) amongst others. However, he is more interested in *using* Wittgenstein's insights to inform a discussion of familial language games than in exploring the implications of these insights for his own ontological position (see Bateson, 1955, for a good example). Had he explored the ontological implications of language, who knows how his thinking might have been transformed.

3.7 The Theory of Autopoiesis

Having discussed the attacks on naive subject/object dualism from Ludwig von Bertalanffy and Gregory Bateson, we can now move on to our third author, Humberto Maturana, who takes quite a different approach from the last two. Maturana's best known work, proposing the *theory of autopoiesis*, was co-authored with Francisco Varela (Maturana and Varela, 1992), but he has also produced extensive, sole-authored writings. In my view, two papers of particular note are Maturana (1988a,b). It is Maturana (1988a,b) and Maturana and Varela (1992) that I shall refer to in my review, but the interested reader may like to search the literature for other contributions. Some of the secondary literature on autopoiesis is also valuable, particularly Mingers (1995) who, in my opinion, has the capacity to present some of Maturana's more difficult-to-follow ideas in a clear and concise manner without losing the subtlety of the arguments.

Maturana (1988a,b) and Maturana and Varela (1992) contrast "objectivity in parentheses" with "objectivity without parentheses". Objectivity without parentheses is the view that independent observation is possible: that 'pure' objectivity, or universal knowledge, can be attained. Taking as his starting point for analysis the lived experience of an observer, Maturana (1988a) argues that this kind of objectivity is impossible: all knowledge is known from a particular standpoint by human beings. Therefore, there is a need to place the word 'objectivity' in parentheses. Maturana acknowledges that we cannot stop using a language of objects, but when we theorise as scientists we can ensure that we do not maintain the illusion that the objects we talk about have a reality independent from (i) the observer, and (ii) the language with which observers co-ordinate their actions and create consensual domains. Here, then, we find a clear challenge to naive subject/object dualism: in the view of Maturana (1988a,b) and Maturana and Varela (1992), there is no 'external' reality that the observer observes. Rather, there are multiple realit*ies* (one for each observer) that have overlapping content because of the use of shared language.

Having asserted that pure objectivity is impossible, Maturana (1988a) argues that his task as a scientist is to provide a biological explanation for why this is the case. The theory of autopoiesis enables this explanation. Essentially, the term 'autopoiesis' means *self-producing*. An autopoietic system is one which acts to maintain its internal organisation and, when it interacts with its environment to maintain itself, the actions it takes are determined by its current structure (Maturana and Varela, 1992). The *structure* of a system is its arrangement of components in such a way that its *organisation* (that which gives it identity) is maintained. The structure of a system changes over time, but within limits laid down by its organisation (which cannot change without the system losing its identity as a self-producing entity—in other words, without it dying). The implications of this are profound, especially the observation that interactions with the environment are determined by a system's structure. While it is common to hear talk about people or organisms being *caused* to act in particular ways, Maturana and Varela (1992) say that the environment cannot be a cause, only a *trigger*. The environment produces perturbations that may or may not be received as meaningful information by the organism. If the perturbation is meaningful, it is because the internal structure of the organism allows it to be received as such. Even if the perturbation is life-threatening (if it will disrupt the organisation of the system), the organism will not be able to react unless its internal structure allows it to be receptive to the perturbation—i.e., if it is meaningful to the organism.

The challenge to naive subject/object dualism is therefore somewhat similar to that proposed by von Bertalanffy, in that observations necessarily result from the particular viewpoints of organisms (rather than originating from outside the material world), but Maturana's theory of autopoiesis differs in one crucial respect from von Bertalanffy's theory of open systems: Maturana explicitly considers the role of language. As social animals, human beings do not only *act*, we also strive to *co-ordinate* our actions. Language helps in this process: it allows us to *co-ordinate our co-ordinations of actions*. The 'co-ordination of co-ordinations of actions' is a rather obscure phrase, but it summarises Maturana's position quite neatly: we act in co-ordination with others, and language supports the co-ordination of these co-ordinations.

Interestingly, when Maturana talks about co-ordinations of actions, he has something very specific in mind. While an organism can only react to outside forces on the basis of its current structure (maintained by, and maintaining, its organisation, or identity as a system), it is organisationally predisposed to identify recurrent patterns of interaction and adapt its structure accordingly, thereby giving rise to habitual responses. When an organism and an aspect of its environment (which may or may not be another organism) have a recurrent relationship, sufficient adaptations occur, and sufficient habitual responses are set up, to allow us to describe the relationship between the organism and the aspect of its environment as *structurally coupled*. Structural coupling, when taking place amongst a group of organisms, allows the working out of co-ordinations of actions in ways that are of mutual benefit to all those concerned. Of course, language may facilitate and strengthen this process.

Language is socially shared only in as much as each individual who participates in its use implicitly understands the relationship of language with the co-ordinations of his or her own actions and those of others. When the use of language gives an unexpected result in terms of a person's perception of the actions of others, it is evidence that the use of language was inappropriate for that event of social co-ordination. This is an unusual understanding of language in two respects: (i) language does not describe a 'real world' external to subjective realities, but only the co-ordination of actions; and (ii) it can never be taken for granted that words mean exactly the same to all people (they are always dependent for meaning on their use by acting subjects appreciating a local context).

Language also forms "rational domains" in which people participate. Over time, a particular use of language to co-ordinate co-ordinations of actions may become more and more elaborated, allowing people to exist in very subtle, well co-ordinated, structurally coupled relationships. Thereby, whole human activity systems are created. People may actually participate in a variety of human activity systems, but the movement of individuals from one to another—and hence from the use of one form of language to another—crucially depends on the invocation of emotion. According to Maturana (1988b), emotions make individuals switch from one 'rationality' to another. All rational arguments are "braided" with emotion (in other words, forms of language come to be associated with emotional states within individuals), so when a particular emotion is experienced, this triggers a switch to the appropriate, associated rational domain (or elaborated system of language). This is why an appeal to the emotions can have such a powerful effect in terms of changing people's ways of thinking (Bilson, 1996, 1997). Of course, this presumes that the relationship between rational domains (forms of language) and emotion is two-way: the use of a particular language game associated with an emotion will give rise to that emotion, altering the set of rational domains that become available to participating individuals at that moment.

Now, talk of 'language games' may remind the reader of Wittgenstein's (1953) radical challenge to Cartesian dualism: the idea that the content of both mind and matter are constructed through language. Like Maturana, Wittgenstein also proposes that we move between a variety of 'language games', but these can be mediated by (translated through) 'everyday' language (the language we understand implicitly from childhood). Unlike Wittgenstein, however, Maturana does not take language as having ontological primacy: he sees the relationship between language and organisms (autopoietic, biological entities) as essentially *co-constructive*:

> "....since we exist in language, the domains of discourse that we generate become part of our domain of existence and constitute part of the environment in which we conserve identity and adaptation.... We humans, as humans, exist in the network of structural couplings that we continually weave through the permanent linguistic trophallaxis of our behavior. Language was never invented by anyone only to take in an outside world. Therefore, it cannot be

> used as a tool to reveal that world. Rather, it is by languaging that the act of knowing, in the behavioral co-ordination which is language, brings forth a world. We work out our lives in a mutual linguistic coupling, not because language permits us to reveal ourselves but because we are constituted in language in a continuous becoming that we bring forth with others. We find ourselves in this co-ontogenic coupling, not as a pre-existing reference nor in reference to an origin, but as an ongoing transformation in the becoming of the linguistic world that we build with other human beings" (Maturana and Varela, 1992, pp.234-235).

Human beings, then, are self-producing organisms which co-construct their realities through language. The biological and linguistic levels interact in a dynamic tension: people co-ordinate their co-ordinations through language, and their very identities are framed in language, but their *essence* (if I may be permitted to use this term, accepting that Maturana would not see essences as external realities) is still biological. Fundamentally, language has a *biological* explanation (Maturana, 1988a). Maturana's challenge to naive subject/object dualism is therefore similar to both von Bertalanffy's and Bateson's, in that the observer is placed *in* the material world (rather than outside it), but for Maturana observation cannot be independent of the organisation and structure of an observer or the forms of language s/he uses to frame the observations that s/he makes.

3.7.1 Critique of the Theory of Autopoiesis

Maturana's theory of autopoiesis pays extraordinary attention to detail in developing a new language to explain the nature of life, which one would expect of a theory that recognises the importance of language to *human* life in particular. Nevertheless, I contend that there remains an ambivalence with regard to the fundamental ontological issue the theory addresses: whether there is *a* reality, or multiple subjective realit*ies*. Certainly, in some of his writings, Maturana appears unequivocal:

> "....outside language nothing (no thing) exists because existence is bound to our distinctions in language.... I am saying that all phenomena.... are cognitive phenomena that arise in observing as the observer operates in language.... Nothing precedes its distinction; existence in any domain, even the existence of the observer themselves, is constituted in the existence of the observer" (Maturana, 1988b, p.79)

If "all phenomena.... are cognitive phenomena" then this certainly indicates the existence of multiple subjective realit*ies*. However, we then have to ask, what is the status of Maturana's biological theory? Is it, as Bilson (1996) believes, no more than Maturana's personal preference of a rational domain which makes emotional sense to him (which it would ultimately have to be for Maturana to be consistent)? Maturana's (1988a) answer to this question is that the theory of autopoiesis is special in one sense: it is *scientific*.

Maturana (1988a) is at great pains to describe what it means to him to generate a scientific theory: it must explain specific phenomena while satisfying some minimum criteria that allow a theory to be called 'scientific' (see Maturana, 1988a, for details of these criteria). He is explicit that, in seeking to *explain* something, a theory does not assume reference to an external reality: it merely requires a reference point in the language games of a consensual community of actors. In other words, a scientific theory can be called 'scientific' because it conforms to certain criteria held to be important to a community of scientists, and it is *meaningful* because it fits into a rational domain of other scientific explanations.

This answer is consistent with Maturana's ontological position already detailed, but a number of authors have questioned its credibility. In particular, Mingers (1992a, 1995) challenges the motivation for producing 'scientific' theories if they really have no meaning beyond the consensual domain of a few scientists. Indeed, he points out the significant similarities between the philosophy of science advanced by Maturana and that proposed by Bhaskar (1975, 1986)—except that the latter does accept that language has a reference point in external reality (even if we cannot know the exact nature of this reference in any particular case). In Mingers's view, there is little point in producing a scientific theory if it can have no external reference: scientific theories can have no advantages over non-scientific theories in terms of co-ordinating co-ordinations of actions if neither contain references to external events.

Nevertheless, whether or not one accepts Mingers's critique, one cannot avoid the irony in Maturana's logic: we explain the phenomenon of multiple realities (pluralism) by the use of a single (unitary) biological theory. In the very act of proposing a unitary theory, it is hard to escape the conclusion that Maturana *wishes* it to be more than just his personal preference (shared by others who engage with the rational domain he brings forth)—otherwise, why not take the phenomenon of pluralism as his starting point? Why seek a unitary explanation at all?

In conclusion, I suggest that Maturana's challenge to naive subject/object dualism is successful to a point, but is ultimately made problematic by the proposal of a unitary biological theory that, *by virtue of being unitary*, presupposes a degree of independent observation. Maturana (1988a) strives to square the circle, arguing that the theory is only meaningful in relation to a consensual scientific domain (and therefore does not make external references), but in my view this makes it hard to justify the effort of creating the theory in the first place: within a group of people who accept the existence of a phenomenon (say, pluralism of viewpoints), isn't this acceptance sufficient to co-ordinate the co-ordinations of actions? A theory to *explain* the phenomenon is only needed if there is something more at stake.

Finally, the pre-supposition of independent observation can be witnessed in the criteria that Maturana (1988a) proposes for an explanation to be thought of as scientific: he explicitly talks about phenomena needing to be apparent to a "standard observer" (p.7)—but, if observation is constructed by individual subjects operating through the use of language, how can the concept of a "standard observer" be justified? There can be no *standard* observers—only *particular* observers whose structure at any point in time enables the use of language to give rise to observations. I suggest that a "standard observer" is an independent observer by another name, and therefore subject/object dualism (albeit a less naive form than the one systems thinkers are wont to criticise) creeps back into Maturana's Universe under a new guise.

3.8 Interpretive Systemology

Let us now move on to discuss the work of the last of the four systems theorists whose challenges to naive subject/object dualism are being dealt with in this chapter—Ramsés Fuenmayor (1991a-c). The work reviewed below provides the philosophical basis for a systems approach called Interpretive Systemology which has been extensively detailed in two journal special issues (volume 4, part 5, of *Systems Practice*; and volume 12, part 1, of *Systemic Practice and Action Research*). However, Interpretive Systemology as a whole will not be discussed here. I recommend consulting these two special issues if you are interested in moving beyond the philosophy to the methodology and practice of Interpretive Systemology.

Fuenmayor (1991a) seeks to identify what is the essential nature of systems thinking. His initial answer is that all concepts, systems thinking included, are defined in the context of that which they are not. Therefore, systems thinking is defined against reductionism. In a similar manner to myself, Fuenmayor (1991a) identifies the connection between reductionism and subject/object dualism: both the Cartesian dualism of "mind" and "matter", and the Aristotelian dualism of "appearances" and "beings" that preceded the thinking of Descartes in Western philosophy. Also important in the philosophy that gives rise to reductionism is the principle of non-contradiction:

> "The principle of non-contradiction, as announced by Aristotle, states that 'The *same thing cannot* at one and the *same time be* and *not be'*.... There are two meanings usually attached to this saying:
>
> (1) Either A is or A is not.
>
> (2) Either A is B or A is not B.
>
> The first case refers to the whole being of A.... The second case refers to a particular determination of 'A.' 'A' cannot have and not have the same determination ('B') at the same time (A cannot be green and not green at the same time). ....the principle of non-contradiction can also be stated in a positive way in the form of the principle of identity (A=A). The latter says that A is identical to A or that A is the same with itself" (Fuenmayor, 1991a, pp.436-437, emphases in the original).

The principle of non-contradiction might be called common-sense logic. It simply asserts that something cannot be and not be at the same time. However, it is this supposedly obvious principle that Fuenmayor challenges in the following manner:

> "Epimenides, a Cretan, claims 'All Cretans are always liars.' Is this a lie?.... If such a proposition is true, then it is false, and vice versa. Another form of the self-referential paradox.... is the following:
>
> Let P1 and P2 be two propositions.
>
> P1: 'P2 is true.'
>
> P2: 'P1 is false.'
>
> Here P1 *is* true if and only if it is not true. The same can be said about P2. The apparent violation of the principle of non-contradiction is plain" (Fuenmayor, 1991a, p.444, emphasis in the original).

It is interesting to note that the principle of non-contradiction has been challenged in a similar manner in physics (Gödel, 1931; Bohr, 1963), mathematics (Spencer Brown, 1972) and philosophy (Hofstadter, 1979). Although apparently common sense, it is a principle that does not stand up to much detailed analysis.

Fuenmayor (1991a) then goes on to propose that the fundamental, ontological concept that can be used to underpin systems thinking is neither "matter" nor "mind"; "being" nor "appearances"—it is a unified form that shows the two poles in an *essentially recursive relationship*. Matter gives rise to mind, which gives rise to matter, which gives rise to mind, etc. Neither is prime, nor for that matter really meaningful without an understanding of the other.

Recursive forms *appear* to be contradictory, but in Fuenmayor's view they represent the essential observation that no concept (including

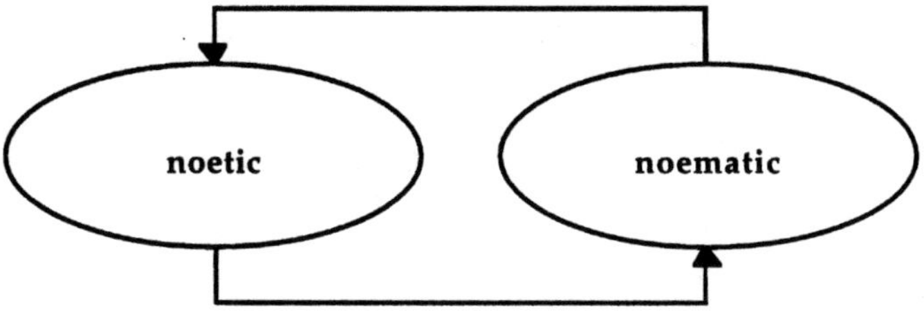

FIGURE 3.3: Noetic/noematic recursive form

any ontological concept) can be meaningful except in relation to that which it is not. Any attempt to take us beyond recursion to an undifferentiated unity simply hides the context in which this unity is grounded. The fact that essential recursion seems contradictory shows us just how far subject/object dualism (in which the elements which should be in a recursive relationship with one another are viewed as independent, thus preserving so-called 'non-contradiction') has penetrated Western consciousness.

From this philosophical foundation, Fuenmayor (1991b,c) goes on to propose an ontology and epistemology for systems thinking—or, in his own words, an "onto-epistemology", because reality and knowledge are related recursively together. I will give a highly summarised account of Fuenmayor's "onto-epistemology", but I also recommend returning to the original literature for more details as the arguments are far more complex than portrayed here. In places I have slightly altered Fuenmayor's terminology to facilitate the presentation of his ideas in truncated form.

Fuenmayor's (1991b) starting point is the recursive form of "noetic" (subject-side) and "noematic" (object-side) (see Figure 3.3). A situation can be thought of in terms of either side of this recursive form, and indeed a holistic analysis requires both sides to be considered as a recursive unity. Below, the "noetic" and "noematic" sides will be detailed in turn.

First, the "noetic" (subject) side. This is characterised by another recursive form (Figure 3.4) of "self" and "otherness". In any situation, there is the self that perceives the situation always in a relationship with that which is not the self—the "other". A situation only has meaning from the noetic side in terms of the intentionality of the self whose project is to explore the other and transform it into "self-history". If my understanding is correct, this means that the self builds its own history by making distinctions in the vast, uncharted territory of the other, thereby making the other meaningful to itself.

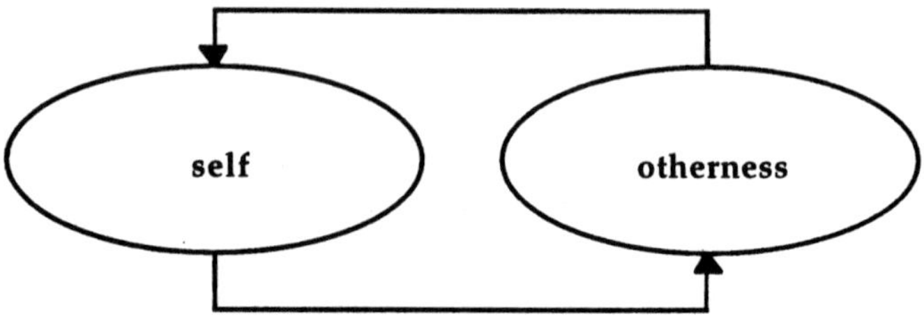

FIGURE 3.4: Self/otherness recursive form underlying the noetic side of Figure 3.3

Self and other are in a recursive (paradoxical) relationship as follows. On the one hand (from the standpoint of the self), the self has limitless possibilities to transform the other into self-history: indeed, it is self-history itself (the previous experiences of the self) that gives rise to these possibilities by providing the ground (categorisation systems) upon which many possible distinctions can be based. On the other hand, from the standpoint of the other, the self only has one possibility to transform the other into self-history: that is, the possibility that is dictated in a deterministic manner by a self-history that has been constructed in a very specific manner and ultimately can only go in one direction. In the first instance there is pure intentionality, and in the second there is none. These, it would seem, are two incommensurable positions. One embraces the idea of free will and the other denies it. However, Fuenmayor (1991b) argues that neither position is tenable on its own. They represent two sides of a recursive form.

Then there is the "noematic" (object) side. This is also characterised by its own recursive form (Figure 3.5) of "distinction" and "scene". On the noematic side there is no 'subject' as such, only distinctions that are made. However, "what has been distinguished has been distinguished from its scene" (Fuenmayor, 1991b, p.464). How exactly the relationship between distinction and scene is recursive is explained by Fuenmayor as follows:

> "What are, in terms of the situation, the distinction and the scene? A first attempt to answer: the distinction is that which has been distinguished from the scene. The scene is that from which the distinction has been distinguished. Such answer shows that the 'whatness' of the distinction is referred to—or, more properly, *transcends* to—the scene and the whatness of the scene moves to the distinction. Scene and distinction.... are, logically speaking, the recursive sides of an essential recursive form...." (Fuenmayor, 1991b, p.466, some emphases removed).

If we are to appreciate the noematic (object) side of understanding

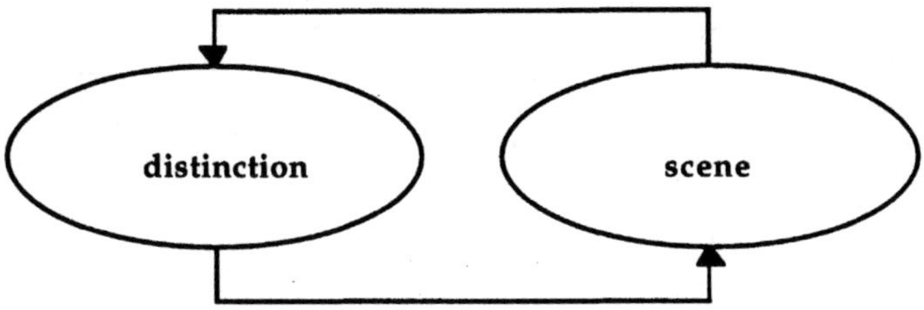

FIGURE 3.5: Distinction/scene recursive form underlying the noematic side of Figure 3.3

situations, it is also important to acknowledge Fuenmayor's (1991b) point that it is possible to identify what has been distinguished, but the "scene" is always illusive. This is because, if any aspect of the scene is identified, it too will have been distinguished and will have moved to the other half of the recursive form "distinction/scene".

Having explored both the "noetic" (subject) and "noematic" (object) sides of understanding what is going on in a situation, Fuenmayor (1991b) returns to the recursive form ("noetic/noematic") that he started with. However, in his conclusion he renames this form "Intentionality/ Distinction" (the capital letters are important, as the term "distinction" was used previously as one part of the noematic recursive form). In Figure 3.6, I have represented the "Intentionality/ Distinction" recursive form along with the other recursive forms (already discussed) that are nested within it.

"Intentionality" is the intentionality of the self that is always distinguished in relation to something else. That "something else" can be seen as both defined by, and defining, the self. "Distinction" is a distinction made in relation to a scene. The scene is not identifiable, but can only be distinguished as either that which is not distinguished or as the general ground that gives rise to a distinction. In Fuenmayor's view, *both* ways of appreciating situations (Intentionality and Distinction) are implied by the other (i.e., they constitute a single, recursive form) and are equally necessary to the construction of holistic understandings.

3.8.1 Critique of Interpretive Systemology

Fuenmayor's approach to the problem of naive subject/object dualism is quite different to that proposed by von Bertalanffy, Bateson and Maturana. Somewhat paradoxically, Fuenmayor believes it is inevitable that situations can be perceived from both a 'subject' and 'object' point of view. Indeed, he says that the two exist in an essentially recursive relationship. Thus, although Fuenmayor would

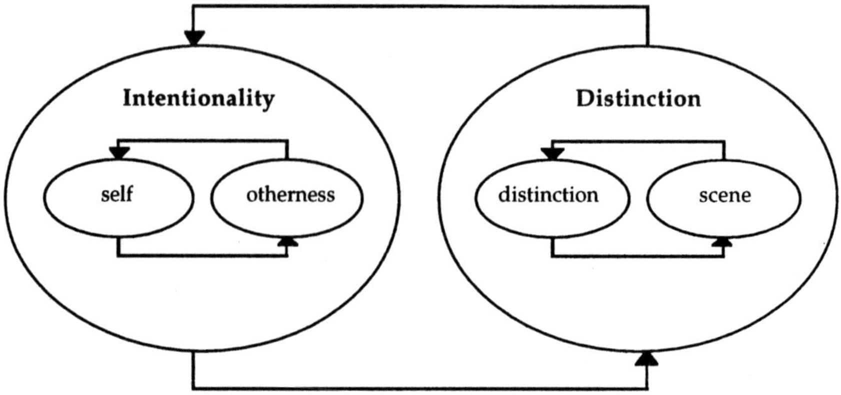

FIGURE 3.6: Intentionality/Distinction recursive form containing other recursive forms

probably disagree with this, I believe it is fair to say that he *embraces* subject/object dualism—he does not really strive to overcome it. However, the kind of subject/object dualism proposed by Fuenmayor is quite different from the naive dualism criticised earlier, which involves the radical separation of the observer from the observed. In Fuenmayor's "onto-epistemology", each side of a recursive form always implies the presence of the other, and an analysis can only be called holistic if both 'subject' and 'object' are explored.

Now, it is important to note Fuenmayor's (1991a) acknowledgement of the influence of Phenomenology (a philosophical movement that emerged in the early-to-mid 20th Century) on his work, particularly the writings of Heidegger (1927) and Merleau-Ponty (1962). The fact that Interpretive Systemology has been so closely identified with Phenomenology has given rise to a critique from Mingers (1992a) who claims that Interpretive Systemology suffers the same ailments as that philosophical movement. One aspect of Mingers's critique is particularly relevant to this discussion of subject/object dualism: his view that, in seeking to challenge naive objectivism, Fuenmayor falls into subjectivism. In other words, the radical separation of subject and object is still present, but instead of claiming pure objectivity Fuenmayor claims pure *subjectivity*. In Mingers' (1992a) words,

> "Phenomenology, by its very nature, is concerned with articulating and expressing the experiencing, the being, of individual people. ....[T]he individual is taken as an already existing and constituted being who is experiencing the world. However, it is argued (by Habermas and Maturana among others) that there is a more fundamental level at which one considers the constitution of human

> subjects.... The distinctive faculty of humans is language, and language is unavoidably shared and therefore social, i.e., intersubjective not individual" (Mingers, 1992a, p.178).

In conversation, Fuenmayor has made it clear to me that he rejects Mingers's critique on the grounds that Interpretive Systemology goes beyond Phenomenology, and the criticisms that can rightly be levelled at Phenomenology are dealt with in Interpretive Systemology. Specifically, Fuenmayor argues that Interpretive Systemology does not root the origin of meaning in the subject, but within the recursive form of "subject/object". In my view this is a reasonable defence, but what is still missing from Fuenmayor's philosophy is any analysis of what gives rise to recursive forms. Mingers (1992a) identifies the importance of language, and it is my contention that "essential recursion" is a feature of the way language operates.

In an earlier work (Midgley, 1992c), I advanced the following argument. Fuenmayor (1991a,b) claims that every concept has an "other" from which it is distinguished. What he fails to acknowledge, however, is that concepts are expressed in language, and the 'otherness' that is present whenever a concept is distinguished is a function of language. 'Black' only has meaning in relation to 'white', 'man' is only meaningful in relation to 'woman', etc. This understanding of language is common to a wide range of writers, from Wittgenstein (1953) to Derrida (1976, 1978) in the discipline of philosophy, and from Bateson (1955) to Maturana (1988a,b) in systems thinking.

It becomes clear that recursion is a function of language when we consider how language can be used. Imagine that you are sitting in an armchair. For a brief period of time you reflect on your own self-understanding, musing that nothing has an existence beyond your own consciousness. However, when somebody calls you from the next room you are shaken out of your reverie and react to their voice without any thought that the voice might just be a product of your mind. Now, if this way of using language (to think about all phenomena as purely subjective) is generalised into a philosophical position (which is often called 'solipsism' in the literature), then all 'truth' quickly becomes '*my* truth'.

However, we can also use language to think in terms of truth without any consideration of the role of the subject, as a scientist might do for a short while when conducting an experiment. However, once the experiment is over, the scientist will move into another context where his or her subjectivity may again come to the fore. Generalised into a philosophical position (which is often called 'naive realism' in the literature), the use of language involved in scientific experimentation hides the 'I' that is appreciating the 'truth', leaving us with pure objectivity.

We have arrived, then, at the two sides of Fuenmayor's recursive form, "self" and "other". I believe that there is therefore a strong case for accepting that recursion can be explained with reference to generalisations of ontological positions from the use of language in time-and-context-specific reflection, thus producing contradictory positions that cannot be resolved. If the logic of these contradictory positions is to be preserved, but there is still the desire for comprehensiveness, then the only option is to express them as recursive forms. What these forms hide, however, is the way that the two sides have been constructed using *language*, where each side represents a concept, or way of speaking, that has its "other".

Also, missing the linguistic nature of recursion results in a second problem with Fuenmayor's (1991a,b) "onto-epistemology". Fuenmayor (1991a) observes that a person, when presented with a visual recursive form that he or she has never seen before, first notices the totality before entering the recursion. This, and the desire to overcome subject/object dualism through the generation of a unified philosophy, leads him to claim that one can describe the *fundamental unity of reality* by depicting the operation of recursive forms. In my view, however, this is not the case.

Recall what I said earlier about how a context-specific use of language focusing solely on truth or subjective understanding can be generalised into a philosophical position. As I see it, it is the unmitigated pursuit of one such position to the exclusion of others that results in a philosophy which, through its desire to suck all existence into itself, becomes one side of a recursive form. Because we know this is unsatisfactory, we end up moving between the two sides. The only thing that recursive forms can describe, then, is the *dis*unity (of which subject/object dualism is one example[33]) promoted by a particular use of language.

In my view, the most we can do to show that an underlying unity exists is to exploit the limitations of language. We can chase our tails around a recursive form until the absurdity of the activity breaks the bonds of language, just for a second. In that moment, we can 'feel' unity. However, this is inevitably a private experience: all we can do using recursive forms is to point in its direction. Recursion does not allow us to describe the fundamental unity of reality, but reveals the impoverished nature of description itself.

---

[33] In Chapter 4 I will discuss recursive forms with more than two sides.

In summary, Fuenmayor (1991a,b) has proposed a fascinating solution to the problem of naive subject/object dualism. However, while he overcomes the naiveté present in pretensions to pure objectivity (see also Fuenmayor, 1993), he nevertheless still accepts that 'subject' and 'object' cannot be reconciled, and says that the best we can achieve is to express them as two sides of a recursive form. Thus, a kind of subject/object dualism is still present. What Fuenmayor misses, however, is the role of language in creating recursive forms, and this insight leads us to the conclusion that recursive forms merely express *dis*unity (of which subject/object dualism is one example) rather than the fundamental unity of reality.

3.9 Conclusion

In this chapter I have argued that the boundary concept lies at the heart of systems thinking. I have also discussed the systems critique of reductionism and the naive dualism that is assumed by reductionist philosophies of science. In all, I reviewed four challenges to naive dualism that have been mounted in the systems literature: from Ludwig von Bertalanffy, Gregory Bateson, Humberto Maturana and Ramsés Fuenmayor. However, while each of these authors has managed to find an alternative to the radical separation of observer and observed, a less naive subject/object dualism is nevertheless still evident in their work. In Chapter 4, I will ask if it is possible to construct an alternative systems philosophy that can offer a different view of subject and object which will take them out of a dualist relationship.

4

Process Philosophy

All through the last chapter, when reviewing the work of other authors, I mentioned the potential that lies in understanding the importance of language for overcoming dualism. This chapter is divided into two parts. In the first part I will explore the 'linguistic turn' that took place in Western philosophy in the latter half of the 20th Century: this is the shift to a new paradigm in which language is given ontological primacy (language is seen as constructing both the subject and the object). At first sight the linguistic turn promises to overcome subject/object dualism, and indeed much of my own early work seeking a solution to this problem was premised on it. However, I will argue that the linguistic turn is not ultimately the answer. While language adds a 'third dimension' to ontology (the first and second dimensions being subject and object), it does not help us move beyond the paradox of Fuenmayor's (1991a,b) recursive form ("subject/object") highlighted in the last chapter. In essence, language complicates rather than resolves subject/object dualism by adding a third aspect without eliminating the element of paradox. Therefore, in the second part of this chapter, I will advocate setting aside the linguistic turn (at least for now), and will propose an alternative set of categories that I will argue can take us beyond the dualism of subject/object *and* the complications introduced by the linguistic turn.

4.1 The Problem of Subject/Object Dualism

Before discussing the linguistic turn, however, I first want to refresh the reader's memory about why the challenge to subject/object dualism is so important. It is important because subject/object dualism underlies the mechanism that has characterised so much scientific theorising in the past three centuries, but which is now being undermined by a variety of research perspectives from across the disciplines (see Chapter 1 for details).

Also, as we saw in Chapter 3, this form of dualism is assumed by reductionist methodologies of science: by focusing solely on supposedly objective, linear, cause-and-effect relationships, scientists have missed the potential for more holistic analyses which allow phenomena to be seen as emergent properties of whole systems. In the past, a great deal of policy making has been based on reductionist analyses, and as a consequence many side-effects have been experienced that could have been anticipated and avoided. In Chapter 3 I mentioned the side-effect (an increase in risk-taking) of making it mandatory to wear seatbelts in cars, but many other examples could be provided. One is the realisation that a largely unrestrained pursuit of economic growth through industrial development has resulted in massively increased emissions into the atmosphere, thereby producing the 'side-effect' of global warming (Meadows *et al*, 1992).

Finally, we find that it is subject/object dualism which underpins the illusion of perfect objectivity—and historically the pursuit of high-quality observation has marginalised any serious consideration of how moral and ethical issues can be explored as part of the mainstream of scientific activity. Lately, however, there has been a resurgence of interest in *intervention* rather than just observation, which is clearly value-*full*, not value-neutral (see Chapters 1 and 6). It is in support of this emerging agenda that I wish to construct an alternative to subject/object dualism.

4.2 The Linguistic Turn

So let me begin by introducing the linguistic turn that many people have heralded as the means by which subject/object dualism can be overcome.

The linguistic turn is often attributed to Wittgenstein (1953) who sought to undermine Cartesian dualism by seeking to show that all subjects and objects are constructed through language. Giddens (1991) puts it like this:

> "Self-consciousness has no primacy over the awareness of others, since language—which is intrinsically public—is the means of access to both. Intersubjectivity does not derive from subjectivity, but the other way round" (Giddens, 1991, p.51).

The essence of the argument is that it is not possible to think of, let alone talk about, either a 'subject' or an 'object' except in words. Therefore, language is ontologically prime. This idea swept through the humanities and social sciences during the latter half of the 20th

Century, and has influenced numerous prominent thinkers of very different persuasions (e.g., Lyotard, 1979; Habermas, 1984a,b; Derrida, 1976, 1978; Douglas, 1986; Luhmann, 1986; Rorty, 1989; and Gergen, 1991). Since then, the basic insight has been elaborated by social psychologists who have, for instance, produced theories of how language plays its part in constructing the identity of the individual (e.g., Middleton and Edwards, 1990; Gergen, 1991; Shotter, 1993; Harré and Gillett, 1994; Simons and Billig, 1994). These theories emphasise the linguistic, and therefore social, nature of phenomena that were previously thought of as 'private'.

Below, I will briefly present my own earlier use of this idea to construct a systems perspective.[34] I will then mount a critique of this, drawing attention to the problem that still remains—specifically, that we are led into a more complicated version of Fuenmayor's philosophy of recursive forms.

4.3 A Linguistic Turn in Systems Thinking

Between 1988 and 1992, I worked on developing a new ontology and methodology for systems inquiry. Various aspects of this work have been published in Midgley (1989a,b, 1990a,b, 1991a, 1992a,c,d, 1996a, 1997a, 1998), and some of the ideas have also been picked up and developed by Mingers (1997a). However, only their bare bones are presented here.

I started by acknowledging that no philosophical position can be described except using language (Midgley, 1992a). In order to understand why this is the case we can follow Heraclitus (approximately 600-500 BC) who declared that the fundamentally interconnected nature of the Universe is simply not accessible to human rationality. It can be accessed only when language is by-passed: "when you have listened, not merely to me (the speaker), but when you maintain yourselves in hearkening attunement, then there is proper hearing".[35]

Heraclitus talked about the *Logos*. At the risk of making a slipshod translation into modern systems jargon, the term *Logos* can be said to refer to the ultimate reality of interrelation and change that

[34] Other systems perspectives focusing on language and communication have also been advanced by, for example, Watzlawick *et al* (1968), von Glasersfeld (1984), Luhmann (1986) and Maturana (1988a,b).

[35] Translated from Greek into German by Heidegger (1954), and from German into English by Krell and Capuzzi (1975).

binds everything together into a dynamic, unfolding process.[36] When we see and think, Heraclitus argued, we can be aware of only a tiny part of the picture, and the image we have of reality is distorted by our static classifications and the bounded nature of our vision. The *Logos*, then, escapes adequate description. The everyday interconnectedness we have empirical knowledge of is not the ultimate interconnectedness of reality. As far as our everyday thinking goes, the *Logos* cannot be known. It can only represent an ideal, reminding us that no boundary is absolute.

If the *Logos* escapes description, it might seem that the only adequate vision of ontology is one which dispenses with language and thought in the exploration of reality. Indeed, this is what some people writing from a spiritual tradition have claimed (e.g., McBurney, 1990). However, in 1992, I argued that ontology is, most basically, concerned with *discourses* about reality. It is not about 'experiencing' reality through spiritual enlightenment. This is not to say that such experiences are invalid—just that ontology consists of *statements and arguments* about reality. Even the writings of spiritual visionaries like Krishnamurti (1991), beautiful though they are, are essentially discursive: they *guide one* down a spiritual path.

In setting out an agenda for my own particular linguistic turn in systems thinking (Midgley, 1992a), I also asked what aspects of previous philosophical positions would need to be accounted for. I came up with a list that included the ability to talk about 'truth' (as discussed by Popper, 1972; and Bhaskar, 1986); the ability to talk in terms of subjective understanding (as discussed in the work of Berkeley, 1710; Kant, 1787; and Kelly, 1955); and the ability to talk about inter-subjective construction and morality (e.g., Foucault, 1972, 1980; Habermas, 1984a,b). I selected these three aspects of philosophy for two reasons: first, as I see it, they represent the three major paradigms of philosophy that have emerged since the birth of Enlightenment thinking; and second, I believe that the systems community (which I was writing for) has also fragmented into paradigms largely along the same lines. I then searched the literature for a theory of language that might explain how it could be possible to talk about truth, subjective understanding and morality without internal contradiction.

[36] This is certainly a commonly accepted view of what *Logos* means. However, Crowe (1996), who returns to the original Greek texts rather than relying on other people's translations, suggests that the word *Logos* simply refers to Heraclitus's argument. He may be right, but this doesn't change the fact that Heraclitus wrote about the world as an ever-changing, dynamically unfolding process, as Crowe acknowledges. I have kept the word '*Logos*' in this text, partly because it is commonly used in this context, and partly to remain faithful to my original 1992 writings which were produced before Crowe made his translation.

The theory I chose was the theory of 'three worlds' advanced by Jürgen Habermas (1976, 1984a,b). However, I found that I had to reconstruct aspects of this theory in order to make it fit my requirements. This theory was touched upon in Chapter 2, and a reminder of it is provided below.

4.4 The Theory of 'Three Worlds'

Habermas (1976, 1984a,b) argues that, in uttering a statement, a speaker automatically claims that it is intelligible; its propositional content is true; the speaker is justified in saying it; and that s/he speaks sincerely. The first of these implicit claims, that the statement is intelligible, is simply a precondition for effective communication. However the other three claims, when made explicit, can all be questioned and justified through argumentation. It is these three claims that refer directly to three 'worlds': the claim that my statement's propositional content is true relates to *the external natural world*; the claim that I am justified in making it relates to *our social world*; and the claim that I speak sincerely relates to *my internal world*. These are not three distinct worlds in the Cartesian sense of the soul and the material world being separate. Rather, it is the nature of language that *allows the differentiation* of the 'natural', 'social' and 'internal' when we enter debate.

It is important to be clear about a key implication of Habermas's notion that all three types of claim are already inherent in any act of communication: although a statement may *appear* to be about just one world (the objective external world, the normative social world, or the speaker's subjective internal world), in fact *a position on the other two is always implied in it*. This allows the hearer to mount one of three types of challenge, regardless of which world the original statement appeared to refer to: a challenge to its propositional content, its normative acceptability, or the sincerity of the speaker. In Habermas's view, this is what good rational argumentation is all about: making distinctions between the objective, normative and subjective, and thus challenging the unseen assumptions of the speaker who is generally only aware that s/he is making a statement about one of the worlds.[37]

[37] As we saw in Chapter 2, Habermas (1984a,b) says that 'good' argumentation is to do with *extricating* the three 'worlds' from one another in any analysis. Some cultures, he maintains, have a prevailing worldview which collapses two or more of the 'worlds' together. For instance, the rights and wrongs of social relationships might be seen as an extension of nature in some cultures because the dominant view of both is governed by some form of myth. What is considered *right* is therefore taken for granted because of what is considered to be *true*, and both are 'solidified' in myth. Habermas believes that such worldviews

In reconstructing Habermas's work, I found it necessary to move away from 'sincerity' as an ideal, and I replaced this with 'subjective understanding' (making a claim about the subjective perspective or motivations of an individual). This was because I found the term 'sincerity' too narrow to represent the ideal of inquiry into the internal worlds of individuals: when we wish to understand the perspective of another, there is much more than sincerity to consider.

Thus, I suggested that it is possible to make, and challenge, *truth* statements (about the objective external world), *rightness* statements (about our normative social world) and statements about an *individual's subjectivity* (a person's subjective internal world). I then went on to argue that all existing methods for intervention prioritise the investigation of one of these kinds of statement (see Midgley, 1992a, 1998, for full details).

In my view, this theory makes meaningful talk of truth, rightness and subjective understanding, which is what I wanted to take from the three major paradigms of philosophy I mentioned earlier. The question I then had to ask was, what assumptions about the relationship between language, subjects and objects was I going to make? I could either follow Habermas (1984a,b) and make language ontologically prime, or I could theorise about the relationship of language with physical reality and subjects. I chose to do the latter (Midgley, 1992a). I argued that the dynamism of language demonstrates that individual subjects bring unique insights to bear and thereby change the use of language, and that it would not be possible to have individually differentiated subjectivities (with the capability of changing language) if there were no physical reality separating individuals from one another.

Although this view of ontology is widely shared (for instance, by Bhaskar, 1986, and Mingers, 1995), I was never completely happy with it. As I acknowledged in a footnote (Midgley, 1992a), the position I ended up with is simply a truth claim about the nature of language, and truth claims (in terms of the theory) relate to the external, natural world of objects. Therefore we are left with a rather paradoxical relationship between language and physical reality, creating the suspicion that I have simply created a new recursive form (or dualism)

---

represent an intrinsic restriction of 'good' rational argumentation: "myth binds the critical potential of communicative action, stops up, so to speak, the source of inner contingencies springing from communication itself" (Habermas, 1984b). In contrast, I believe that what constitutes 'good' argumentation has to be defined in the context of other discourses we regard as important. It is therefore possible for us to claim that, in some contexts, extricating the three 'worlds' from each other might be necessary, while in others it might not be. We therefore escape Habermas's inevitable conclusion that forms of rationality other than the most 'advanced' Western rationality are in some sense poorer.

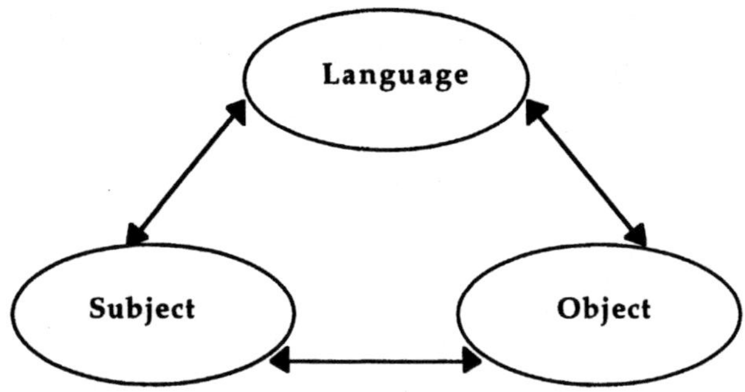

FIGURE 4.1: Language/subject/object recursive form

of 'language/object', with the subject marginalised. This leads me now to propose a critique of the linguistic turn.

4.5 A Critique of the Linguistic Turn

It is my contention that all theories of language are, by definition, truth claims. Indeed, in a critique of his own previous work, Gergen (1994) reached the same conclusion. He said it was paradoxical that, in a desperate rush to escape naive objectivity, he created a new object—language. For Gergen, the new dualism is one of "subject" and "language". If Gergen has produced a dualism of 'subject/language', and I have produced a dualism of 'language/object', I believe we need to take a more critical look at what we are doing.

The problem with saying that language is prime is that language can also be shown to have a recursive relationship with both the subject and object. Language can be seen as either 'language from my point of view' (subjective)[38] or 'a feature of the real world' (objective).[39] Thereby, we generate a new recursive form with three sides: 'language/subject/object' (see Figure 4.1). It is because of my argument in Chapter 3 (that recursion does not really provide us with an adequate ontology, but reveals the impoverished nature of description itself), that I suggest we abandon this line of philosophical inquiry, at least for now, to see if a more satisfactory alternative might present itself. My

[38] See von Glasersfeld (1999) for a strong argument for this position.

[39] Bhaskar (1986) views language in this way.

own belief is that the new philosophical position offered below provides such an alternative.

4.6 *The Origins of Knowledge*

Let us start by taking the works of the authors reviewed in Chapter 3 and look at what they have in common. I suggest that von Bertalanffy, Bateson and Maturana[40] all believe that it is possible to create an epistemological theory by specifying a prime originator of knowledge—or *knowledge generating system*—whether this is a biologically-situated observer (von Bertalanffy, 1968); an autopoietic organism (Maturana, 1988a,b); or a circular information pathway (Bateson, 1970).[41] I use the term 'knowledge generating system' to mean something that gives rise to the existence of knowledge through its own activity. The list of knowledge generating systems that it is possible to define is certainly not exhausted by the above theorists: for example, Douglas (1986) and Luhmann (1986) both view self-organising linguistic systems as generating knowledge.[42]

As I see it, however, there is a problem with the way in which von Bertalanffy, Bateson and Maturana all try to identify one specific knowledge generating system as the centrepiece of an epistemological theory. If we have a view of the Universe as a continually unfolding, interconnected entity—which is the view (traceable to Heraclitus, 600-500 BC) that lies behind most systems theories—then the idea that any one type of organism or system, acting autonomously, can generate knowledge must be open to question. Any such organism is dependent on interactions with its environment, and if this environment needs to be responded to, then the knowledge and actions of the organism must logically arise from the organism-environment pairing in interaction. The boundary of the knowledge generating system is therefore wider than it first appears. Indeed, theoretically, the boundary of the knowledge generating system can be seen as the boundary of the

[40] Fuenmayor's work will be discussed later.

[41] Another thing they have in common is a neglect of the issue of power—or, in Bateson's (1979) work, an active *hostility* to using the concept. Bateson (1979) calls power a "mythical abstraction" (p.223), which to me is highly problematic.

[42] In my view, the term 'knowledge generating system' is useful because it provides a general category describing systems which produce knowledge, implying that rivalries between particular epistemological theories exist (e.g., the theory of open systems and the theory of autopoiesis), and suggesting that no particular theory is all encompassing.

Universe—assuming that the Universe is a closed system (which is, of course, an assumption that is open to question).[43]

Let us take Maturana's autopoietic organism as an example. Is Maturana *really* saying that an autopoietic organism immersed in a linguistic environment is all that is needed for knowledge generation? Could such an organism survive without its animate and inanimate surroundings? I am sure that Maturana would never make such a claim, and indeed he talks in a very lucid manner about how the organism structurally couples with aspects of its environment (Maturana and Varela, 1992). Maturana argues that there is organisational closure at the boundary of an organism, with external factors being mere triggers, but isn't his identification of the boundary as the point of closure a mere convenience to aid explanation of the ways in which individual organisms function? What about Bateson's (1970) argument that information transmission can just as easily be seen as cutting across the boundary of the organism (the distinction between 'inside' and 'outside' being to do with the nature of transmission)?

What this points to is something that is commonly recognised by many late 20th Century writers in the philosophy of science (e.g., Bhaskar, 1986): all theories are partial, and their partiality is a function of the purposes and values of their creators and their communities of users—including epistemological theories which try to specify knowledge generating systems. Von Bertalanffy, Maturana and Bateson can all make claims to having coherent epistemological positions, but each of them leave out considerations that are important to the theories of others.

If partiality is inevitable, then it seems to me that we have a choice: we can either identify one preferred theory from the multiplicity available, and defend this against those who choose other ideas, or we can recognise the possibility of working with a variety of theories in the knowledge that each privileges particular insights, values and purposes. My own preference is for the latter, because a plurality of theories ultimately yields more insights for intervention than if we work from one position alone (for more details see Bernstein, 1983, 1991; Morgan, 1986; Francescato, 1992; Gregory, 1992, 1996a,b; Romm, 1996; and Chapter 8 in this book). Of course, this raises two thorny issues: how to justify moving between theories that make contradictory assumptions; and how, practically speaking, to exercise choice between theories in the context of intervention. These matters will be dealt with more fully in Chapters 8-11. In proposing an

---

[43] A 'closed system', as compared with an 'open system', is one where nothing crosses its boundary (von Bertalanffy, 1950). A closed system is therefore autonomous.

alternative philosophical position (below), I am simply establishing the groundwork that will allow for greater theoretical pluralism.

4.7 From Content to Process Philosophy

If the common assumption of Bateson, von Bertalanffy and Maturana is the specification of a prime originator of knowledge, let us ask if there is anything other than a knowledge generating system that could be treated as analytically prime. My answer is that we can view as prime the *process* of bringing knowledge into being. Bateson, von Bertalanffy and Maturana all offer a *content* philosophy. They try to make some propositions (specify some content) about what the knowledge generating system must be like. In contrast, we can switch analytical primacy to the *process* of specifying that content.[44]

Now, in saying this I should acknowledge that I am using the term 'process' in a related, but subtly different, manner to others who have talked about 'process philosophy' (e.g., Bergson, 1911; Whitehead, 1929; Pols, 1967; Capek, 1971; Leclerc, 1972, 1986; Mathews, 1991; and Gare, 1996). Tracing the origins of process philosophy, Gare (1996) cites the ancient Greek philosopher Heraclitus who says, "nothing *is*, everything is *becoming*" (p.310, my italics). However, as I understand it, 20th Century process philosophers do not assume that 'nothing is'. Rather, they take as analytically prime those 'objects' (or systems) that provide the *means of becoming*. Thus, Gare aligns von Bertalanffy with process philosophy because, in general systems theory, the activities of open systems give rise to change: inputs are transformed into outputs, and properties of whole systems emerge. For von Bertalanffy (1968), *open systems* are therefore the means of becoming. In contrast, I wish to avoid the identification of any one type of object or system as analytically prime—as I see it, a process should not be logically reliant on the prior identification of just one type of object or system, otherwise we have merely generated another content philosophy (albeit one which is slightly more sophisticated than content philosophies that disregard process altogether). It is for this reason that I cannot accept von Bertalanffy as a process philosopher: he is primarily interested in specifying the nature of systems (i.e., content) giving rise to process.

[44] *Analytical* primacy is not the same as *ontological* primacy. Something is analytically prime if it is advisable to look at it first, but this does not necessarily mean that it has a more fundamental reality.

So, for me, process philosophy involves identifying a process that is not dependent on the further identification of a single type of system giving rise to that process. Fuenmayor (1991a,b) goes quite a long way towards such a position. As we saw in Chapter 3, he proposes a recursive form relating together the intentional subject and distinctions of its other (which also serve to delineate the subject). Essentially, making distinctions is *process* and the subject is *content*. So, while Fuenmayor takes a step toward process philosophy, he still hangs on to an aspect of content. It seems to me that the subject has to be expressed as content because of the assumption that Fuenmayor inherits from Phenomenology that the starting point for building a philosophical position should be lived experience. From an experiential point of view, it would be inconceivable not to have a subject (or self) in a semi-pivotal position. Of course, when the self is placed in relation to its other to create a vision of epistemology, this generates the paradoxes expressed in Fuenmayor's recursive forms (and indeed, these can be made even more paradoxical through the introduction of language, as we saw earlier in this chapter). So, although Fuenmayor distances himself from the tendency of biological epistemologists to try to root everything in one prime originator of knowledge, there are still problems with his position (which I believe can be overcome).

In switching analytical primacy from content to process, the particular process I have in mind is making boundary judgements (which are similar to Fuenmayor's distinctions).[45] If we regard the process of making boundary judgements as analytically prime, rather than a particular kind of knowledge generating system, then *subjects come to be defined in exactly the same way as objects—by a boundary judgement.*

That this is the case can be demonstrated as follows. Churchman (1970) suggests that a boundary defines what is the focus of attention—what is to be taken as pertinent at any moment in an analysis. In other words, a boundary delineates the *object* of attention. Where there are multiple objects in relationship with one another, there are multiple boundaries—and the set of objects is delineated by a wider boundary that defines that set in relation to everything that is excluded from attention (the invisible "scene" to use Fuenmayor's term). I would hope that this can be accepted as uncontroversial, so I will not dwell on it. Where controversy might surface, however, is when we ask, who or what is drawing the boundary? What gives rise to the boundary's existence? What gives rise to the possibility that an object appears the way it does? I will spend some time addressing these questions, but this

[45] See Chapter 3 for an introduction to the idea of boundary judgements, and Chapter 7 for a more detailed discussion.

should not be taken as an indication that the process of identifying knowledge generating systems is more important than identifying other phenomena: I have simply focused on this aspect because of the need to provide support for a position that is significantly different to the usual content philosophies.

Von Bertalanffy's answer to the question 'what gives rise to the boundary's existence?' would no doubt be to say that only open, living systems can draw boundaries. Maturana's (1988a,b) answer would be to refer to the theory of autopoiesis, arguing that boundaries are drawn by autopoietic organisms using language. Bateson's (1970) answer would be that the identity of a knower cannot be delimited by the skin of an organism, so it is the organism as part of a larger circular information pathway that gives rise to the drawing of boundaries. Fuenmayor's (1991a,b) answer would almost certainly be to relate the drawing of boundaries to the intentional self, defined in turn in relation to distinctions of its other. My answer, in contrast, is to say that *it depends on where the boundaries are drawn.*

It is possible to make a variety of boundary judgements when looking 'outward' towards the world, and a variety of judgements when looking 'back' at the knowledge generating system which produces these 'outward' judgements. Borrowing from the language of cybernetics (e.g., von Foerster, 1984), I will call a boundary drawn when looking 'outward' a *first-order* distinction. In contrast, I will use the term *second-order* to denote the distinction of the identity of a knowledge generating system which is instrumental in making a first-order distinction.

When operating with process philosophy (as I have described it), any number of possible second-order boundaries might be used: one which identifies an individual human being in isolation; a person using a particular language to construct their understanding; a solitary animal; a group of animals; a group of people; a group of people acting within the constraints of their culture; an organisation or institution which constrains the possible actions of its members; a group of people viewed as representative of their social class; a group of people shaping, and being shaped by, the ecosystem they are immersed in; etc.

In each case, different theories might throw light on the ways in which the knowledge generating system has been instrumental in producing first-order knowledge: for example, Freud (1915) and Kelly (1955) both have different theories of individual motivation; Campbell *et al* (1994) discuss equilibrium theories of small group and organisational behaviour; Douglas (1986) and Luhmann (1986) talk in different ways about how social institutions restrict the thinking and actions of their members; Marx (1887) and Mandel (1975) discuss how the "working class" may be subject to a "false consciousness" within the capitalist system; Roszak (1993) considers how people think and act as part of ecosystems; and both Lovelock (1988) and Abram (1988) talk about how people can be regarded as 'organs' of the 'body' of the Earth. There are, of course, a variety of other relevant theories too, giving rise to many possibilities for theoretical pluralism.

4.8 Defining Knowledge

In talking about 'knowledge generating systems' I should be explicit that I am using the term 'knowledge' in a wide sense to mean *any* understanding, whether this is phrased in language (giving the potential for inter-subjective communication), or whether it takes the form of imagery (visual, auditory, olfactory, etc.) in the absence of language. Thus, in my understanding of 'knowledge', what is commonly thought of as 'perception' may also be included within it. Perception is not simply recorded images of an external reality, but is a complex construction by a sentient being in interaction with its environment (and what counts as 'environment' in any one case is defined through a local boundary judgement).[46] Clearly, this assumes that there can indeed be knowledge without language—but this does not in turn imply that, in humans, language maps onto non-linguistic knowledge in a simple fashion. The relationship between language and imagery is no doubt complex and co-constructive too.

It is important to be clear about this definition of knowledge because it takes us away from an understanding of knowledge as already inscribed in language—and therefore both human and theoretical in nature. Knowledge is not necessarily the property of academic discourse or the subject of erudite books: it may be seen as the fleeting perceptions of a sentient being (whether human or non-human); the theory-in-use of an organisation; the ideology of a political group; or a scientific theory. All are forms of knowledge which may be explained with reference to many possible knowledge generating systems.

---

[46] See Wanner (1975) for a review of some empirical evidence surrounding perception-as-construction—although this deals with the phenomenon at an individual level and not a social or ecological one. For some ecological considerations, see Roszak (1993).

4.9 Sentient Beings

In the above, I have been careful to talk about knowledge generating systems as being, or containing, individuals or groups of humans or animals (in the case of humans, they can be seen in relation to languages, institutions, economies, social classes, networks of power relations, ecosystems, etc.). The implication of this is that the identification of one or more sentient beings (human or animal) is necessary as part of a second-order boundary judgement, but the boundaries of the knowledge generating system containing the sentient being(s) may be much wider than the skin (or skins) of the organism(s) concerned.[47] In any second-order analysis, when using process philosophy, at least two boundary judgements need to be identified: one specifying the extent of the knowledge generating system; and one specifying the nature of the sentient being(s) who are part of it.[48]

The placing of these boundaries is always dependent on the purposes being pursued, and the theoretical ideas employed, in a local situation—so, as purposes change (allowing a different point of view to be taken), there is always the possibility of identifying a second knowledge generating system and associated sentient being(s); and a third one beyond that, etc. Also, every time the question is asked, 'what gives rise to the purposes motivating this second-order boundary judgement?' another second-order boundary judgement needs to be made—and, in theory, this can go on *ad infinitum*.

Of course, mention of sentient beings raises the question of how a 'sentient being' should be defined, given that I do not regard the traditional boundary used to do so (the skin) as an absolute dividing line between an organism and its environment. My answer is that, for the process philosophy I am evolving to be consistent, *I should not actually propose a universal definition*. To do so would be to say that there is an aspect of content—a sentient being with a single, set, theoretical definition—that has analytical primacy over boundary judgements. This would be a return to content philosophy. Rather, what counts as a sentient being from a process point of view must depend on the particular second-order boundary judgements being made in any local situation. For some situations it may be necessary to use a biological theory of living systems (e.g., the theory of autopoiesis) to understand the nature of a sentient being (although, in my view, the theory of

[47] I would not want to suggest that non-sentient beings can generate knowledge, except in their interactions with sentient beings.

[48] In Chapter 6 I discuss the nature of human agency in relation to this understanding of second-order boundary judgements.

autopoiesis has a fairly restricted view of the environmental factors that can contribute to knowledge generation[49]). In another context, it may be sufficient to simply specify the name of a person or the identity of a group. Sometimes the use of language may need to be very precise, while at other times it could be looser or even quite creative. Exactly how the presence of sentient being(s) should be interpreted is very much context-dependent, relying on judgements about what is appropriate for particular purposes.

The question may also be asked, why do I choose to talk in terms of 'sentient' rather than 'human' beings? The answer is that I do not want to restrict knowledge to *human* knowledge. Non-human animals can be seen as sentient beings too. As far as I can see, non-human animals have more or less the same relationship with their environment as human beings, in that what gives rise to an animal's understanding can be seen as having a wider origin (using whatever boundary judgement appears correct in the local circumstances). A non-human animal interacting with a specified aspect (or aspects) of its environment can therefore be considered a knowledge generating system.

However, the key difference between non-human animals and human beings is in the latter's use of language. As far as we are currently aware, in the vast majority of cases, non-human animals only have command of 'basic' language: verbal and non-verbal signals that carry some meaning-in-context. However, these signals do not provide *contexts for one another* that change their meanings. Words used by humans seem to be unique in this respect: they can be combined into sentences which have a meaning that only makes sense as an emergent property of the whole sentence-in-context. The meaning of a sentence-in-context is different to the sum of the meanings of its contributory words in the same context. An important implication of this is that, if we are looking to explain the emergence of knowledge that has a linguistic expression, the sentient being that needs to be identified as part of the knowledge generating system will almost always be human. Non-human animals may be included within the boundary as well, but not exclusively.[50]

---

[49] Maturana and Varela (1992) talk about language, and aspects of the environment that the organism may structurally couple with, but the boundary of a knowledge generating system may actually be pushed out much further using my process philosophy.

[50] Williams (2000) argues that experiments in teaching primates sign language suggest that some animals are as capable as humans of using words in this more complex manner. However, the use of sign language by primates is a special case brought about by human intervention. Therefore, while I acknowledge that there may be exceptions to the rule that knowledge with a linguistic expression arises out of knowledge generating systems containing human beings, these exceptions are only very rarely encountered. Also, I would be reluctant, without seeing some evidence, to accept the idea that primates can use

There are also three other important consequences of acknowledging that humans may use language to frame boundary judgements. The first is that second-order inquiries of the type discussed earlier (i.e., using explicit theories to analyse the production of first-order knowledge) can only be conducted by human beings (embedded in wider knowledge generating systems). This is because explicit theories are expressed in human language. The second consequence of acknowledging the human use of language stems from the fact that language can express *values*. In the human creation of boundary judgements, *value* judgements are also implicated: values direct the drawing of boundaries about what is pertinent to the purposes being pursued, and particular boundaries necessarily constrain the values and purposes that can emerge (Ulrich, 1983, and see Chapter 7 of this book). The third consequence of the human use of language in making boundary judgements is that people are not only able to distinguish what actually exists or is happening (or what is pertinent) using boundary judgements—they can also distinguish what *might possibly* happen under different circumstances, or what *ought* to happen. It may be possible for some non-human animals to use visual imagery to anticipate simple future scenarios, but language enables a far more elaborate expression of possibilities—and is certainly necessary for the framing of moral injunctions (expressions of what *ought* to be the case).

In discussing the similarities and differences between humans and other animals in their generation of boundary judgements, it may appear that I am labouring an obvious point. However, I believe it is important because there is a tendency in much 'humanist' literature to treat all knowledge as human knowledge—as if non-human animals are mere objects rather than sentient beings. This is an assumption which obviously gives rise to the exploitation of non-human animals as 'natural resources' no different from, say, the iron ore we dig from the ground. It seems to me that the humanist focus solely on human knowledge comes about for two reasons: first, the importance of linguistic knowledge to all human endeavours that involve more than just individual perception; and second, the tendency for humanist discourses to marginalise the non-human, giving rise to an ideology of human supremacy (Midgley, 1994). The process philosophy I am proposing takes us away from this ideology, but without slipping into the trap of anthropomorphism (treating animals as if they are human):

explicit theory in the same way as human beings do when conducting second-order inquiries. In addition, while primates may have community norms, these do not come to be expressed as values (which, as far as I can see, requires human language).

the key difference between humans and other animals, the use of linguistic systems, is explicitly recognised.

4.9.1 Shorthand Expressions of Boundary Judgements

Now, having discussed the need for two boundaries to be used when making a second-order judgement [one specifying the relevant sentient being(s) and one specifying the wider environment that should be seen as part of the knowledge generating system], I feel that I should pass comment on a common practice that might conceivably undermine this. Often, a knowledge generating system is identified without mentioning the individuals and groups within it. Arguably, the classic case is when there is talk about 'organisational learning' as if it is the organisation as a whole, rather than the people within it, who learn. My own view is that this is justified only in so far as we recognise that omitting the human element is a convenient shorthand. The organisation can be viewed as a self-organising system, and the human parts can be replaced by others without necessarily disturbing the function of the whole (Beer, 1985), but it is not meaningful to talk about organisational learning unless there are actually people involved.

For the vast majority of authors and intervention practitioners, this will not be a problem. Even a writer like Luhmann (1986), who has been criticised for consciously excluding the biological level of human existence from his analysis of "autopoietic" linguistic systems (Teubner, 1993; Mingers, 1995; Brier, 1999), is not claiming that language operates independently from human use. Luhmann has analytical reasons for marginalising human beings as *biological* entities because he wishes to show that language games, pivoted around the practice of institutions in modern society (e.g., the economic, legal and educational systems), are self-producing. Essentially, he believes that language games, human beings as biological entities, and individual consciousnesses are *all* autopoietic (organisationally closed), but structurally coupled with one another. He is not claiming that language games could even exist, let alone be autopoietic, in the absence of human beings. Therefore even Luhmann, who consciously chooses not to discuss the involvement of sentient beings in knowledge production, does not deny the necessity of their presence.

4.10 Second-Order Reflections on the Nature of the Self

Apart from the wide variety of second-order distinctions that it is possible to make (as described earlier), there is also one kind of second-order distinction worthy of special note: the distinction of the self as one particular knowledge generating system.[51] It is commonly noted by philosophers exploring the relationship between 'self' and 'other' that, when first-order investigations are going on, the identity of the self doing the investigations is hidden from its own view (see, for example, the arguments of Fuenmayor, 1991a,b). However, when attention passes to this identity—when the investigator starts to investigate him/herself—the external world falls into darkness instead. It is not possible to conduct first- and second-order investigations simultaneously when the second-order investigation is into the self, although it is possible to move rapidly between them. Of course, in common with other second-order distinctions, the self need not be defined as an autonomous individual: any aspect of the environment of the body may contribute to the knowledge-producing capacities of the self.

This idea is similar to Roszak's (1993) "ecopsychology", where he argues that the self is not a discrete entity bounded at the skin. On the contrary, the self is an interactive part of its ecosystem, the planet, and ultimately the whole cosmos. Therefore, the boundary we place around the self is arbitrary—except that using the word 'arbitrary' suggests it is accidental, and I do not believe this to be the case. Elsewhere (Midgley, 1994), I argue that the tendency to gravitate towards the use of boundaries around human systems (individuals, groups, organisations, linguistic systems, economies, societies, etc.), excluding the ecosystems of which they are a part, is a function of a humanist discourse that results in the marginalisation of ecological concerns and ultimately produces environmental degradation that rebounds on human society. Similarly, Roszak (1993) and Hillman (1995) both identify the ecopsychological view of the self as providing a key concept for a new ecological paradigm. In the words of Hillman:

> "Since the cut between the self and the natural world is arbitrary, we can make it at the skin or we can take it as far out as you like—to the deep oceans and distant stars. But the cut is far less important than the recognition of uncertainty about making the cut at all. This uncertainty opens the mind to wonder once again...." (Hillman, 1995, p.xix).

Humanist discourses tend to be *anthropocentric*. That is, they centre human beings in analyses, and root knowledge in human systems only. Reacting against this, some environmentalists argue for prioritisation of the planetary boundary (e.g., Allaby, 1989): we might call this attitude 'Gaiacentric' (following Lovelock's, 1979, 1988, theory that the Earth,

[51] I am, of course, talking about a *human* self here: language is necessary for the kind of self-reflection that involves theorising about the self as a knowledge generating system.

which he calls "Gaia", is a living system). However, the process philosophy I am advocating in this chapter contrasts with both these positions: it allows the centring and decentring of human beings and all other possible knowledge generating systems as and when it is considered appropriate by the knowledge generating systems making the boundary judgements (see also Midgley, 1994). Ideally, when it come to intervention, the consequences of using a *variety* of boundaries should be considered (Churchman, 1970; Ulrich, 1983; Midgley, 1992b), and in Chapters 7 and 14 some of the practicalities of this will be discussed.

4.11 The Importance of Time

Earlier, when I discussed Fuenmayor's (1991a,b) version of process philosophy, I noted that he accepts the assumption inherited from Phenomenology that a philosophical position should be built up from the starting point of lived experience. This results in the need to place the self in a semi-pivotal position, generating the paradoxes expressed in Fuenmayor's recursive forms (which can be made even more paradoxical by introducing language into the equation). These paradoxes arise because, from one side of the recursive form, it looks like all distinctions are in the mind of the self, while from the other side it looks like whatever is in the mind of the self is determined externally. Introducing language simply adds a third side in which both the identity of the self and distinctions of the external world are given in language.

In my own version of process philosophy, when a knowledge generating system external to the self is being identified, there is no problem: as I have argued, knowledge generating systems containing sentient beings are delineated through boundary judgements in exactly the same way as non-sentient objects. However, introduction of the self as a special case of a knowledge generating system (even a self with a variety of possible boundaries) introduces the spectre of a similar kind of recursion that I have claimed is an issue for Fuenmayor's position. It is not quite the same recursion, because in this case it arises when a boundary judgement is made about the nature of the self and we then ask, what is the identity of the self making this boundary judgement? When this is answered, the question can be asked again *ad infinitum*.

In my view, the means of resolving this problem is to introduce the concept of time. Instead of seeing one self as *simultaneously* giving rise to boundary judgements about another self, we need to view this as an activity happening over time. Witness the following hypothetical scenario. At one moment the self feels the need to define its boundaries.

Having done so, the very next moment the question is asked, what is the self that gave rise to this definition? Reflection may produce a second, different definition of the self (or possibly the same one). If we see this as a process happening over time, then there is no recursion: rather, there is a spiral of reflection involving movement from questioning the nature of the self, to defining the self, to questioning the self, etc. Theoretically, the spiral can go on indefinitely—but this never happens in practice because the need to make boundary judgements concerning matters other than the self inevitably intrudes. Indeed, I suggest that relatively little time is spent in self-reflection of this kind compared with time spent on making other first- and second-order boundary judgements. Also, moments of reflection on self-identity are interspersed amongst the many other moments of 'outward' looking inquiry [at which times there can be no appreciation of the knowledge generating system(s) giving rise to these inquiries], and experiencing two moments of self-analysis following directly on from one another is a particular rarity.[52]

Having demonstrated the importance of time to overcoming the problem of recursion, I should note that the idea of movement between moments of inquiry is also vital to the practice of drawing upon multiple methods that I describe in Chapter 10.

4.12 The Indeterminacy of Process

The distinction between process and content philosophy should now be clear. Content philosophy presents a theory specifying exactly what counts as a knowledge generating system, while process philosophy allows for a variety of possible knowledge generating systems (with the proviso that there are sentient beings identified as part of them). Also, content philosophy is mono-theoretical (proposing a single theory to account for the existence of knowledge), while process philosophy allows for theoretical pluralism in relation to the many different possible first- and second-order boundary judgements that can be made.

However, the reader may be left wondering why I have only talked in broad terms about the process of making boundary judgements, and have not specified exactly how these are generated. The answer is

[52] I should note that there appears to be a consensus right across the 'natural' and 'social' sciences on the importance of time for solving this kind of problem. Spencer Brown (1972) makes note of the role of time in casting a fresh light on mathematical paradoxes, and this is a theme that is still being discussed in mathematics today (see, for example, Kauffman, 1999). Also, some quantum physicists argue that time prefigures the existence of matter (Prigogine, 1989).

that, as soon as we move from discussing boundaries in general to a generative mechanism, we have moved away from process to content![53] In this sense, it would be contradictory to create a supposedly universal theory of what generates boundary judgements.

However, this does not mean that we should not theorise about generative mechanisms at all—just that these theories should not be regarded as universal, or true in an absolute sense. For example, in Chapter 7, I detail a theory of the relationship that can be found in many social situations between boundary and value judgements, and I show how social processes involving multiple boundary judgements generated by different stakeholders can operate to stabilise a situation characterised by value conflicts. This is a theory that I have found to be very useful for informing intervention (see Chapter 14), but I would not wish to claim that boundary judgements should *always* be seen as arising from stakeholder groups in this way. The theory can be said to generate useful insights about how knowledge is shaped in situations of value conflict, but it is not a general theory, and should not be regarded as universal. The only theory of boundaries that could conceivably have the label 'general' applied to it (although I would not say 'universal'[54]) is the process philosophy already outlined, where the origins of boundary judgements are left unspecified. They will, however, be specified in a variety of different ways during other activities of theory production which will have meaning in local contexts, such as in Chapter 7 (the local context being production of a methodology for systemic intervention that has specific uses).

4.13 *Some Consequences of Process Philosophy for Speaking about Reality*

There is one final act to perform before closing this chapter, and that is to reflect on the consequences of process philosophy for what it is possible to say about reality. You will recall that, earlier, I claimed that there are at least three major paradigms of ontological thought

---

[53] See the earlier discussion of Gare's (1996) alignment of von Bertalanffy with process philosophy, which does not fit with the way I use the term.

[54] I tend to resist claims to universality because of the possible interpretation that no other way of seeing or acting could have any validity. When people propose so-called universal theories they may not mean to imply that they have found an absolute truth, but the claim to universality is often taken to mean this. However much I might value my own vision of process philosophy at the present time, I would not wish to close myself off to the possibility that there might be a better philosophical position for my purposes. Such a closed attitude is fundamentally uncritical.

that have emerged since the birth of the Enlightenment. I also said that, in my earlier work (1992a), I wanted to preserve the best from these: the ability to talk in terms of truth, subjective understanding and rightness (morality). The three paradigms I was referring to are *realism, idealism* and *social constructionism.* Below, I will give very brief definitions of these, acknowledging that my words will, to some extent, produce caricatures: the most sophisticated proponents of each position have inevitably tried to take account of the strengths of the other two (as Willmott, 1993, shows, distinctions between paradigms can often be rather artificial). I will then demonstrate that, through the process philosophy outlined here, it is possible to make *all* the kinds of statements associated with valid practice in these paradigms. Process philosophy therefore allows us to talk about a real world, social construction and subjective understanding without contradicting ourselves.

4.13.1 Realism

The first of the three philosophical paradigms is *realism*: the idea that there is a real world which knowledge refers to. The more sophisticated proponents of realism (e.g., Popper, 1972; Bhaskar, 1986; Mingers, 1995) argue that, although knowledge does refer to a real world, we cannot know the exact nature of that reference (see the discussion of Popper's critical fallibalism in Chapter 2 of this book). Therefore, everything we are aware of is actually knowledge (models) of reality, not reality itself. Nevertheless, realists insist that it would be counter-intuitive to say that there is no reality beyond knowledge. They argue that we should pursue an *ideal* of truth. Although we can never be sure that what we know really *is* the truth, we can still make propositional statements (explanations) that can be subject to argumentation, with evidence being provided to support the case of those mounting an argument for or against a truth claim. It is widely taken for granted that any theories which assume the existence of material entities (e.g., the planet, biological organisms or economic conditions) must be based on a realist philosophy—hence the controversy over Maturana's (1988a,b) claim that it is possible to have a non-realist, biological theory (see, for instance, Mingers, 1995).

4.13.2 Idealism

The next major paradigm is *idealism*. Berkeley (1710) and Kant (1787) were two of the earliest post-medieval thinkers to argue that, because it is not possible to know the real world, we should not try to talk about it. Kant (1787) distinguishes between the "phenomenal" world (of knowledge) and the "noumenal" world (of reality), but says that the latter is merely a 'limit concept': a concept that must be proposed because if we say that knowledge exists, then the concept of knowledge must be distinguished against something it is not. Kant's philosophy is wide-ranging, but ultimately he roots meaning in the "transcendental subject": a conscious being who is able to make choices between true and false, right and wrong. Similarly, in the 20th Century, Kelly (1955, 1970) talks about the active subject constructing his or her own reality, and Maturana (1988a,b) talks about individuals bringing forth their own realities (albeit using language). While Maturana (1988a) believes that his own idealist position grants validity to scientific attempts to explain phenomena (on the understanding that those explanations are relative to the language games being played by other scientists), Kelly (1955) insists that the only valid form of inquiry is into the subjective understanding of individuals.

4.13.3 Social Constructionism

Finally, there is the third paradigm that I have called *social constructionism*. I have taken this term from Gergen (1991), but I am using it more broadly to denote any position that talks about the inter-subjective (usually linguistic) construction of reality. The roots of this paradigm are most often traced back to Wittgenstein's (1953) argument that nothing can be said about either the external world or a subjective position, except using language. Therefore, like Kant (1787) before him, Wittgenstein insisted that we should not try to talk about the 'real world' at all. This paradigm actually embraces a wide range of ideas, including Rorty's (1989) discursive construction of truths; Habermas's (1984a,b) theory of argumentation (and the systematic distortion of debate in modern societies); Foucault's (1980) history of the construction of human identities through power/knowledge dynamics; and Gergen's (1991) theory of the dynamic relationship between language and individual selves. What constitutes appropriate inquiry for these authors obviously differs, but they all refute the possibility of perfect knowledge of either external reality or human subjectivity.

4.13.4 What can be Said using Process Philosophy?

Because it is possible to make any number of conceivable boundary judgements in both first-order inquiry (looking 'outward' on the world) and second-order inquiry (looking 'back' at knowledge generating systems engaging in first-order judgements), I argue that the vast majority of ideas from all of the above paradigms are compatible with the process point of view. However, this does not mean that all the authors who produced these ideas would agree with everything I have said. Process philosophy cannot take us beyond the debate between paradigms because it makes assumptions (about the analytical importance of process compared with content, and the need for theoretical pluralism) that are different from the assumptions made by others. Therefore, process philosophy provides the basis for establishing a *new* paradigm.[55] Nevertheless, I argue that the process paradigm is *preferable* to realism, idealism and social constructionism because it can offer an alternative grounding for most of the work from these other paradigms without the need to alter their most important contributions. What *is* altered, however, once this work is seen from a process point of view, is the degree of certainty that can be ascribed to the boundary judgements that are implicit in the various theories. Crucially, process philosophy undermines the dogmatic attitude that can accompany blind confidence that there is only one correct boundary to work with.

4.13.5 From Realism to Process

First, I will demonstrate that materialist theories (those which make truth claims about the world, and which are generally assumed to be realist in orientation) are compatible with process philosophy. Two examples will illustrate: Marx's (1887) theory about class struggle in capitalist societies, and Lovelock's (1979, 1988) theory of the Earth as a living system (Gaia).

Marx is primarily concerned with the workings of the capitalist economy and the resulting stratification of society into social classes (those who own the means of production, and those who, through necessity, are forced to sell their labour at less than its market value, generating a profit for the owners). His boundary therefore includes both the economy and the human beings which operate according to its dictates (and who, in the view of Marx and Engels, 1888, can only alter this state of affairs through revolution). The states of consciousness of the classes are also included, in that the 'working class' (those forced to sell their labour) tend to accept the capitalist system as inevitable because the pressures of poverty and work leave them little time for the collective design of an alternative form of society (i.e., in Marx's terms,

[55] See Chapters 10 and 11 for a more detailed discussion of paradigms, including (in Chapter 11) a justification for why I believe that it *is* possible, contrary to the thinking of some authors (e.g., Kuhn, 1962; Tsoukas, 1993a), for an individual to propose a new paradigm.

they suffer from a "false consciousness"). From a process point of view, what Marx is doing is drawing boundaries around the economic and social bodies, effectively excluding, for example, the ecosystems of which societies are a part. Marx's materialism is therefore compatible with process philosophy—which is not to say that everyone will agree with his boundary judgements. For example, Marx measures "value" in terms of the labour that goes into producing commodities rather than, say, the energy from the environment that is consumed in the production process. In my view, environmental thinkers such as Daly and Cobb (1989) are right to criticise earlier economists (including Marx) for excluding environmental concerns from calculations of value.

A similar argument can be made that Lovelock's (1979, 1988) Gaia theory, which is also materialist in outlook, is likewise compatible with process philosophy. Lovelock chooses to place his boundary around the planet, and argues that the Earth is a self-regulating system. In a second-order reflection on this, Abram (1988) argues that "the things around us.... are our co-participants in the evolution of a knowledge and a science that belongs to humankind no more, and no less, than it belongs to the Earth" (Abram, 1988, p.128). In other words, the knowledge generating system that has produced Gaia theory can be seen as Gaia itself. Of course, there are problems with this choice of the planetary boundary, not the least of which is explaining how Lovelock can reconcile his view that human beings are part of the self-regulating processes of the planet with his belief that we can change the course of the Earth's ecology through our actions (Lovelock, 1990). Nevertheless, because particular boundary judgements are open to challenge doesn't affect the basic argument that materialist theories are compatible with process philosophy—indeed, the fact that all single uses of boundary judgements bring problems to light actually supports the process view that theoretical pluralism is needed, together with caution about setting boundaries in stone.

What is particularly interesting for me about this reflection on materialist theories is that the need to be cautious and critical about the use of boundaries is entirely consistent with the view of most realists writing in the second half of the 20th Century that knowledge can never be perfect (see, for example, Popper, 1972, and Bhaskar, 1986).

4.13.6 From Idealism to Process

Having shown that ideas usually associated with the realist paradigm are compatible with process philosophy, let us now do the same for idealism (which, as I have defined it, prioritises the "transcendental subject"). In this case, let us take Kelly's (1955) personal construct theory (briefly reviewed in Chapter 2) as an example. To remind the reader, Kelly (1955, 1970) suggests that there are as many worlds, or realities, as there are human beings. While he does not rule out a social or linguistic dimension to the construction of realities, he regards this as relatively unimportant compared with the bringing forth of markedly different realities at the level of the individual. Kelly also claims that individuals work to construct their realities primarily in terms of activities. It is the things that are perceived as impacting on decision making for action that become part of a reality. In terms of methodology, Kelly argues that it is only valid to explore individual subjectivities—and then only in a way that is action-focused. The method he advocates is to represent decision making by asking subjects for a scenario in which they have to make a simple choice between options. He then asks them to generate the "personal constructs" (variables) that they will take into account in making the choice. By cross-tabulating the options with the personal constructs, and entering numbers into the boxes to represent the preferred option in terms of each personal construct, Kelly is able to produce a visual representation of the factors impacting on the decision to be taken. Other writers since Kelly (1955) have built on this method to include weightings of the constructs as well as multiple contexts of decision making, and have validated this means of representation by predicting simple, well-structured decisions (see McKnight, 1976, for a review).

Essentially, what Kelly and his followers are doing in terms of process philosophy is conducting a second-order reflection in which the knowledge generating system is bounded around individual consciousness. Then all subsequent first-order inquiries (into the decision making of individuals) are constrained by the assumptions made in this second-order reflection. Of course, the use of such a narrow boundary can obviously be subject to criticism. I would argue that factors outside the conscious awareness of individuals may affect decision making (which is why Schön, 1983, claims that what people say and do can be quite divergent). Also, there are many situations in which individuals believe their actions are determined by forces beyond their control, suggesting that there is something existing outside the mind which is influencing behaviour (Munlo, 1997). Nevertheless, because the work of Kelly and his followers can be described in terms of the use of a boundary placed around individual consciousness, it is perfectly compatible with process philosophy.

4.13.7 From Social Constructionism to Process

Finally, let us look at the paradigm of social constructionism (which contains theorists with quite different opinions on some issues, as well as a common focus on social or linguistic construction). Like realism and idealism, I nevertheless believe it is still possible to show that process philosophy is compatible with most of the ideas and practices from this paradigm. Let us use the contrasting works of Habermas (1976, 1984a,b) and Foucault (1980, 1984a) to illustrate (although I should note that both authors have proposed elaborate theoretical networks of ideas that cannot be summarised adequately in a couple of paragraphs).

Let us begin with Habermas who, as we saw in Chapter 2 and earlier in the current chapter, proposes a theory of language. Language gives participants in debate the freedom to question the intelligibility, truth, rightness and/or sincerity of any statement. Habermas also proposes a normative ideal for debate: we should pursue "undistorted communication" where everybody is completely free to question intelligibility, truth, rightness and sincerity—and this freedom implies that forces of power are neutralised at both the micro and macro levels. At the macro level, Habermas (1984a,b) argues that economic forces have created a situation where pressures for instrumental reasoning (reasoning about *how* to do something rather than why people would want to do it) are creating a distortion of local speech situations so that arguments around truth claims are still possible, but arguments about rightness have become marginalised. This is a systematic distortion in society which is resulting in what Habermas calls the 'colonisation of the life-world by the system' (my paraphrase). The "life-world" is the sum total of social practices, inscribed in language, which makes life meaningful to human beings within society. The "system" is made up of the steering mechanisms in society—money, law and power. Therefore, the system is colonising the life-world in the sense that concerns about, for example, efficiency and cost-effectiveness (and the paraphernalia of measurement that accompanies these concerns) are intruding into all aspects of social life. The antidote, according to Habermas, is the rejuvenation of civil society to create space for debates about rightness. This will restore balance once again to local speech situations and remove the macro level distortion from communications.

In terms of process philosophy, what Habermas is doing is conducting a second-order analysis of communication, bounding it in terms of language. He views the life-world—the reservoir of meaning in society—as being inscribed in language, but subject to the influence of the steering mechanisms. Because of the unavoidability of people interfacing with and discussing the steering mechanisms as part of their everyday activities (e.g., when using money), a whole set of language games are elaborated which come to dominate other understandings in the life-world. What is excluded from Habermas's (1984a,b) analysis, however, is any consideration of the biological roots of human beings within ecosystems.[56] Nevertheless, the main point in terms of this chapter is to point out that, because we can show that Habermas takes the primary knowledge generation system as a 'linguistic community', and all his prescriptions for change (revitalising civil society to create space for debates on rightness) flow from this assumption, his work can be shown to be based on one particular second-order boundary judgement, and therefore all his ideas are perfectly compatible with process philosophy.

Now let us move on to the work of Foucault (1980, 1984a) which, while sharing many similarities with that of Habermas, diverges from the latter in its view of power (see Fay, 1975; Smart, 1983; Couzens Hoy, 1994; Kelly, 1994; and Ashenden and Owen, 1999, for comparisons). Foucault argues that there is a complex, interactive relationship between knowledge, power and identity. Knowledges (especially, in the modern era, those generated by the applied sciences, such as psychology) provide the basis for both defining the identities of people and for the exercise of power in the form of social control. For example, knowledge about rationality, insanity and the confinement of those labelled 'mentally ill' all go hand in hand (Foucault, 1961; and also see Rose, 1990, and Dávila, 1993).

In contrast with the work of Habermas (1984a,b), for Foucault power is not 'owned' by anybody: it resides in the development of forms of knowledge which people use to order their social relationships. What appears on the surface to be one person exercising power over another is actually the *end result* of a process of knowledge formation in which certain social practices come to be legitimated. Hollway (1991) provides an interesting example of this: what is perceived as the 'power' of the manager over the worker is a result of the formation of knowledge about what 'management' actually is. Foucault talks in terms of "power-knowledge" because of the intimate relationship between these two concepts: once knowledge has been used, on the one hand, to define the identity of subjects, and on the other hand to

[56] This is interesting because, in his earlier work, Habermas (1972) proposed just such a theory which he later abandoned because of criticisms from both social constructionists (e.g., Foucault, 1980), who smelled a whiff of universalism in Habermas's work, and environmentalists (e.g., Eckersley, 1992) who highlighted Habermas's questionable assumption that our relationship with nature is one of domination and control.

legitimate actions of social control, the resulting exercise of 'power' (in the Habermasian sense of one person determining what happens to another) simply reinforces the total power-knowledge system. A good example can be seen in Foucault's (1977) analysis of the legal system, where the whole knowledge base (supported through applied disciplines like criminology) enables the labelling of 'criminals' and the construction of apparatuses of social control (such as the building of prisons). Historically, once the system reached the point where it was seen as legitimate to incarcerate 'criminals', the action of incarceration simply reinforced what had become the status quo.

Once again, Foucault's work is compatible with process philosophy. Just about all his writings have as their primary focus the production of second-order theories (about the nature of knowledge generating systems). The role of his first-order inquiries—e.g., his examinations of mental illness (1961); scientific understandings of the natural world (1970); the establishment of hospitals (1973); the criminal justice system (1977); and human sexuality (1976, 1984b,c)—all support his second-order reflections. Foucault bounds knowledge generating systems to include within them knowledges produced through disciplines, power, and the identities of human beings. While some might criticise him for excluding biological reality (see Levy, 1999, for a discussion of this issue), he justifies this on the grounds that biological theories are part of the power-knowledge networks he writes about: such theories therefore cannot specify *a priori* truths (Foucault, 1980). Indeed, all truths (from Foucault's, 1980, point of view) can only be accepted as such because of their resonance with previously established power-knowledge. Hence, Foucault's almost exclusive focus throughout his life on *critique*—defined as the historical demonstration of the construction of knowledge walking hand in hand with power (Foucault, 1984a).

Therefore, while Foucault's method of critique is certainly compatible with process philosophy (in that it involves the setting of particular first- and second-order boundaries), if he were still alive and reading this it is unlikely that he would accept my call for theoretical pluralism. This is because it involves seeing truths as actual statements about the world at some moments (when second-order boundaries around material entities are employed) and as socially constructed devices for purposes of control at other moments (when second-order boundaries are confined to power-knowledge and discourse). If I have understood his work correctly, Foucault was only willing to talk about things in terms of the latter (even though, as Habermas, 1985, demonstrates, he still had to make truth claims about the nature of truth itself).

It is because there are still clear differences between my own more pluralistic perspective (based on process philosophy) and the positions of those (like Foucault) within the three paradigmatic traditions of philosophy, that I do not claim to be *subsuming* those positions. To reiterate an earlier point, I am setting out a *new* philosophical agenda that I argue allows us to make all the kinds of statements associated with valid practice in the three paradigms without slipping into a dogmatic insistence (sometimes found within these paradigms) that there is only one correct boundary to work with.

4.14 Conclusion

In this chapter, I have described how process philosophy treats subjects and objects in the same way: in each case they are identified through a process of making boundary judgements [but in the case of subjects, it is often necessary to use two boundaries—one defining the relevant sentient being(s) involved, whatever theory is used to do this[57], and the other defining a wider knowledge generating system]. Because subjects and objects are defined through the same process (of first- and second-order boundary judgements), I believe we can claim that this overcomes the subject/object dualism we identified in the work of the systems writers reviewed in Chapter 3.

This claim is based on the observation that a new analytical (rather than ontological) dualism has been established between process and content. Both subjects and objects are viewed as *content* defined through the *process* of making boundary judgements. It is therefore important for me to be explicit that dualism itself has not been swept away. However, if the assumptions of process philosophy are accepted, then ontological *subject/object* dualism can be removed from the centre stage of philosophy.

I also believe that process philosophy can provide the grounds for a new theoretical pluralism that will allow human beings to be centred or decentred in analyses, depending on the purposes and values being pursued. Thereby, a much more flexible, critically aware form of systemic intervention than many of those currently practised can actually be developed—if we are prepared to put the time and energy

[57] This need not be a theory of sentient beings as physical systems, but might focus only on the contents of consciousness (e.g., Kelly, 1955).

into the necessary research. The second and third sections of this book seek to make a contribution to this endeavour.

II

Methodology

5

Why Methodology?

Having outlined my own version of process philosophy in the first section of this book, in this second section I discuss methodology—starting, after these introductory few paragraphs, with an answer to the basic question, why methodology? I then move on, in Chapter 6, to consider the meaning of the term 'systemic intervention'. I argue that all uses of method are interventionary, including scientific methods for structuring observations. However, *systemic* intervention is something more specific: it refers to intervention that embodies pursuit of the ideal of comprehensiveness. As *absolute* comprehensiveness is impossible (see the argument in Chapter 3), an adequate methodology for systemic intervention must facilitate *considerations of issues of inclusion, exclusion and marginalisation* by promoting reflection on boundaries. It should also allow for theoretical and methodological pluralism. An outline methodology for systemic intervention is presented at the end of Chapter 6, which is then fleshed out in subsequent chapters.

Following this analysis, in Chapter 7 I build on the boundary idea already introduced in the first section on philosophy in order to propose a normative (prescriptive) theory of *boundary critique*. This describes the essential relationship between boundary and value judgements made by human agents, and a systemic model of human conflict is presented that I have found particularly useful to inform reflections during intervention (see also Chapter 14).

Boundary critique gives rise to the possibility of embracing theoretical pluralism. This is because different theories imply different boundaries of analysis, meaning that choice between boundaries also involves choice between theories. The idea of theoretical pluralism is explored in Chapter 8, and it will be argued that, while universal standards for choice between theories cannot be devised, this doesn't imply the deterioration of standards and a descent into absolute relativism. Chapter 8 will lay the foundations for a normative (prescriptive) model of interventionist learning that will be presented in Chapter 11.

In my view, although boundary critique and theoretical pluralism are both necessary if we are to call intervention 'systemic', they are not sufficient by themselves. It will usually be important to develop an intervention using appropriate methods of engagement with other participants in the situation, beyond the discussion of boundaries and theories. In Chapter 9, the need for pluralism in the use of methods is discussed. This need is grounded in the observation that there is no one method, or set of methods, that can deal with all eventualities. Fortunately, a massive literature on intervention methods and methodologies has been produced during the 20th Century, providing a substantial resource for practitioners willing to embrace methodological pluralism. A short (and inevitably incomplete) review of this literature will be undertaken in Chapter 9, focusing in particular (but not exclusively) on methodologies and methods developed by the management systems community, and references will be provided to key texts to enable interested readers to conduct their own explorations and develop an appropriate armoury of methods of intervention.

Having established the need for methodological pluralism, Chapter 10 then goes into more detail, explaining a strategy for mixing methods during systemic intervention. Finally, in Chapter 11, I address three important arguments that have been raised against methodological pluralism: (i) that it is not theoretically coherent because different methods embody the contradictory assumptions of different paradigms; (ii) that it is not culturally feasible because academic research communities have vested interests in promoting single methodologies and methods; and (iii) it is not psychologically feasible because it requires too much intellectual effort from interveners. Interveners are said to have psychologically ingrained preferences and too little time to become proficient practitioners of more than a narrow range of methods (Mingers and Brocklesby, 1996; Brocklesby, 1997). In answer to these criticisms, I propose a model of interventionist learning about theory, methodology and methods.

However, let us start this section with a very basic question: why methodology?

5.1 *Why Methodology?*

In Chapter 2 I gave an answer to the question, why philosophy? That chapter was motivated by the need, as I see it, to counter the arguments of interveners who look down on philosophy and declare it irrelevant to systemic intervention. I argued that it is *very* relevant: both in substantive terms (philosophical analysis can reveal hidden assumptions embedded in methodologies and methods) and strategically (we should not accede the philosophical high ground to those who frown on intervention and favour supposedly 'value-neutral' science). In a similar manner to Chapter 2, this chapter answers the question, why methodology? It is aimed at three kinds of reader: those who like to keep their systems philosophy 'pure' (untainted by discussions of methodology and practice); those who believe that a focus on methodology encourages purely 'instrumental' thinking (forcing thought into a strait-jacket dominated by a concern with the means to reach pre-defined ends); and those who believe it is acceptable to follow an atheoretical line, using methods simply as tools without any explicit methodology at all.

However, before entering into the argument in favour of methodology, let me clarify some terminology—in particular, the meaning I ascribe to the term 'methodology' itself. I will then argue against both philosophical purists and those who believe that methodology enforces instrumental thinking. Finally, I will tackle the arguments of those who are prepared to separate methods from methodology, and use the former while discarding the latter.

5.2 Method and Methodology

Many authors use the terms 'method' and 'methodology' interchangeably, especially in the management science and operational research communities. In my view, this is rather unfortunate: in writings on the philosophy of science, and also in some of the systems literature (see, for example, Checkland, 1981), 'method' and 'methodology' have a distinctive meaning that can be most useful. A 'method' is a set of techniques operated in a sequence (or sometimes iteratively) to achieve a given purpose. A 'methodology' is the set of theoretical ideas that justifies the use of a particular method or methods. When an operational researcher says "I designed a new methodology to deal with this circumstance", s/he is usually talking about a method, not a methodology (at least in the terms that both Checkland and I use). If one wanted to be cynical, one could say that this degraded use of the term 'methodology' is a symptom of the 'dumbing down' of operational research: treating methodology as method places the theoretical and political assumptions made in the construction of methods beyond critique.

Of course, methodology is not a wholly discrete area of study. There is often a blurring of the boundary between methodology and philosophy: some philosophical ideas may feed into methodology (and vice versa). There can also be a blurring of the boundary between methodology and practice, in that practice is very often interpreted by interveners in the light of a particular methodology (Romm, 1995a), and an intervention methodology that is not informed by practice would be strangely contradictory (practice, as I am using the term, is the practice of intervention). This blurring of boundaries is not a problem: it is partly why I believe it is necessary to cover all three fields of inquiry (philosophy, methodology and practice) in a book such as this, and show their inter-dependence.

One thing that all methodologies have in common, however, is a concern with the validity and/or legitimacy of methods. The term 'validity' is generally used by proponents of observational science: if a method is valid, it yields knowledge that reflects reality without known distortions or intervention by the observer. However, those (like myself) who believe that truly independent observation is impossible (see Chapter 6) tend to avoid the word 'validity' and talk about *legitimacy*. If a method is legitimate, it is viewed (by the researcher, stakeholders and/or other interested parties) as *appropriate in the circumstances*.

5.3 The Argument against Philosophical Purism

The argument for talking about methodology, and against philosophical purism, is essentially a moral argument. While I find issues of ontology and epistemology fascinating in their own right, I am also moved by my feelings when I encounter what I see as injustice and destructive greed[58]: hunger in the midst of plenty; victims of preventable disease; the atrocities of war; abuses of human rights[59]; unsustainable economic growth; the plunder of the rain forests;

58 The link I have made in this sentence between morality and feelings is not meant to indicate that morality should be seen solely in terms of the emotion of the individual. MacIntyre (1985) offers a strong argument against what he calls "emotivism": this is the use of a degraded understanding of morality, brought about in modern societies by liberal individualism. There are two aspects to the degradation: (i) moral decision making is seen solely as an individual rather than a community affair; and (ii) morals are seen as emotional commitments only, so the possibility of *considering* moral issues (through personal reflection and/or debate) becomes unthinkable. Nevertheless, in my experience, it *is* the case that feelings are involved in moral understanding—but that does not mean that the value of reflection and debate on moral issues should be neglected.

59 My own view of human rights is that they are not absolute, 'natural' or God-given. Rather, they are relative and subject to debate. However, minimally acceptable standards for the treatment of human beings can still be defined with sufficient rigour to enable legislation to be framed, and it is this legislation that gives meaning to the concept of 'rights'.

violation of tribal lands and cultures; unnecessary cruelty to both humans and animals; the abuse or neglect of children; discrimination on the grounds of gender, race, disability, age, sexuality, religion, etc. The list is long, and I am sure that the vast majority of readers share these feelings.

Given the scale of injustice, cruelty and greed in the world, and the complexities of defining them anywhere near adequately and in a manner with which others can agree, we inevitably ask ourselves, what should we do? We each have just one pair of hands, and limited time on this Earth. As distressing events are presented to us in discrete packages by the media (hiding the systemic links between issues), we tend to make our own priorities among 'worthy causes'. To an extent, systems methodology can provide an antidote to this kind of piecemeal thinking, although an issue-based practice is difficult to avoid given the complexities and sheer scale of some of the problems we face, the bounded nature of our understandings, and the need to keep intervention on a 'human scale' (giant projects tend to have many unforeseen side-effects).[60]

Perhaps more importantly, however, we also have to prioritise moral action for the wider social good (beyond personal or family benefit) in relation to other forms of action, including action in pursuit of personal advancement, knowledge for its own sake, and pleasure more generally (these forms of action may be seen as moral or amoral depending on the context). It is in making these kinds of choices that I find it hard to justify philosophy purely 'for its own sake', turning one's back on issues of injustice and cruelty. Indeed, as I see it, the pursuit of personal pleasure (including that gained from pure philosophy) is hollow if no heed is paid to the needs of others: ultimately, we are connected through natural and social systemic relationships with those others, so can expect a negative reaction to purely selfish action. This reaction may not be direct, but may come in the form of 'systemic readjustments' which take place within the wider systems in which we are embedded. An example of a systemic readjustment is the latest phase of capitalist development in the West, which is requiring many workers to spend more and more time at work, and away from their families, so that industry can produce the material goods (beyond those

[60] An example of a giant project which has ignored the side-effects of human misery and environmental destruction in the name of 'industrial progress' is the decision to build the Three Gorge Dam in China (Zich, 1997). This is now under construction despite the fact that the Chinese government commissioned an evaluation from a group of systems practitioners of the likely social and ecological effects of the dam. The evaluation recommended that the project should not go ahead, but this finding was set aside by the government and was never made public (Midgley *et al*, 2000).

that are needed to maintain a sustainable and reasonably comfortable existence) that these workers and their families wish to acquire: in this case, short-term selfish acquisition (and pressure from employers when acquisitive motivation breaks down) leads to an impoverished quality of life when these workers eventually realise that they have little leisure time left to enjoy the fruits of their labour (Sachs, 1999). One does not need a mystical idea like Karma to understand this phenomenon: there are many systems theories that can help to explain these effects (e.g., Bogdanov, 1913-17; von Bertalanffy, 1968; Miller, 1978).

It seems to me that the philosopher who refuses to engage in *applied* philosophy is making a moral choice, even if s/he is unaware that this is the case: s/he is choosing intervention in narrow philosophical discourse over intervention in wider discourses with significant life-consequences—and personal gratification over all notions of the wider good. Philosophy becomes *applied* philosophy when consideration is given to the consequences for both discourse *about* change, and action *for* change. Methodology is one particularly important vehicle through which philosophers can apply their ideas: it is through methodology, which sweeps in philosophical reflection, that we can better understand how methods of intervention can be used to create and sustain valued personal, social and ecological change.

Lying behind this view is a theoretical understanding (explored in more detail in Chapter 7) that, in the case of human agents (together with the knowledge generating systems of which they are a part), boundary and value judgements are intimately connected. In other words, if excessive attention is paid to a narrow boundary of intellectual inquiry (philosophy), marginalising everything to do with methodology and practice, then (unsurprisingly) the values pursued by the philosopher are likely to reinforce this narrow boundary. Conversely, if a wider boundary is used, admitting issues of injustice, cruelty and greed (defined in terms other than the purely philosophical), then the values that it is possible to pursue will also be widened.

This moral stance is certainly not new to philosophy: for example, it was a cornerstone of the Pragmatist movement at the turn of the 20th Century. Authors like James (e.g., 1904), Pierce (e.g., 1934), Dewey (e.g., 1946) and Singer (1959) argued for a morally committed philosophy which, instead of pursuing a Grand Truth, viewed 'truth' as 'what works in practice'. However, theirs was not a naive notion of 'working in practice', but one which required a significant effort of inquiry to tease out the assumptions underlying what it means to say that something 'works'. While some (in my view justifiable) scepticism has surrounded a few of the claims of the Pragmatists—particularly the desire of Pierce to find a universal basis for validating knowledge in action (Rorty, 1989) and Singer's over-emphasis of the power of mathematics to solve problems (Churchman, 1987)—their basic argument that philosophy should have a practical face in a morally challenging world still stands. One such practical face is indeed methodology, where the meanings of methods of intervention can be explored.

5.4 The Spectre of Instrumental Rationality

One argument against a focus on methodology and methods is that it encourages purely instrumental thinking. A number of authors writing in the 20th Century (e.g., Marcuse, 1964; Habermas, 1984a,b) have contrasted *instrumental* rationality (rationality harnessed in the service of meeting some pre-defined end) with *practical* rationality (which enables moral reflection and the pursuit of mutual understanding). Marcuse, for example, argues that a significant problem in modern capitalist societies is that instrumental rationality has become a dominant force, and practical reason has become marginalised. Thus, people are able to think seriously about developing the best means to meet their ends, but meaningful discussion of the ends themselves is downplayed or degraded (or even, in the discourses of traditional science, labelled 'unscientific' and put to one side).

Churchman (1970) strongly criticises mainstream writers in operational research and management science who are almost exclusively concerned with developing techniques for applying mathematics to the solution of discrete problems. Essentially, these operational researchers and management scientists provide managers with the *means* for solving problems without subjecting the *ends* they are pursuing to any scrutiny. Thus, they serve the political and organisational status quo, regardless of whether or not this can be morally justified—which is indeed the practice of instrumentality. However, Mvula (1999) aims the same argument at me. He suggests that, because I champion a focus on methodology, I encourage the reader to neglect philosophy and theory, which are equally important, and which take us beyond merely instrumental reason.

I have three answers for Mvula, and any others who might share his concerns. First, contrary to Mvula's assertion, I am explicitly interested in the development and use of theory (see Chapters 8 and 11). Second, because I argue in favour of methodological pluralism (Chapters 9-11), I am equally interested in methods for critiquing ends as I am in methods for meeting those ends. Therefore, the methodology I am proposing in this book is specifically designed to enable change agents to *transcend* purely instrumental rationality. Of course, this is somewhat paradoxical in that I could be accused of *instrumentally* transcending instrumentality! This is where my third answer comes in. It should be obvious throughout this book that I am *not* only concerned with methodology, but wish to see it as an essential part of the trinity of philosophy, methodology and practice: like many other authors, I argue that we should indeed look "beyond method" (Morgan, 1983), but this should not imply the abandonment of methodology and method. Seen in relation to philosophy and practice, and incorporating a focus on the critique of ends as well as the development of means, I believe that methodology cannot easily slide into the instrumentality that Mvula (1999) and I both wish to challenge.

5.5 *Arguments against the Atheoretical Use of Methods*

Having addressed philosophical purists and those wishing to avoid instrumentality, we can now move to the other extreme and ask why we should care about methodology and not just look, in an atheoretical manner, at which methods 'work' in practice. A superficial interpretation of the writings of the Pragmatists might lead one to do just this, but of course the Pragmatists were very keen to interrogate the meaning of any claim that something 'works', and present-day interveners can still learn a great deal from them (Brauer, 1995). The Pragmatists were in no sense anti-philosophy, let alone anti-methodology: they merely believed that philosophy should have practical relevance. In my view, it is a shame that the term 'pragmatism' has been degraded over the course of the 20th Century: in common use it now means practical *as opposed* to theoretical, whereas the original Pragmatists celebrated the fact that *a good theory has significant practical implications.*

Using the common, degraded understanding of 'pragmatism', several authors writing in the management systems literature (Jackson, 1987a; Flood, 1989a,b, 1990; Midgley, 1989b) have argued against atheoretical 'pragmatism' and in favour of a theoretically-informed approach to methodology (I will keep the word 'pragmatism' in parentheses to indicate that this is the degraded use of the term). 'Pragmatism' is defined by Jackson (1987a), building upon previous work by Reed (1985), as follows:

> "The pragmatist strategy is to develop management science by bringing together the best elements of what may appear to be opposing strands [of management and systems thought] on the

> criterion of what 'works' in practice. Pragmatists are distrustful of theory, believing that the wranglings to which it gives rise distract attention away from management science practice.... Pragmatists, therefore, do not worry about 'artificial' theoretical distinctions. They concentrate on building up a 'tool kit'.... Proven techniques from different strands of management science are employed together in the course of problem-solving if the situation warrants it. The choice of techniques and the whole procedure is justified to the extent that it brings results in practice" (Jackson, 1987a, p.462).

Flood (1989a) adds the following:

> "The pragmatist may be seen as someone who has a systems tool bag....which....is used in an analogous way to cathedral building of old. The craftsmen were able to build complex structures using their own tool kit but had no idea why the thing stood up, why a beam fixed one way cracked but fixed another way did not. They only knew how to do it from the practice of trial and error...." (Flood, 1989a, pp.78-79).

These atheoretical 'pragmatists' have been criticised on a number of grounds. The following points have been distilled from the works of Jackson (1987a) and Flood (1989a). First, the trial and error approach means extensive and costly experimentation in the social domain. Theory is needed to develop understandings of *why* methods sometimes work and sometimes do not, so that people can learn more effectively from their mistakes. Second, 'pragmatists' without a common theoretical language find it difficult to pass their knowledge on to others—theory enables communication between practitioners and even across disciplines. Third, what may appear to 'work' in the short term might have disastrous consequences in the longer term: theory is needed to expand our understandings of what it means for a method to 'work'. Finally, because 'pragmatists' are not concerned with *the terms in which* methods 'work', their activities may unwittingly lend support to authoritarian practices—after all, methods often work, "not because they are the most suitable for the situation in which they are employed, but because they reinforce the position of the powerful, and implementation is therefore enforced" (Jackson, 1987a, p.464).

Of course, all these uses of theory—to examine the strengths and weaknesses of methods; to interrogate what it means for a method to 'work'; to differentiate between the application of methods and the effects of authoritarian power relations; and to communicate insights to others—are all essentially methodological. I therefore suggest that engaging in methodological discourse is vital if a superficial and potentially dangerous form of intervention is to be avoided.

5.6 Conclusion

In this short chapter I have defined methodology as a set of theoretical ideas that justify the use of a particular method or methods. I have argued against philosophical purism on the grounds that turning one's back on the suffering of others in favour of 'pure' philosophical reflection is a moral choice, taken either wittingly or unwittingly—and one that is hard to defend given the embeddedness of all people (including philosophers) in wider social and ecological systems. I have also argued against so-called 'pragmatists' who advocate the use of methods as simple tools without methodological reflection: it is methodology that allows us to examine the strengths and weaknesses of methods, and to ask what it means for a method to 'work'. Having made the case for methodological inquiry, I will start my own methodological reflections in the next chapter with an examination of the meaning of the term 'systemic intervention'.

6

Systemic Intervention

Having answered the basic question 'why methodology?', I can now begin to lay out my own methodological ideas, starting with a definition of the term 'intervention'. This will provide a foundation for bringing together, at the end of this chapter, the systems idea (the meaning of which was touched upon in Chapter 3) and 'intervention' in order to produce a definition of *systemic intervention*. Finally, it will be possible to present an outline of the systems methodology I have developed, that will be fleshed out in the coming chapters.

To give an initial definition of intervention, it simply means *purposeful action by a human agent to create change*. In discussing process philosophy in Chapter 4, I was keen to include non-human sentient beings as knowledge generators. However, in moving on to discuss methodology, I intend to set aside the activities of non-human animals and focus on human agency alone. This is not to say that the activities of non-human animals cannot be described as interventionary (they most certainly can), but it would be pointless to produce a *methodology* for non-human use. Methodologies are constructed using language. Therefore, the definition of intervention provided above should not be considered a *general* definition. Rather, it is a *methodological* one—specifically relating to human action alone.

Of course, even though we have now eliminated non-human animals from discussion, it should be acknowledged that what constitutes a human agent is not necessarily a simple matter to identify. Actions can be ascribed to a variety of possible agents: e.g., an individual person; a group; a team; a family; an organisation; a community; a nation; etc. Note that this list is very similar, but not identical, to the list of knowledge generating systems it is possible to define (see Chapter 4).[61] It is therefore necessary to discuss the meaning

[61] The difference is that each of the above are exclusively human systems, albeit ones which interact closely with non-human environments.

of the term 'agent' in relation to the understanding of sentient beings and knowledge generating systems produced in that chapter.

To refresh our memories, I argued in Chapter 4 that a sentient being or beings (which can be defined in many possible ways), embedded within a knowledge generating system (which can also be bounded in a variety of ways), can be seen as giving rise to knowledge. Where the boundaries are placed around the sentient being(s) and the wider system depends on the theories being employed in local contexts [in some cases the sentient being(s) and knowledge generating system will be regarded as synonymous, in which case the sentient being(s) will be viewed as autonomous]. Here the concept of 'agent' can be introduced. I suggest that an agent is either a single human being, or an identifiable group of human beings in interaction (e.g., a family, team or organisation), that have purposes ascribed to them. In the case of a group, this definition does *not* assume that all participating individuals are seen as sharing the purpose of the whole (indeed, some sub-agents may be seen as acting in opposition to the dominant purpose), but a group can be called an agent when it (or its representatives) are perceived as acting to realise a dominant purpose at the group level regardless of the actions or views of sub-agents. The word 'dominant' here is crucial, as it indicates that the group purpose is a function of whatever mechanisms of legitimation exist within and beyond the group (whether autocratic, democratic or bureaucratic) that allow the group as a whole to be perceived as moving in one particular direction, regardless of any counter-arguments being produced by internal opponents. Therefore, when a government minister declares war on behalf of a nation, it is generally accepted that the nation is at war even if half of its citizens do not wish to support this.

Because the precise definition of a sentient being has been left open (allowing theoretical pluralism and multiple possibilities for boundary judgements), the exact specification of an agent needs to be just as open. *Exactly* what constitutes a human being, or a group of human beings in interaction, or a mechanism of legitimation (allowing a group purpose to be visible), can only be determined in local contexts using particular theoretical understandings.[62] Of course, the action of an agent is taken on the basis of knowledge (defined widely to include perceptions, implicit understandings, unconscious motivations, behavioural habits, etc.), so *action can be said to be undertaken by an agent under the influence of the*

[62] Here, the term 'local' does not necessarily indicate *geographical* locality. A 'local context' is one where particular (non-universal) conditions apply. All contexts are local (including ones where global issues are being addressed) in the sense that a limited set of agents develop their understandings and take action in relation to the particular conditions that appear to obtain.

knowledge generating system in which s/he is embedded (and which, as we saw in Chapter 4, has to be defined using a boundary judgement).

Now, if the identity of the *agent* can be variously defined in the light of process philosophy, the concept of 'action' must be variable too. While 'action' assumes an effect, how the action and effect are understood will depend on the theories being used and the boundaries being drawn in any analysis. An effect could be on the material world; on a subjective interpretation; on a system of knowledge; on language; on others' actions; on ecological equilibrium; etc. As it is possible to use any from a huge variety of boundary judgements, and it is consequently legitimate to draw upon any corresponding theory (in the knowledge that all theories imply boundary judgements and are therefore partial), it would be contradictory to establish a theory of action in this book—other than to welcome a variety of theories of action that people might want to use in a pluralistic manner (in a manner that does not imply only one such theory is valid in an absolute sense).

Having produced an initial definition of intervention, and having highlighted the plurality of potential understandings of particular interventions implied in it, I will now go on to explore how the concept has been used by others (not necessarily in the same way that I use it). Traditionally, ideas about intervention have been contrasted with those about 'observation', and this will be the starting point for my analysis. However, after comparing these two concepts, I will seek to show that the distinction between observation and intervention is not as simple as it might at first appear. Indeed, I will argue that observation should be viewed as a 'special case' of intervention. As we shall see, this has profound consequences for understanding the relationship between 'science' (which has observation as its traditional focus) and other activities that are more obviously concerned with intervention (e.g., policy making, personal and/or group decision-making, management and community development).

6.1 Observation versus Intervention

Many writers contrast observation and intervention: it appears that both scientists (who champion observation) and action researchers[63] (who champion intervention) have an interest in

[63] There are others in the 'intervention camp' too, such as operational researchers, management scientists, evaluators and systems practitioners. These labels refer to people in a variety of semi-independent research communities who have similar interests, but slightly different emphases.

maintaining this pair of concepts in opposition to one another. Let us start with the views of the scientific camp.

6.1.1 Observation as the Basis of Science

While many philosophers of science have discussed observation, Popper (1959, 1972) is arguably the best known. Popper claims that, to be worthy of scientific attention, "[an] event must be an 'observable' event; that is to say, basic statements must be testable, inter-subjectively, by 'observation'" (1959, p.102). The assumption is that information provided by the senses (observations) will naturally provide a platform for consensus, thereby guaranteeing inter-subjective agreement. Hence, traditional science seeks to place all statements that cannot be tested by observation outside its remit.[64]

The reliance of Popper's (1959) understanding of science on observation leads him to propose the idea that, fundamentally, science can be differentiated from non-science by the methods that are used: if an approach is to be called scientific, it must use methods that enable high-quality, independent observation. Hence the emphasis in most traditional scientific methodologies on quantitative comparisons between 'experimental' and 'control' conditions. For example, a psychologist may wish to test the hypothesis that, in a simple memory test, grouping numbers in sets of three will improve recall compared with numbers left ungrouped. Two comparable cohorts of people (in terms of gender, age, occupation, etc.) can be given the same numbers, either grouped into sets (the experimental condition) or not (the control condition). Only if the experimental cohort display superior recall (on average) compared with the control cohort can the hypothesis be confirmed in a manner that can be called scientific.[65] Of course, statistical tests of significance may also be used (assessing the probability that the result may have happened by chance), and any number of additional variables may be introduced (for example, a scientist may wish to test whether the difference between recall of grouped versus ungrouped numbers is affected by gender, age, alcohol consumption, etc.). However, what I have described is the scientific

[64] Of course, the validity of this understanding of science rests upon the further assumption that the 'observer' and the 'observed' are independent of one another. If they are actually *inter*dependent, then observations would be just as much a property of observers (with all their peculiar interests, idiosyncratic cultures and power relations) as of the observed. See Chapters 1 and 3 for further discussions of this theme.

[65] This hypothetical example has been influenced by my reading of an experiment conducted by Miller (1956), but is not directly based on it.

method at its most basic. See Wright *et al* (1970) for a more detailed introductory discussion of the need for controlled observations.

Arguably, one of the most important aspects of controlling observation, as far as many scientists are concerned, is the need to prevent *intervention*. The observer should not influence the observed, other than by establishing the required difference between the experimental and control conditions, otherwise the results of the observation could be due to the activities of the scientist rather than the variable(s) under investigation.

6.1.2 Intervention as the Basis of Action Research

In marked contrast with Popperian science, action research is concerned primarily with intervention and not observation (some action researchers, such as Reason, 1996, disagree with using the term 'intervention', but I will deal with this argument later). The birth of action research is widely attributed to Lewin (1947, 1948), who argues that the focus of the philosophy of science on independent observation creates a divorce of the scientific method (especially as it is used in the social sciences) from social practice. He stresses that science should be harnessed for the benefit of human society, and this requires a very different set of philosophical and methodological ideas from those traditionally associated with the sciences.

To appreciate why action research emerged in the mid-20th Century, and gained a great deal of popularity very quickly amongst many people working outside academia (even though it only occupied a marginal position in the academic scientific community), it is necessary to understand the orthodoxy that was being propounded at the time. Popper had been writing about the importance of experiment and observation since the 1930s, and his work built on previous philosophies of science that also placed independent observation at the centre of scientific practice. While there were strong debates about the extent to which human knowledge is fallible, the orthodox view was that the need for independent observation was not in question. It began to appear to many people that the reasons or purposes for undertaking scientific research were secondary to the robustness of the methods used (this was certainly my own perception as a student of psychology graduating as late as 1982). Some scientists advocated a radical denial of purpose, saying that all organisms, including human beings, are deterministic 'learning machines' (e.g., Skinner, 1971). Even if the existence of purposes was accepted, such purposes could not be considered 'scientific' in the same sense as observations; they were generally omitted from reports of experimental practice, and could often only be deduced by reading between the lines of hypotheses. In this way, the purposes and debates that made the hypotheses meaningful were largely hidden from view.[66]

It was in this atmosphere that Lewin (1947, 1948) mounted a strong critique of 'pure' science in favour of action research. Lewin's argument is that the institutions of science invest massive resources into research that has largely become divorced from the goals of meeting human need and satisfying human desires (that is, the desires of those outside the scientific community—the latter tends to value knowledge for its own sake). In Lewin's view, it is generally a matter of accident whether this research is relevant to people working in industrial and welfare organisations. Of course, there are the 'applied' natural sciences[67], like medicine, but really nothing comparable for the worlds of industry and human welfare where it is much more difficult to control observations.

Essentially, Lewin (1948) advocates the harnessing of science in the service of *intervention* rather than observation. That is, science should be undertaken in organisations for social benefit. He believes that scientists have a choice: they can either conduct research for the sake of pure curiosity, or help themselves and others improve the social conditions that surround them. When a problem is encountered in an organisation, research may be undertaken to help define a way forward. However, social purposes should not be subordinated to methodological purity: in Lewin's view, if research is being conducted in support of action, it makes little sense to subvert the purposes that guide that action in the name of scientific rigour. This means, for Lewin, 'adapting'

---

[66] Given the dominance of this way of thinking, it is possible to see why the work of von Bertalanffy (1968) appeared so radical at the time. By claiming that organisms are purposeful, he was flying in the face of orthodoxy. In the light of more recent systems theories, however, von Bertalanffy's work appears to be quite strongly influenced by the traditional scientific focus on experiment and observation (Midgley, 1998): he was really only concerned with the production of an organismic, general systems theory—which would have to be validated by scientific means. He also made many of the same philosophical assumptions as Popper (see Chapter 3). It appears that the emphasis of our interest has shifted in the last thirty years: what was once a radical proposition—that organisms can be said to have purposes—is now a generally accepted truth, making the differences between von Bertalanffy and his opponents seem less meaningful to us now than they were in 1968.

[67] The terms 'pure' and 'applied' science refer to whether or not observation is being undertaken for some social purpose. Microbiology is a 'pure' science, insofar as observations of organisms at the microscopic level are conducted just to gather knowledge for its own sake or to prove a point in a debate (which does not have any immediately obvious wider significance). Microbiology becomes 'applied' when observations are undertaken, say, to identify a particular virus that causes a disease in order that research may be started on how to treat it. Of course, 'pure' science may inadvertently give rise to applications, and 'applied' science may give rise to 'pure' knowledge, but the two kinds of science can nevertheless be distinguished by the primary purposes that motivate them.

the scientific method to make it more meaningful in social situations: instead of testing hypotheses, scientists should identify questions that need answering. Likewise, if it is impossible to set up perfectly controlled conditions, they should not call research 'invalid', but should still generate data in a manner that supports decision making—even if *strongly* scientific conclusions cannot be reached. After all, organisational decisions will have to be taken anyway, and it is preferable to take them on the basis of imperfect data than using no data at all.

Of course, embedding scientific practice in social situations, and adapting it in the service of intervention, will affect the stance of independent observation. Far from keeping one's distance from the observed, in Lewin's (1948) action research the observer is encouraged to eliminate socially undesirable phenomena and promote desirable ones. What counts as desirable or undesirable obviously needs to be defined by participants in the local situation, which is why Lewin (1952) produced his "field theory"—a "field" is a set of phenomena that can be seen as directly interacting with an object (person, group or organisation) of concern. The boundaries of the "field" demarcate what is and is not relevant in an analysis.[68] We see that, in Lewin's perspective, observation is not independent of the values of the observer (these values determine what initial question is asked), but is nevertheless 'factual' in the sense that a realist ontology is assumed—so observations reflect the real world (albeit imperfectly through our fallible perceptions). Also, because of the context of action which takes place over time, observations tend to be most meaningful as a sequence which constitutes feedback to actor(s) who are required to make judgements about the success, or otherwise, of their actions.

It appears that, while Lewin (1948) is primarily concerned with intervention, he does not entirely abandon observation—but it is harnessed into the service of the former. Also, where controlled observation is impossible, other means of supporting intervention through research are explored.

This work has since been developed by a variety of different authors, both in the action research and other communities. One of the most notable examples is Seidman (1988) who, following Dewey (1946) as well as Lewin (1947), advocates a much stronger opposition between observation and intervention. Instead of arguing that science should be *harnessed* into the cause of intervention, Seidman suggests that the two

[68] Lewin's (1952) field theory bears some comparison with Churchman's (1970, 1979) theory of boundaries, but I suggest that Churchman is much clearer about the need to consider the ethics of drawing system boundaries.

concepts are mutually exclusive because they are differentiated by the involvement of action: science requires the *exclusion* of action on the grounds that changing the phenomenon of interest corrupts the purity of observation, while intervention is *founded upon* action.

6.1.3 Summary of the Distinction between Observation and Intervention

At this point we have made a clear distinction between observation (as used in science) and intervention, the former being about seeing things in a manner that is not 'contaminated' by the actions of the observer, and the latter being about the actions of agents to promote change. However, it should already be apparent from the discussion of Lewin's (1948) work (above) that observation and intervention do not have to be regarded as oppositional concepts (although they often are)—observation can be undertaken in the service of intervention. Later in this Chapter, I will pick up this point and develop it further. As a result, a more detailed theory of intervention will emerge.

6.2 Arguments against Intervention from within Action Research

First, however, I wish to deal with the argument put forward by Anderson and Goolishian (1992) and Reason (1996) against using the term 'intervention'. The crux of their view is that 'intervention' has unwelcome connotations of expert consultancy, and it is preferable to initiate an unplanned, agendaless dialogue (Anderson and Goolishian, 1992) or cycles of Co-operative Inquiry (Reason, 1988a; Reason and Heron, 1995; Heron, 1996).

It is worth highlighting the fact that Reason and Heron (1995) advance much the same strongly-oppositional argument against observational science as Seidman (1988), but favour the word 'action' to 'intervention'. This is because they are part of a new movement in action research (see Reason, 1988b, for some edited readings) that is critical of the expert, neo-scientific role of the researcher in Lewin's original model. Reason (1996) is also critical of more recent participative approaches (e.g., Midgley, 1996b) where the researcher acts as a facilitator of participant-led change (here, the researcher is still in some sense an 'outsider'). In contrast, he advocates a method of Co-operative Inquiry in which the participants themselves are co-researchers: there is no need to invite someone in from outside. Participants work through cycles of group and individual reflection on issues of mutual concern. In Reason's view, the term 'intervention' refers to situations where the researcher acts as an outsider who, because s/he has something to offer that the participant group feels they need, inevitably gains the status of expert. Similarly, Anderson and Goolishian (1992) insist that an 'intervention' is a planned change where the outcome is predetermined by the intervener, placing him or her in an authoritative position in relation to others in the situation.

I am prepared to defend my use of the term 'intervention' against these arguments on three grounds. The first, in answer to Reason, is that his Co-operative Inquiry is just one of a number of useful methods available to people who wish to deal with problematic issues. Its use may be appropriate in some circumstances and not in others. I would certainly question the value of Co-operative Inquiry if people are coerced into participation, or if they participate voluntarily but do not feel that they can talk openly about some issues. This might be the case, for example, if all the participants work together in a strongly hierarchical organisation where some people routinely exercise authority over others with little participation, or even consultation. In such a situation it might be more appropriate to invite somebody in from outside to facilitate debate so that these authority relationships can be taken into account by the facilitator in the way debate is organised. Indeed, there may also be occasions when seeing the researcher as an expert is of *positive value*: a good example is provided by Flood and Romm (1995a) and Midgley (1997b) who discuss Flood and Zambuni's (1990) intervention with an African tourism company: Flood and Zambuni used their status as experts to expose corruption, thereby creating possibilities for dialogue and change that might not otherwise have existed. If it is acceptable to use a plurality of methods, some of which might be facilitated by organisational 'insiders' and others by 'outsiders', then the term 'intervention' is quite appropriate. Indeed, use of the word highlights the fact that, whether change is facilitated by an 'insider' or an 'outsider', whether it is owned by an individual or a participative group, there is still purposeful action by an agent to create change.

The second reason for keeping the word 'intervention' follows on from this. If it is legitimate to choose between a variety of methods in pursuit of change, and Co-operative Inquiry represents just one possible choice that can be made, then the act of choice itself, and the implementation of the chosen method, should be seen as an intervention. If Co-operative Inquiry is chosen, then it will have been chosen by an agent (whether a group or individual) who will have done so because, in the agent's view, it will bring about desired learning and change. This is most definitely an intervention (in the way I conceive of the term).

Indeed, a similar argument has been used by Larner (1995) in reply to Anderson and Goolishian (1992): the trajectory of their agendaless dialogues might be unplanned (which is why they say they are not interventions), but the decision taken by a group to enter a dialogue in any given situation is most certainly planned—or at least purposeful. It is possible that Anderson and Goolishian missed this point because they made the assumption that planning must involve predicting every aspect of the outcome. However, in my view, this is a rather unrealistic definition of planning. Not only is it usually impossible to predict the detailed course of an intervention (an intervener can rarely do more than predict a general direction, based upon his or her methodological knowledge and experience), but a wise intervener plans for the unexpected by making it clear in advance to all those involved that, as the situation changes and people become aware of different aspects that they want to deal with, the direction of the intervention will almost certainly need to change too. It is for this reason that I now prefer to talk about 'purposeful' rather than 'planned' action: the word 'purposeful' allows us to bypass the rather peculiar definition of planning (action based on accurate prediction[69]) that seems to have been accepted by some writers in the action research community, making meaningful communication about planning difficult.

Finally, I would wish to preserve the term 'intervention' because, without it, the actions of agents who initiate co-operative processes, and their many actions within these processes, become largely invisible. Co-operative Inquiry is based on dialogue, but each moment an agent intentionally contributes to this dialogue s/he can be seen as making an intervention. Certainly, the dialogue may have emergent properties that cannot be understood in terms of the sum total of individual interventions, but to obscure the contributions made by these interventions allows the possibility of seeing dialogue as some kind of mystical, harmonious process—thereby hiding the struggles and power plays that contribute to its emergence. The result could be that individual participants on the 'losing end' of these power plays never have their negative experiences addressed by the group—or even acknowledged as meaningful.

[69] This understanding of planning as involving accurate prediction is arguably employing a rationality of scientific management, where plans can only be considered legitimate if they are based on valid data (i.e., deriving from independent observation), allowing predictions of the future to be made. The whole paradigm of scientific management has recently been brought into question by chaos and complexity theorists who argue that it is simply not possible to plan with such certainty (see, for example, Stacey, 1992).

6.3 *The Impossibility of Independent Observation*

So, I have sought to justify talking in terms of intervention, not only to make it clear that research should be seen as an integral part of social practice, but also to keep visible the purposeful actions of all the agents involved (not just the researcher).

Having said this, I made clear at the beginning of this chapter that 'observation' and 'intervention' might have been treated as opposites by some authors (e.g., Seidman, 1988), but that a deeper analysis will reveal that we do not *have* to treat them in this way. Some of my argument has already been prefigured by Lewin's (1948) preparedness to use techniques of observation *in the service* of intervention. However, there is still more to be said.

We have already seen (Chapters 3 and 4) that the systems view of the Universe, that sees everything as interconnected, precludes the possibility that an observer can be truly independent of the observed (making naive dualism a non-starter for systems approaches, even if more sophisticated forms of subject/object dualism are still a part of some systems theories). However, there is another argument that, in my view, is even more persuasive in countering claims to independent observation. This is the argument, which is related to my discussion of knowledge generating systems (Chapter 4), that it is not possible to have knowledge without the presence of a knowledge generating system (a system containing a sentient being or beings) giving rise to it. I suggest that this sometimes escapes people's notice because, when first-order observations are being undertaken ('outward' looking), the nature of the knower (observer) is hidden. It is only when second-order investigations are undertaken to clarify the identity of the observer within its knowledge generating system that awareness of the contingency of knowledge comes to the fore, and the impossibility of truly independent observation becomes apparent.

A variety of authors from quite different disciplines have reached the same conclusion with regard to observation: that observation without the influencing presence of an observer (or knowledge without a corresponding knowledge generating system) is impossible. Let us take a small sample of disciplines to illustrate, starting with physics.

In the discipline of physics, Einstein (1934) claims that our inability to know the world 'as it really is' means that non-empirical "speculation" has to be an integral part of physics (see the quotation from Einstein reproduced on p.43 of this book). Of course, Einstein suggests that the origin of speculation is essentially human, and in Chapter 4 I made clear my own view that identifying a knowledge generating system with the boundaries of the human body is only one amongst many possible options (although a knowledge generating system giving rise to non-empirical speculation will inevitably *contain* one or more human beings). Be this as it may, Einstein's essential point still stands: that what is observed cannot be independent of the observer [also see Bohr (1963), Bohm (1980) and Prigogine (1989) in the discipline of physics].

Similar ideas have been explored in biology too. Northrop (1967) focuses on the inevitability of metaphysics (non-empirical ideas introduced into observation by the observer). If biological theories are about the identification of patterns in empirical data, then an understanding of metaphysics reveals that human beings, in *looking* for patterns, must employ ideas that have their origins outside the empirical data itself.[70]

Likewise, in psychology there have been theorists who have stood out against the philosophy of independent observation (e.g., Kelly, 1955; Weimer, 1979; and Hollway, 1989), as there have in sociology (e.g., Brown, 1977) and systems thinking (e.g., de Zeeuw, 1992). Even in chaos theory (see Gleick, 1987, and Capra, 1996, for reviews), which has permeated many disciplines and, in my opinion, has been peculiarly blind to the critiques of independent observation that have abounded in the latter half of the 20th century, there are now authors who acknowledge its subjective side (Fitzgerald, 1999), or who talk about how something can only be seen as 'simple' or 'complex' (chaotic) once a limiting boundary judgement has been applied (Flood, 1999a,b).

Of course, this lightening review of a variety of disciplines has ignored the often substantial differences between the opinions of the cited authors. Rather, I have focused on what the authors have in common: a critical attitude to the idea that it is possible to have genuinely independent observation. The substance of the critique of independent observation is that there are always sentient beings embedded in a knowledge generating system making the observations. Hence a first-order observation (looking 'outward') can always be supplemented by a second-order observation (looking at the particular identity of the knowledge generating system constructing the first-order observation). Add to this the assumption of systems approaches that everything is ultimately interconnected, and it makes the notion of independent observation quite untenable.

Of course we are now left with the question, if truly independent observation is impossible, where does this leave the practice of

---

[70] If we replace Northrop's focus on human beings as the sole origin of metaphysics with any bounded knowledge generating system that includes a human being, then this position is in line with the philosophical arguments explored in Chapter 4.

observational science? My argument, to be developed below, is that the construction of scientific observation should be regarded as *a* form, but by no means the only valid or useful form, of intervention.

6.4 *Observation* as *Intervention*

A key word in the above sentence is 'construction'. Scientific observation is not just any observation, but a moment in which the situation is constructed to facilitate observation under controlled conditions. There are two levels at which this kind of observation is dependent on the involvement of agents within knowledge generating systems: first, in establishing the goals and parameters of the observation; and second, in actually undertaking the observation.

In talking about setting the goals of the observation, I mean asking the question, what exactly should be observed? This is a moral question as much as a practical one, as scientists may just as easily look in one direction at the effectiveness of a new irrigation system for food cultivation or in another direction at a new weapon of mass destruction. Of course this is over-simplistic: the skills of weapons and agricultural scientists are not necessarily that interchangeable. Nevertheless, the principle applies to all scientific observations: there is a value judgement, whether consciously recognised or not, involved in every decision about what to study. It is precisely these value judgements that Popper (1959) places outside the remit of science—which he has to do if any semblance of independent observation is to be preserved. However, from a systems point of view, this absolute separation of moral decision making from the act of observation cannot be sustained: because the two interact, in principle they should both be available for critical analysis. Of course, in practical situations, boundaries have to be drawn around the inquiry process, but it seems to me that there can be no *general* case for excluding value judgements from inquiry—only local cases for *momentary* exclusions while observations are being undertaken. In other words, moral inquiry can be suspended temporarily while an act of observation is carried out, simply because the agent cannot do two things at once, and it can be resumed once again in the light of the observation and previous moral inquiries.

One possible argument against this is that there is a difference between 'pure' and 'applied' science. Some might say that those conducting applied science should indeed undertake moral inquiry, but pure science is curiosity-driven; many of its ethical implications are uncertain; and it is less obviously interventionary. My answer to this is that even pure science is interventionary, in the sense that it is designed to produce knowledge that will make a difference in scientific debates (which, incidentally, may be engaged in by pure and applied scientists alike). There may be similarities and differences between the ethical issues impacting on pure and applied scientific projects, but in choosing to undertake a particular piece of pure, curiosity-driven research, the scientist is still making a value judgement that this is the right thing to do (rather than taking on some other research project, for instance). This kind of judgement is therefore just as amenable to moral inquiry as that made by the applied scientist—it just means acknowledging that factors other than curiosity can and should be considered in forming pure research agendas.

The above is more or less the same as the view of Lewin (1948) who argues for a value-*full* scientific inquiry. I depart from Lewin, however, when I also argue that there is a second level at which scientific observation is dependent on the involvement of knowledge generating systems: in actually undertaking the observation. I cannot accept the view of Popper (e.g., 1959) that good control yields observations that are necessarily identical for all observers, simply because observation is a biological process. I argue that interpretation is integral to the act of observation itself.

What the scientist is able to see will in part be determined by his or her expectations, which in turn will be coloured by the language s/he uses and the values flowing into the act of observation. To illustrate, in experiments in which people are asked to look into a machine which feeds one picture into one eye and another into the other, some interesting effects occur. If people are fed two faces, one upside-down and the other the right way up, they invariably only see the one that is the right way up (Engel, 1956; Hastorf and Myro, 1959). Similarly, Bagby (1957) took US and Mexican citizens and fed them the same two images: one a US landscape and the other a Mexican one. In almost every case, people only saw the one that was culturally familiar to them. This indicates that the brain, linked to its environment, is actively *constructing* the observation, not simply reflecting what enters the eye. Observation is clearly not 'pure': it is mediated by conceptual and emotional frameworks of *interpretation*.

So, observation is in no sense independent from knowledge generating systems which give rise to agents' interpretations (knowledge). This is clear from empirical studies of observation (e.g., Engel, 1956; Bagby, 1957; Hastorf and Myro, 1959) as well as from moral arguments about the uses to which science can be put (Lewin, 1948) and metaphysical reflections on systemic interconnectedness. But there is one final nail to be hammered into the coffin of independent observation, which is revealed when we understand that science involves *communicating* about, not just conducting, observations. Therefore, one particular kind of knowledge generating system—the scientific community using language—is clearly implicated in the construction of scientific observations.

Scientific observation is undertaken to confirm or falsify hypotheses, which are of course constructed using language. A hypothesis is produced in the light of *debates* in scientific communities, conducted partly through the medium of writing (in books and journals, and on the internet) and partly using the spoken word (in meetings, conferences, and over lunch in the University canteen). The idea that language is a transparent medium of communication, with each word meaning the same to all its users, is simply untenable. Of course, there is enough commonality for meaning to be taken for granted most of the time in common conversation, but everyone still finds that they are strangely misunderstood from time to time (or finds that they have misunderstood another person). In scientific debates, not only are the same words used with different implications in different paradigms (Kuhn, 1962), but individuals may also have developed particular understandings of concepts that they are unaware are not universally shared. It is only when the intelligibility of concepts are interrogated during debate that people realise that the words they are using might not be as transparent as they initially appeared.

Given this situation, observation is constructed by language in two ways. First, the decision about what should be observed (and how) is taken based upon the scientist's understanding of the use of concepts in the relevant scientific debates (this level of constructing the observed can either be seen as connected to, or separate from, the moral decision making mentioned earlier). An obvious example of this is when sociologists talk about the effects of social class on attitudes. There are many ways in which social class can be defined, and which particular definition is selected will structure the act of observation and its outcome. Second, observations are affected by language when they are communicated to others: the particular choice of words may have one *intended* meaning (in the mind of the writer) and several *received* meanings (in the eyes of different readers). Meaning is not located in either the writer, the reader, or the text—but a *variety* of meanings may be produced through their interaction (Belsey, 1980).

In many different ways we have seen that agents embedded in knowledge generating systems are implicated in constructing observations: through their indirect interactions with the observed; through moral decision making about what to observe; through their selection of concepts to guide observation; through their interpretations of sense data; and through their selection of words to produce meaning(s). It should be clear from this that observation, as a purposeful act (involving interpretation), can only be isolated from its context by artificially ignoring what flows into it and the consequences it gives rise to. In my view, it is hard to justify placing this artificial boundary around it—especially as the choice of what to observe and how to observe it has unavoidable moral consequences for action (which may sometimes be anticipated and sometimes not). Given this state of affairs, I argue that it is more appropriate to *take account* of the construction of observation than to turn one's back on it. Once the moral, subjective, linguistic and other influences on observation are opened to critical reflection—in other words, once it becomes legitimate to reflect on the identity of the agent(s) and wider knowledge generating system involved in any given observation—*scientific observation has to be seen as a form of intervention*. Observation is undertaken *purposefully*, by an *agent*, to *create change* (in knowledge and/or practice).

Of course, the methods of scientific observation provide a set of techniques for intervention which can be seen to have significant uses *and* limitations. Scientific methods have been given pride of place in the last 300 years of Western intellectual history, largely because of the focus of philosophers of science on maintaining the shibboleth of independent observation and thereby denigrating methods of intervention. As the interventionary nature of observation itself has been demonstrated, I argue that scientists should welcome in all those other methods that are more self-consciously interventionary (see also Chapter 1). Of course, there are many communities of writers, including several with an interest in systems thinking, which have been developing methodologies and methods of intervention despite the disinterest, or even disapproval, of the scientific establishment. It is mainly to this work that I will refer in Chapter 9 when I argue in favour of methodological pluralism: the use of a wide variety of intervention methods to pursue a correspondingly wide variety of purposes.

6.5 Systemic Intervention

Having defined intervention in terms of purposeful action by an agent to create change, we can now integrate this with the initial understanding of 'system' provided in Chapter 3 to produce an understanding of *systemic intervention*.

To refresh our memories, I argued in Chapter 3 that the boundary concept lies at the heart of systems thinking: because of the fact that everything in the Universe is directly or indirectly connected to everything else, where the boundaries are placed in any analysis becomes crucial. The 'cut-off point' for analysis will make some things visible and others invisible. Systems thinking pursues the ideal of comprehensiveness, but knows that this is unattainable. However, reflection on the boundaries of knowledge at least enables us to consider options for inclusion and exclusion. It also reminds us that all understandings are incomplete: there is a need for humility and openness to the perspectives of others (Churchman, 1979).

If intervention is purposeful action by an agent to create change, then *systemic* intervention is purposeful action by an agent to create change *in relation to reflection on boundaries*. This statement embodies the core concern of the methodology of systemic intervention that I will be introducing in the next five chapters, and which is briefly summarised below.

6.6 Towards a Methodology for Systemic Intervention

At minimum, I suggest that an adequate methodology for systemic intervention should be explicit about three things. The first is the need for agents to reflect critically upon, and make choices between, boundaries (as mentioned above). The meaning of this will be expanded upon in Chapter 7, but for the purposes of this section it is important to be clear that it is only possible for agents to make boundary judgements through the use of (implicit or explicit) theories and methods, and that reflection leading to the making of boundary judgements is an *activity* (it is intervention to shape the agent's understanding, which may in turn influence future action). Critical reflection upon boundary judgements is vital because it is only by way of boundary critique that the ethical consequences of different possible actions (and the ways of seeing they are based upon) can be subject to analysis.

The second aspect that should be made explicit is the need for agents to make choices between theories and methods to guide action, which requires a focus on theoretical and methodological pluralism. These two forms of pluralism have meaning in terms of process philosophy (Chapter 4) and the focus on boundary judgements mentioned above: if understandings can be bounded in many different ways, then each of these boundaries may suggest the use of a different theory (and conversely, each theory implies particular boundary judgements). Methodological pluralism then also becomes meaningful because methods and methodologies embody different theoretical assumptions: choices between boundaries and theories suggest which methodological options are most appropriate (and conversely, methodological choices imply particular theoretical and boundary judgements). Choice between theories and methods is also a form of action, in the same way as reflection on, and choice between, boundary judgements can be seen as action: it is intervention in the present to shape a strategy for future intervention. Theoretical and methodological pluralism will be explored further in Chapters 8 to 11.

Finally, an adequate methodology for systemic intervention should be explicit about taking *action for improvement* (action for the better, which cannot, of course, be defined in an absolutely objective manner). There are two key words to consider here: 'action' and 'improvement'. Let us look at 'action' first. As we saw earlier in this chapter, it is not possible to create a *general* definition of action: rather, the meaning of action can only be determined in local contexts (though a 'local context' may be wide in scope, for instance when dealing with international relations or global environmental problems). This is because the use of different boundaries, theories and methods will give rise to correspondingly different understandings of what it means for an agent to take action.

Similarly, 'improvement' needs to be defined temporarily and locally: as different agents may use different boundary judgements, what looks like an improvement through one pair of eyes may look like the very opposite through another (Churchman, 1970).[71] Also, even if there is widespread agreement between all those directly affected by an intervention that it constitutes an improvement, this agreement may not stretch to future generations. The temporary nature of all improvements makes the concept of *sustainable* improvement particularly important: while even sustainable improvements cannot last forever, gearing improvement to long-term stability is essential if future generations are to be accounted for. We can say that an improvement has been made when a desired consequence has been realised through intervention, and a *sustainable* improvement has been achieved when this looks like it will last into the indefinite future

---

[71] An example is logging a stretch of rain forest, which may bring about an improvement in the eyes of the logging company's employees and those who consume the wood that is generated, but may be considered as damaging by tribal people who are displaced from their ancestral lands, and by conservationists concerned with the preservation of species diversity. As Churchman (1970) says, every improvement assumes boundaries defining what consequences of intervention are to be taken into account, and what are to be ignored or regarded as peripheral. In the above example, the logging will only be viewed as bringing about an improvement if the displacement of tribal people and the reduction of species diversity are excluded from the boundaries of analysis. Clearly, *what* is included in the boundaries of analysis and *who* conducts this analysis are both vital issues in defining improvement.

without the appearance of undesired consequences (or a redefinition of the original consequences as undesirable).

The notion of improvement is important because agents are restricted in the number of interventions they can undertake, and must therefore make judgements about what they should and should not do. The extent to which various interventions look like they may or may not bring about improvements, or may bring about improvements that have greater or lesser priority, is a useful criterion for making these judgements.

Of course, I should say why I have used the term 'improvement' rather than, say, the creation of beauty, pleasure, emancipation,[72] or spiritual enlightenment. The answer is that, if we value any of these things, the creation of these *represents* an improvement. The term 'improvement' is therefore general enough to have meaning in relation to almost any value system: it simply indicates the purposeful action of an agent to create a change for the better (even if, in the case of 'pure' science, this is simply a change in our knowledge base).[73]

These three activities—reflecting on boundary judgements; making choices concerning theory and method; and taking action for improvement—are clearly inseparable. Doing one always implies doing

[72] Several writers in the systems community (e.g., Mingers, 1980, 1984; Jackson, 1982, 1985a,b, 1988, 1991; Ulrich, 1983; Flood, 1990; Flood and Jackson, 1991a,b; Schecter, 1991) have talked about human emancipation being the focus of systemic intervention. This strand of thinking is based on the importation of Habermas's (1972) theory of knowledge-constitutive interests into systems practice (see Chapter 10 of this book for details). However, in previous writings (e.g., Midgley, 1996c), I have criticised this focus on two grounds: (i) the discourse of human emancipation tends to concentrate on issues of power and social justice entirely separately from their ecological contexts, risking the perpetuation of unsustainable means of social organisation; and (ii) Habermas (1972) believes in the rather dubious notion of a human 'march of progress', and there is a suspicion that this is being imported into systems practice along with the notion of emancipation. I regard the idea of a 'march of progress' as dubious because what looks like progress from one point of view may appear to be just the opposite from another. Also, what seems like progress to us now may be the ruin of future generations (Churchman, 1970).

[73] It should be noted that there is a counter-argument to this. According to Rorty (1989), using a term like improvement (or truth, legitimacy, ontology, morality, etc.) suggests a belief in absolute facts or values. Rorty believes that such words are tainted. To talk of improvement is to talk about the attainment of a state that *everybody* would agree is better. Rorty, along with other writers who have been labelled 'post-modern', have launched a fierce critique of the apparent certainties of the modern world, and the attempt to discredit talk of improvement is central to this. Rorty offers a powerful argument, but why *abandon* words like truth, morality and improvement? As will become clear in Chapter 7, if we are prepared to be critical about the business of making boundary judgements, there is no need to assume that understandings of improvement are universal. To abandon words like truth, morality and improvement is to risk slipping into negativity and inaction. To tear away the modernist certainties surrounding their use and clothe them with an awareness of the frailty of human understanding is to preserve the possibility of positive action while facing the complexities of this head on.

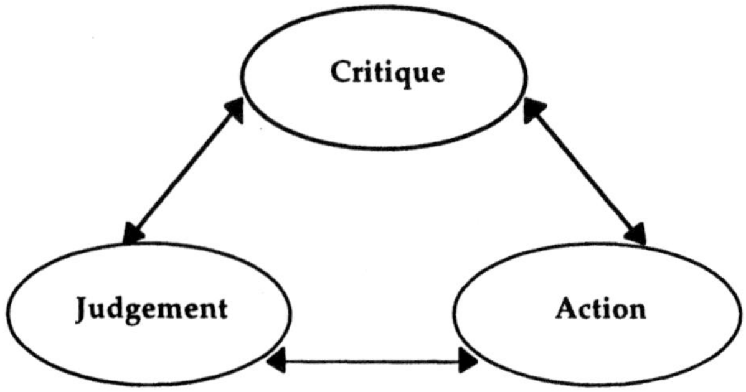

FIGURE 6.1: Three aspects of a methodology for systemic intervention

the other two as well, although the focus of attention may shift from one to another aspect of this trinity so that none remain implicit and thereby escape critical analysis. The separation between the three is therefore analytical rather than factual: it ensures a proper consideration of a minimum set of three 'angles' on possible paths for intervention. Making all of them a specific focus of a methodology for systemic intervention guides the reflections of the agent, ensuring that boundaries, theories, methods, and action for improvement all receive explicit consideration. The three activities, diagrammed in relation to one another, are presented in Figure 6.1. *Critique* specifically means boundary critique (reflection on, and choice between, boundaries); *judgement* means judgement about which theories and methods might be most appropriate; and *action* means the implementation of methods to create improvement (however this is to be understood in the local context).

6.7 Conclusion

In this chapter I have defined intervention as purposeful action by an agent to create change, and have contrasted this with the concept of observation. While some authors suggest that intervention and observation are opposites, I have argued that observation (as undertaken in science) is a 'special case' of intervention. We should therefore welcome scientific techniques of observation into our pluralistic armoury of intervention methods. Next, I related the systems idea to intervention, and suggested that systemic intervention is purposeful action by an agent to create change in relation to reflection on boundaries. Finally, this led to a presentation of an outline of a methodology for systemic intervention incorporating explicit foci on boundary critique (reflection on, and choice between, boundaries); judgement concerning appropriate theories and methods; and action for improvement. The first two of these concepts will be explored in more detail in the coming chapters, starting with the theory of boundary critique in Chapter 7. No specific chapter is devoted to the concept of improvement, but it should be apparent in the discussion of boundary critique (Chapter 7) and the design/choice of methods (Chapters 9 and 10) that the pursuit of improvement (defined in a non-absolute manner, as indicated earlier) underlies all my arguments.

7

Boundary Critique

In the last chapter, I discussed the concept of 'systemic intervention'. Now it is time to go into more depth concerning what it means to subject the boundaries of intervention to critical scrutiny. In this chapter I will propose a theory of *boundary critique*. This is a normative theory (prescribing a course of action rather than simply describing an aspect of the world) about the need for reflection on boundaries during interventions. If the boundaries of analysis are crucial to the generation of knowledge, then a capacity to reflect on different possible boundaries is essential if we are not to simply take for granted assumptions flowing into intervention. However, there is also more at stake: in Chapter 4 we saw that boundary judgements made by human beings are closely tied to value judgements (and this idea will be expanded upon shortly). Therefore, an adequate theory of boundary critique will not only help us reflect on understandings flowing into intervention, but will also support moral reflection on the purposes of intervention itself.

In Section One of this book, I talked about how boundary judgements are made by sentient beings as part of wider knowledge generating systems (the boundaries of which are only established during second-order reflections). However, in this chapter (like the last), I will be dealing with methodology, and will therefore only be concerned with boundary judgements made by human agents (as a sub-set of sentient beings) under the influence of the knowledge generating system(s) in which they are embedded (and which are only identified explicitly in second-order reflections). The rationale for this focus on human agents has already been provided (see Chapter 6).

Also, rather than keep repeating the point that agents are not autonomous knowers and actors, but are embedded in a whole Universe of activities (which is differentiated into knowledge generating systems through the use of boundary judgements), from now on I will simply talk about 'the agent', assuming that this embeddedness is accepted. Of course, it *is* both possible and legitimate to view an agent as an autonomous actor—when the boundaries of the agent and the knowledge generating system are seen as identical—but this is only one possible boundary judgement that can be made during second-order reflections, and needs to be kept open to critical analysis.

As far as I am aware, the term 'boundary critique' was first introduced into the literature by Ulrich (1996b), and was then used again by Midgley *et al* (1998) to enable the consolidation of a research programme advanced in various different ways by C. West Churchman, Werner Ulrich and myself over a number of years. I will discuss the ideas of all three of these authors, plus those of Maurice Yolles (1999a, 2000) who has further developed the theory of boundary critique since Midgley *et al* wrote their paper. The works of each of these authors will be discussed in chronological order.[74] However, when introducing my own contribution, I will elaborate somewhat on the ideas I have previously presented (primarily to draw out some of the key distinctions between my own thoughts and those of Churchman and Ulrich).

One final point of clarification should be made before starting the review. A core idea in the theory of boundary critique is that boundary judgements and values are intimately connected. It should be remembered that, because this is a methodological rather than a philosophical theory, it is not a *general* theory of boundaries like the one explored in Chapter 4.[75] It is directly linked to the context of trying to improve systemic intervention, so already assumes some contextually relevant boundary judgements: i.e., a focus on human agents (discussed above) and the human use of language, including the language of morality (in Chapter 6 I argued that it is the use of linguistic systems that primarily differentiates human from non-human animals).

This review and critical analysis of the theory of boundary critique will start with an examination of the work of C. West Churchman (1968a,b, 1970, 1971, 1979), who is widely acknowledged to be a major contributor to the development of systems thinking and operational research.

---

[74] Consequently, what will emerge is a kind of 'history' of the theory of boundary critique. I should note that all histories are partial: they are driven by the purposes and limited understandings of those who produce them (Carr, 1961; Flood and Gregory, 1988). I am sure that this history will be no different, so I encourage the reader to reflect critically on my inclusions, exclusions and marginalisations.

[75] Although even that general theory should not be regarded as universal (see footnote 54, Chapter 4, for details).

7.1 The Bounded Nature of Improvement

Churchman (1970) is interested in the concept of *improvement*, and if a change is to be justifiably called an improvement then reflecting on the boundary of analysis is crucial. What is to be included or excluded is a vital consideration: something that appears to be an improvement given a narrowly defined boundary may not be seen as an improvement at all if the boundaries are pushed out. Essentially, defining the boundaries of improvement is an *ethical* issue, requiring the exercise of value judgements. For this reason, Churchman argues that as much information as possible should be "swept in" to definitions of improvement, allowing the most inclusive, and therefore most ethical, position on improvement to emerge.

As we saw in Chapter 3, in comparison with the earlier ideas of Bogdanov (1913-17), von Bertalanffy (1968) and other general systems theorists, this way of thinking involves a fundamental shift in our understanding of the nature of a system. Prior to the work of Churchman, many people assumed that the boundaries of a system are 'given' by the structure of reality. In contrast, for Churchman, boundaries are social or personal constructs that define the limits of the knowledge that is to be taken as pertinent in an analysis. Also, when it comes to social systems, pushing out the boundaries of analysis may involve pushing out the boundaries of who may legitimately be considered a decision maker (Churchman, 1970). Thus, the business of setting boundaries defines both the knowledge to be considered pertinent *and* the people who generate that knowledge (and who also have a stake in the results of any attempts to improve the system). This means that there are no 'experts' in Churchman's systems approach, at least in the traditional sense of expertise where all relevant knowledge is seen as emanating from just one group or class of people: wide-spread stakeholder involvement is required, sweeping in a variety of relevant perspectives.

Not only did Churchman introduce this fundamental change in our understanding of 'system', but he also discussed critique. In examining how improvement should be defined, Churchman (1979) followed Hegel (1807), who stressed the need for rigorous self-reflection, exposing our most cherished assumptions to the possibility of overthrow. To be as sure as we can that we are defining improvement adequately, we should, in the words of Churchman (1979), pursue a "dialectical process": this involves seeking out the strongest possible "enemies" of our ideas and entering into a process of rational argumentation with them. Only if we listen closely to their views and our arguments survive should we pursue the improvement.

Here, then, we have the foundations for the theory of boundary critique: boundaries are constructs, and may therefore be placed in a variety of different places, bringing forth markedly different 'realities'; they are associated with values, in that different values (associated with different ideas of improvement) may result in boundaries being constructed in different places; participation from a variety of stakeholders is important, because different stakeholders may bring different insights to bear; and even our most cherished ideas should be subject to critique from time to time to test their worth in the light of other value systems.

7.2 *Critical Systems Heuristics*

Churchman produced a great deal of highly influential work in the 1960s and 1970s, and in the 1980s several other authors began to build upon it in significant new ways. One of these authors was Werner Ulrich (1983, 1988, 1994, 1996a,b).[76] Ulrich (1983) agrees that Churchman's desire to sweep the maximum amount of information into understandings of improvement is *theoretically* sound, but also acknowledges that the need to take practical action will inevitably limit the sweep-in process. He therefore poses the question, how can people rationally justify the boundaries they use? His answer is to develop a methodology, Critical Systems Heuristics, which can be used to explore and justify boundaries through debate between stakeholders.

In producing his methodology, Ulrich draws upon the later writings of Jürgen Habermas (1976, 1984a,b) concerning the nature of rationality. Habermas regards rationality as dialogical—and the tool of dialogue is language, which allows us to question. The basis of dialogue is therefore open and free questioning between human beings. However, Habermas does not take a naive line concerning dialogue: he acknowledges that it may be distorted through the effects of power. This may happen directly, when one participant coerces another, or indirectly, when participants make unquestioned assumptions about the

[76] Other authors who have developed Churchman's thinking include Mason and Mitroff (1981) and Mitroff and Linstone (1993). Mason and Mitroff's work is reviewed in Chapter 9, so I will not discuss it here. I will confine my comments to Mitroff and Linstone (1993). These authors talk in terms of "unbounded" systems thinking to emphasise the potential for creativity that can come about through the use of systems methods. However, my own preference is not to use the term "unbounded" because it can lead people into the trap of thinking that it is possible to transcend all limiting assumptions. Of course, this is not what Mitroff and Linstone mean by "unbounded"—they have a similar understanding of boundaries to Churchman (1970)—but in my view the danger of this misinterpretation is always present if we use such terminology.

absolute necessity for, or inevitable future existence of, particular social systems. To overcome these effects of power, we need to establish what Habermas calls an "ideal speech situation": a situation where any assumption can be subject to critique and all viewpoints can be heard.

However, while Ulrich (1983) accepts the *principle* of Habermas's understanding of critique, he nevertheless criticises him for being utopian. For all viewpoints to be heard, the ideal speech situation would have to extend debate to every citizen of the world, both present and future. This is quite simply impossible. Ulrich sees his task as the *pragmatisation* of the ideal speech situation, and a marriage between 'critical' and 'systems' thinking is the means by which this can be achieved. Truly rational inquiry is said to be *critical*, in that no assumption held by participants in inquiry should be beyond question.[77] It is also *systemic*, however, in that boundaries always have to be established within which critique can be conducted. Indeed, Ulrich claims that both ideas are inadequate without the other. Critical thinking without system boundaries will inevitably fall into the trap of continual expansion and eventual loss of meaning (as everything can be seen to have a context with which it interacts, questioning becomes infinite). However, systems thinking without the critical idea may result in a 'hardening of the boundaries' where destructive assumptions remain unquestioned because the system boundaries are regarded as absolute.

An important aspect of Ulrich's (1983) thinking about boundaries is that boundary judgements and value judgements are intimately linked: the values adopted will direct the drawing of boundaries that define the knowledge accepted as pertinent. Similarly, the inevitable process of drawing boundaries constrains the ethical stance taken and the values pursued. Debating boundaries is therefore an ethical process, and a priority for Ulrich is to evolve practical guidelines that can help people steer the process of critical reflection on the ethics of drawing system boundaries. For this purpose, Ulrich (1983) developed a list of twelve questions which can be used heuristically to interrogate what the system currently *is* and what it *ought* to be.[78] It is important to note

---

[77] The meaning of the word 'critical' is hotly contested. Popper (1972) uses it to mean subjecting truth claims to empirical test (i.e., looking at facts from different theoretical angles and asking which theory is best supported by the empirical evidence), and reflection on values is explicitly excluded from 'being critical' (Popper, 1966). In contrast, the 'critical theorists' (e.g., Adorno and Horkheimer, 1944; Adorno, 1951; Arendt, 1958; Wellmer, 1970; Apel, 1977; Habermas, 1984a,b; and Fay, 1975, 1987) say that to be critical means to reflect on facts *and* values. Ulrich (1983), and all subsequent writers (myself included) who have worked under the banner of 'Critical Systems Thinking' (see Chapter 9 for details), follow the latter understanding of being critical.

[78] These twelve questions are derived from Kant's (1788) 'categorical imperatives' (or *moral*

that some of these questions relate to who should participate in discussing boundary judgements in the first place, meaning that there is always the possibility for people to enter or leave discussions. The original questions are presented in Ulrich (1986), and reproduced here in Figure 7.1.[79]

There are two key guiding ideals embedded in Ulrich's work. First, if rationality is dialogical, plans for improvement should, in principle, be normatively acceptable to all those participating in a given dialogue. In practice, this means (if at all possible) securing agreement between those designing an improvement and those affected by it (of course, judging who or what is actually involved and/or affected already involves making a boundary judgement). When agreement is not secured, citizens who disagree with implementing the 'improvement', and who are affected by it, may legitimately use Ulrich's twelve questions in a 'polemical' mode to build an argument with which to embarrass planners in future public debate by exposing the limited nature of the expertise they lay claim to.

The second guiding ideal is that participants in dialogue should respect the principle of *universalisation*. This is the idea, inherited from Kant (1788) and Habermas (1976) amongst others, that moral judgements should be regarded as equally applicable to everyone. Thus, if it is wrong to kill one person, it is wrong to kill anyone. By seeking to justify the universality of their moral judgements, Ulrich argues (following Habermas, 1976, 1984a,b) that participants in debate are driven to look beyond the narrow boundaries of local stakeholder groups. In this way, Ulrich (1983) avoids a potentially problematic implication of Critical Systems Heuristics: by accepting the freedom of communities of stakeholders to set boundaries of participation and the inclusion and exclusion of issues, Ulrich could be accused of making all morality relative to locally defined boundaries. Of course no judgement is flawless, so even universalised moral judgements will assume implicit

---

imperatives). However, in tune with the dialogical turn he proposes, Ulrich has turned the categorical imperatives into questions for use in debate.

[79] In my view these questions include some jargon that might not be immediately transparent to lay participants in dialogue (and terms like 'emancipation' have a negative connotation for some people, being associated with Eastern European Marxist ideology in the writings of the popular press). Several writers have therefore made attempts to translate the questions into plain English (Cohen and Midgley, 1994; Gregory *et al*, 1994; Midgley *et al*, 1997). However, the originals are provided here because they list the questions in generic form, while the translations are tailored to people in specific circumstances (people with mental health problems caught up in the criminal justice system; blind and partially-sighted people evaluating the quality of health care; and older people receiving housing services).

The 12 critically heuristic boundary questions in the *is* mode	The 12 critically heuristic boundary questions in the *ought* mode
(1) Who is the actual *client* of the system design, i.e. who belongs to the group of those whose purposes (interests and values) are served, in distinction to those who do not benefit but may have to bear the costs or other disadvantages?	(1) Who ought to be the *client* (beneficiary) of the system to be designed or improved?
(2) What is the actual *purpose* of the system design, as being measured not in terms of declared intentions of the involved but in terms of the actual consequences?	(2) What ought to be the *purpose* of the system, i.e. what goal states ought the system be able to achieve so as to serve the client?
(3) What, judged by the design's consequences, is its built in *measure of success*?	(3) What ought to be the system's *measure of success* (or improvement)?
(4) Who is actually the *decision taker*, i.e. who can actually change the measure of success?	(4) Who ought to be the *decision taker*, i.e. have the power to change the system's measure of improvement?
(5) What *conditions* of successful planning and implementation of the system are really controlled by the decision taker?	(5) What *components* (resources and constraints) of the system ought to be controlled by the decision taker?
(6) What conditions are *not* controlled by the decision taker, i.e. what represents *"environment"* to him?	(6) What resources and conditions ought to be part of the system's *environment*, i.e. not be controlled by the system's decision taker?
(7) Who is actually involved as *planner*?	(7) Who ought to be involved as *designer* of the system?
(8) Who is involved as *"expert"*, of what kind is his expertise, what role does he actually play?	(8) What kind of *expertise* ought to flow into the design of the system, i.e. who ought to be considered an expert and what should be his role?
(9) Where do the involved see the *guarantee* that their planning will be successful? (E.g. in the theoretical competence of experts? In consensus among experts? In the validity of empirical data? In the relevance of mathematical models or computer simulations? In political support on the part of interest groups? In the experience and intuition of the involved?, etc.). Can these assumed guarantors secure the design's success, or are they false guarantors?	(9) Who ought to be the *guarantor* of the system, i.e. where ought the designer seek the guarantee that his design will be implemented and will prove successful, judged by the system's measure of success (or improvement)?
(10) Who among the involved *witnesses* represents the concerns of the affected? Who is or may be affected without being involved?	(10) Who ought to belong to the *witnesses* representing the concerns of the citizens that will or might be affected by the design of the system? That is to say, who among the affected ought to get involved?
(11) Are the affected given an opportunity to *emancipate* themselves from the experts and to take their fate into their own hands, or do the experts determine what is right for them, what quality of life means to them, etc? That is to say, are the affected used merely as means for the purposes of others, or are they also treated as "ends in themselves" (Kant), as belonging to the client?	(11) To what degree and in what way ought the affected be given the chance of *emancipation* from the premises and promises of the involved?
(12) What *world view* is actually underlying the design of the system? Is it the world view of (some of) the involved or (some of) the affected?	(12) Upon what *world views* of either the involved or the affected ought the system's design be based?

***FIGURE 7.1**: The 12 critically heuristic boundary questions in the* is *and* ought *modes (after Ulrich, 1986)*

boundaries, but by encouraging *consideration* of the universalisation of morality amongst participants in debate, Ulrich claims that people are able to look beyond their local contexts.[80]

7.2.1 Key Differences between Churchman and Ulrich

In many ways, Churchman and Ulrich have similar ideas, but there is a key difference worthy of note. Churchman (1979) advocates a strongly Kantian position, assuming that individuals are relatively autonomous moral decision makers. In contrast, Ulrich (1983) advocates a dialogical position (following Habermas, 1976) which assumes language and debate to be the origin of meaning rather than individual consciousness. This theoretical difference has consequences for their focus of attention when it comes to methodology. Churchman's (1970) prime concern is to argue that *interveners*, who he assumes will have a professional identity (e.g., they may be called 'operational researchers'), should take moral responsibility for their decisions, and this should involve widening their boundaries of analysis and opening up their projects to stakeholder participation (in Churchman's words, they should adopt a 'whole systems methodology'). In contrast, because of his focus on debate, Ulrich (1990) says that moral responsibility should be lifted from the shoulders of the professional and placed in the hands of the participative group. Ulrich therefore seeks to develop "critical systems thinking *for citizens*" (Ulrich, 1996b, p.1, my emphasis). However, both Churchman and Ulrich agree on the principle of universalisation: neither wishes to see morality become wholly relative to the practice of making boundary judgements—for both authors, it is important that participants think about the *general* morality of their actions, not just what might be locally acceptable.

7.3 Processes of Marginalisation

We have seen how Ulrich has built on and developed the work of Churchman, changing the emphasis in certain key respects. In a similar fashion, when I first started publishing as part of this research programme at the beginning of the 1990s (Midgley, 1991b, 1992b), I set out to extend the work of Ulrich. For both Churchman and Ulrich, the

---

[80] At least thirteen papers have been written criticising Ulrich's (1983) work. See Jackson (1985c, 1991), Willmott (1989), Ivanov (1991), Flood and Jackson (1991b,c), Mingers (1992b), Romm (1994, 1995a,b), Brown (1996), Midgley (1997c) and Vega (1999). In my view, this volume of critical literature demonstrates the importance many people attach to Ulrich's ideas: it is a sign of his influence rather than an indication of intellectual weakness.

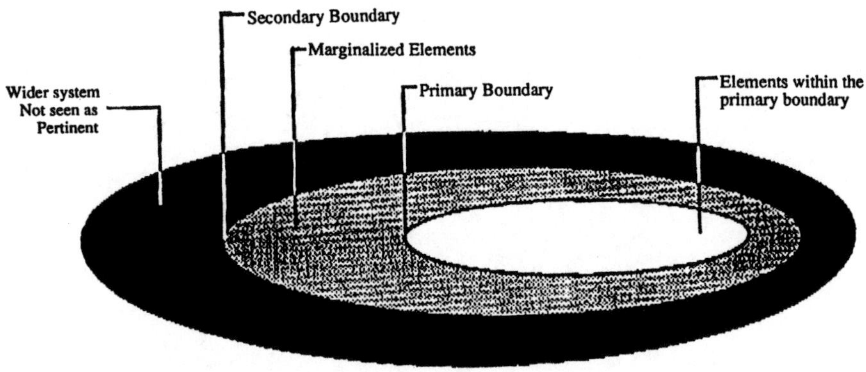

FIGURE 7.2: Marginalisation

question of which system boundaries are to be used in an analysis is essentially an ethical question because value and boundary judgements are intimately related. I used this insight as a starting point to ask what happens when there is a conflict between different groups of people who have different ethics (values in purposive action) relating to the same issue, and thereby make different boundary judgements.

I argued that, if one group makes a narrow boundary judgement and another makes a wider one, there will be a *marginal* area between the two boundaries. This marginal area will contain elements that are excluded by the group making the narrow boundary judgement, but are included in the wider analysis undertaken by the second group. We can call the two boundaries the *primary* and *secondary* boundaries (the primary boundary being the narrower one). This is represented visually in Figure 7.2.

I then went on to argue that, when two ethical boundary judgements come into conflict, the situation tends to be stabilised by the imposition of either a *sacred* or a *profane* status on marginal elements. The words 'sacred' and 'profane' mean valued and devalued respectively. This terminology was borrowed from the tradition of anthropology, exemplified by the work of Douglas (1966), and it should be stressed that it is not meant in an exclusively religious sense, but refers to the special status of a marginalised element. The imposition of either a sacred or a profane status on marginal elements stabilises a conflictual situation in the following manner. When marginal elements become profane, the primary boundary and its associated ethic is focused upon and reinforced as the main reference for decision making. People or issues relegated to the margins are disparaged, allowing the secondary

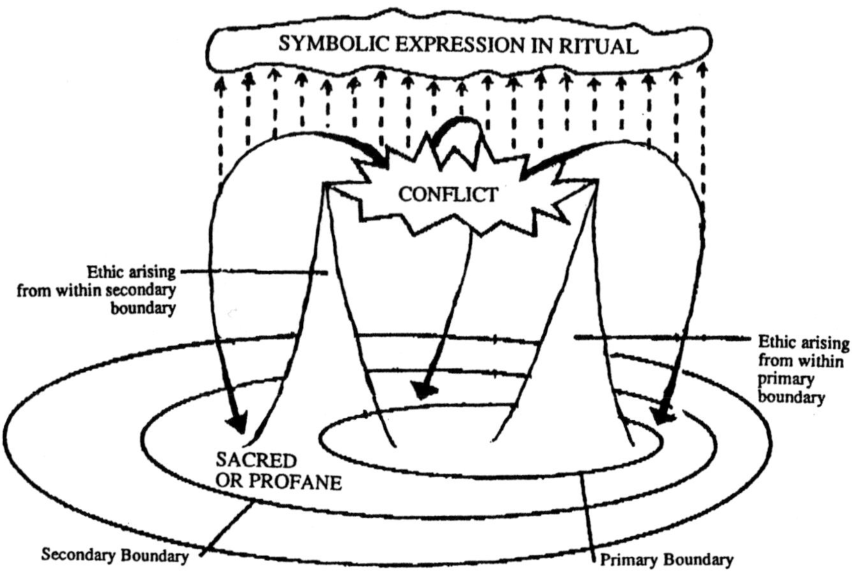

FIGURE 7.3: Margins, ethics, sacredness, profanity and ritual

boundary to be ignored. Conversely, when marginal elements are made sacred (and thereby assume a special importance), the secondary boundary and its associated ethic is focused upon and reinforced.

However, this is not the end of the story. Not only do ethical tensions give rise to sacredness and profanity, but this whole process comes to be overlaid with social ritual. Ritual is behaviour, in whatever context, that contains certain stereotypical elements that involve the symbolic expression of wider social concerns (also see Douglas, 1966, and Leach, 1976, for further thoughts on the relationship between ritual, sacredness and profanity). An observation of the presence of ritual can tell us where sacredness and profanity might lie, and hence where ethical conflicts related to marginalisation might be found. In order to make this clearer, the whole process has been represented diagrammatically in Figure 7.3.

To explain, in Figure 7.3 we see one ethic arising from within the primary boundary, and another from within the secondary. These come into conflict—a conflict that can only be dealt with by making one or other of the two boundaries dominant. This dominance is achieved by making elements in the margin (between the primary and secondary boundaries) either sacred or profane. The whole process is symbolically expressed in ritual which, in turn, helps to support the total system.

Of course, the 'system' represented in Figure 7.3 is a model, and like all models it does not fully express the complexity of the many value and boundary judgements that interact dynamically in social situations. Some discussions of this complexity, together with practical examples that further clarify the theory, can be found in Midgley (1992b, 1994) and Midgley *et al* (1998).

One particularly important point about the complexity lying beyond the model should be borne in mind, however: this kind of 'system' does not exist in isolation. It is 'held in place', or granted integrity, by virtue of the fact that it expresses wider struggles between competing discourses. These competing discourses can develop in families or small groups; at an organisational or community level; or they can be society-wide. Indeed, there may be multiple levels of mutually-supportive discursive conflict, and a boundary judgement needs to be made about which level(s) of analysis will be most appropriate for the purposes being pursued.

To make clear what I mean by competing discourses holding processes of marginalisation in place, I will provide an example of a society-wide conflict which results in the marginalisation of people who are unemployed. There is a conflict in many Western societies between the liberal discourse of citizenship (where all people are seen as having equal value because of their status as rational beings), and the capitalist discourse of good employment practice (which limits the responsibility of organisations to their employees alone). This conflict is not stabilised by either the inclusion *or* exclusion of the unemployed, but by their marginalisation. If unemployed people were to be fully included along with employees in the primary boundary of industrial organisation, 'good employment practice' (indeed, the whole capitalist system of organisation) would become untenable. However, if they were fully excluded, the liberal ideal of equal citizenship would become untenable instead. Both the liberal and capitalist discourses have long histories in the West, and have come to be institutionalised throughout the economic and legal systems of our societies. While on the whole the two discourses are mutually supportive (Booth Fowler, 1991; Midgley and Ochoa-Arias, 1999), there are still significant tensions, and the phenomenon of unemployment points to one of them. The key to understanding the status of the unemployed is to realise that it is only possible to maintain the dual commitment to liberalism and capitalism if people who are unemployed are neither fully included nor excluded. People who are unemployed therefore become marginalised, but the conflict is finally stabilised when a sacred or profane status is imposed on them: when they are regarded as profane, it justifies thinking in terms of narrow organisational boundaries; when they are regarded as sacred, this justifies programmes to support social inclusion. There is rarely a consensus on whether a marginal group or issue should be viewed as sacred or profane, but there are dominant patterns of social action which come to be solidified in rituals. In the case of the unemployed, a typical example is 'signing on' which many people view as an exercise in ritual humiliation.

7.3.1 Is there a Need for a Sociological Theory?

Charlton (1993) has argued that the above theory of marginalisation is useful for informing intervention, but it lacks an explanation of the social mechanisms through which marginalisation comes about. He proposes that these ideas should be seen in relation to Luhmann's (1986) theory of social autopoiesis. Essentially, Luhmann argues that, in modern capitalist societies, institutions (e.g., the legal, financial, educational, scientific and health systems) take on a self-producing character, and can only be subject to change by individuals and groups who are prepared to phrase their discourses in terms which 'resonate' with the ways in which the institutions currently function. For example, the legal system is solely concerned with argumentation around whether particular activities are legal or illegal: any other way of talking about activities will not be meaningful within the system. Similarly the health system is concerned with the distinction between health and illness, and scientific institutions are concerned with what is scientific or not scientific. Discourses that do not use these distinctions are simply not 'heard' within the relevant institutions. Charlton did not expand on the precise connections between Luhmann's work and my own, so I cannot really respond to this idea in a meaningful manner. What I *can* respond to, however, is the basic criticism that I should be proposing a theory of a social mechanism (beyond simply talking about competing discourses) that explains the phenomenon of marginalisation. Essentially, Charlton believes that I should be engaging in sociological investigation to complement my work on intervention.

I have been thinking about this issue for some time, and can see two views of the argument. In favour of producing a sociological theory, *from the point of view of supporting intervention* (I am not really interested in a theory for its own sake[81]), is the need to understand why it is so

[81] Of course, even a theory 'for its own sake'—i.e., a purely explanatory theory that does not make any comment about the social consequences of adopting that mode of explanation—is an intervention of sorts: it is an intervention into our knowledge of the mechanisms of modern societies. While I find such theories intriguing, my primary interest in them is in their consequences for social action rather than their explanatory potential alone. In other words, the key question from my point of view is: if we see society in this way, what does it

difficult to challenge oppressive forms of marginalisation through intervention practice; and why when we are successful in making the challenge at a local level, there is rarely a domino effect (change does not usually generalise beyond the local context to wider society). Perhaps a sociological theory would help us identify alternative intervention practices that would make more of a difference.

However, the argument against producing a sociological theory is as follows. If we think of writers like Foucault (e.g., 1980, 1984a), Habermas (e.g., 1984a,b), Luhmann (e.g., 1986), Douglas (e.g., 1986, 1992), Beck (1986) and Giddens (1991), we realise that this is exactly their agenda. It has taken a life-time for each of these writers to elaborate their ideas in a sufficiently comprehensive form for others to use—the issues are so complex. And of course, while there are significant commonalities, they disagree on major issues, such as (in the case of Foucault and Habermas) how to conceptualise the operation of power (see Fay, 1975; Smart, 1983; Couzens Hoy, 1994; Kelly, 1994; and Ashenden and Owen, 1999, for comparisons). I think one mistake that some previous writers on intervention have made is to import just part of one such theory of society into their own work without sufficient consideration of the wider debates. They have therefore opened themselves up to criticism for over-simplifying the issues and restricting their methodology of intervention by channelling all observation through a simplistic sociological meta-theory (see Midgley, 1996c, for a critique of one such attempt at using a sociological theory). At this stage in my inquiries, I tend to agree with Bernstein (1983) and Gregory (1992) that it is more useful to look at the variety of theories available and learn from these, rather than imposing an overly restrictive meta-theory on ourselves. In any case, fixing on just one sociological theory in an absolute manner would almost certainly contradict my proposal for embracing theoretical pluralism (Chapter 8).

Therefore, I suggest that a variety of theories may be used to partner my work on marginalisation. So far, several different attempts have been made: Charlton (1993) has suggested Luhmann's (1986) theory of social autopoiesis; Yolles (1999a) has synthesised complexity theory (Nicolis and Prigogine, 1989; Cohen and Stewart, 1994) and viable system theory (Beer, 1975, 1979, 1981; Schwarz, 1994, 1995) to explain marginalisation; and Cordoba *et al* (2000) have partnered the theory of boundary critique with Maturana's (1988a,b) theories of biological autopoiesis and language. I am content to accept this kind of

---

mean we have to do to make desirable changes? And related to this, what does this way of seeing mean for defining desirability itself?

experimentation for now, and allow the strengths and weaknesses of these various ideas (in terms of their meanings for intervention) to surface through academic debate.

In the longer term, once the current debate about theoretical underpinnings for understanding marginalisation has reached maturity, my intention is indeed to develop a new political/sociological theory to complement my work on boundary critique (see Chapter 18 for details). However, this will be a theory advanced in the knowledge that it is just one perspective amongst many: I will argue that it should be seen as a contribution to a wider debate as well as a support for intervention practice. Also, in terms of supporting intervention, it will certainly not be universally applicable: there will still be a need for theoretical and methodological pluralism.

7.3.2 Making Critical Boundary Judgements

Having outlined the theory of marginalisation itself, and having considered some of the attendant issues this work raises, let us now examine the implications of this way of thinking for making critical boundary judgements during interventions. Ulrich (1983) stresses the importance of stakeholder involvement in the process of making such judgements: he suggests that a boundary judgement should, in principle, be normatively acceptable to all "concerned citizens". My work problematises this because it is possible to have a consensus between a relatively diverse group of stakeholders on the boundary that they think should be adopted, yet this may be the result of processes of conflict and marginalisation that remain invisible. An example I have discussed elsewhere (Midgley, 1994) is the tendency to focus uncritically on boundaries around human systems (especially in the industrial, economic and political arenas) while marginalising the non-human environment. Elements of the non-human environment are made profane by some actors, thereby justifying the narrow focus, yet the abuse these elements are then subject to may result in damage to human and non-human alike (note that an issue or any aspect of perceived reality may become marginalised, not just stakeholders). When all identified stakeholders share a commitment to a boundary judgement, it is still important to consider what is marginalised by this. On occasion, it may be necessary for the intervener to introduce a different perspective by widening dialogue beyond the boundary of those who are immediately identified as affected or involved. The new participants may argue for the use of a different boundary in the intervention, challenging the consensus and making visible the marginalisation that supports it.

This has implications for stakeholder theory. A good stakeholder theory defines a 'stakeholder' in a broad manner: first, as someone who benefits from (or is harmed by) a particular social situation in the present (the usual definition); and second, someone who can throw new light on 'insider' understandings. The sweeping in of 'outsider' perspectives creates new relationships, meaning that a stakeholder comes to be defined as someone who is, or *ought to be*, involved in or affected by a social situation—in the present or the ideal future.[82]

7.3.3 Key Differences between Churchman, Ulrich and Myself

One of the tensions between Churchman and Ulrich concerns whether the individual should be viewed as an autonomous moral decision maker, or whether moral responsibility should be seen as resting with 'discourse communities' (groups of people who participate in given discourses). A related tension is whether the theory of boundary critique is useful because it supports the professional intervener, or because it provides a language of debate for citizens more generally. I have used it both ways in my own practice: as a device for the intervener to gain insight into a problematic situation (see Chapter 14 for an example), and as a theory that I have communicated to a group so they can generate their own insights.[83] This dual usage reflects my own view that responsibility rests with *agents*—and as I made clear in Chapter 5, an agent may be an individual, group, family, organisation or community. In short, any individual or group who can be seen as acting with purpose.

Churchman's and Ulrich's ascription of moral responsibility to the individual or the group reflects their use of philosophy: Churchman takes a more traditional Kantian view (seeing the individual as morally autonomous) and Ulrich takes a Habermasian line (seeing the

---

[82] In terms of practising stakeholder analysis, this way of thinking can be operationalised through the modified use of methods from Critical Systems Heuristics (Ulrich, 1983) and Interactive Planning (Ackoff, 1981) which surface visions of the ideal that an organisation or group should move towards. An ideal can be used to surface new stakeholders who should be swept in, even though they are not involved at present (see Cohen and Midgley, 1994, and Midgley *et al*, 1997, for practical examples of this kind of stakeholder analysis).

[83] An example of the latter is a workshop I held with nursing students. The students used the model to reflect on disciplinary divisions in the hospital setting; gender relations; relationships between patients and health professionals; relationships between people with dementia and others in the community; etc. The group generated a wide range of rich and subtly textured analyses, and their comments upon debriefing made it clear that they found it of great value. I must say that I was surprised by the extent to which they were able to use it, not just to examine their own relationships with patients (which was the original focus of the workshop), but to interpret a wide range of other relationships.

individual as a participant in dialogue). In terms of the three major paradigms of philosophy reviewed in Chapter 4, Churchman is an idealist and Ulrich is a social constructionist.[84] I have suggested that we should be able to accept the best from both these paradigms (plus the realist paradigm), without getting caught in the rationality of either—and as I have already argued in Chapter 4, process philosophy is the means to achieve this. Therefore, there is a difference in the philosophical bases of our ideas. While I accept that it is human agents who participate in competing discourses (with their associated value and boundary judgements), and these same agents maintain patterns of marginalisation through their actions, they nevertheless interact with other human and non-human phenomena which may be included or excluded from second-order reflections on what gives rise to particular systems of discourse and marginalisation (depending on the second-order boundary judgements being made).

This philosophical difference has implications for the way in which we view systems. As explained earlier in this chapter, Churchman made a paradigm shift from a view of system boundaries as 'given' in the real world, to a view of boundaries as personal or social constructs defining what is pertinent in an analysis. Clearly, the counterposing of these two views reflects the subject/object dualism I argued against in Chapters 3 and 4. Ulrich's introduction of social constructionism promises to overcome this dualism but, like all views of epistemology which stress the primacy of language and dialogue, I argue that it ultimately complicates matters by establishing further oppositions between 'language/subject' and 'language/object' (refer back to Chapter 4 for a more detailed argument). In Ulrich's (1983) case, as I understand it, he is primarily concerned with the 'language/object' problem: he argues against the view that system boundaries are objective properties of the world, and in favour of the view that they derive from dialogue and are expressed in language. My own view is that we need not get trapped into any of these 'either/or' positions. It is perfectly possible, using the process philosophy outlined in Chapter 4, to accept understandings of systems and their boundaries as real world entities, personal constructs *and/or* dialogical phenomena—depending on where the boundaries of analysis are drawn and which theory or

[84] As I said when I reviewed those paradigms in Chapter 4, the boundaries between them are fuzzy because the more sophisticated authors try to take into account arguments from across the board. Therefore, this categorisation of Churchman as an idealist and Ulrich as a social constructionist has only partial legitimacy: Ulrich (1983) in particular uses a number of Kantian ideas, giving him a foot in the idealist camp. Nevertheless, I believe it is fair to say that Ulrich's key departure from Churchman is in his use of a dialogical theory of rationality.

theories are used in first- and second-order analyses. While (like Churchman and Ulrich) I accept that all of these understandings of systems are expressed in language, and are therefore the product of human agents, these agents can either be seen as autonomous or as parts of wider physical and/or linguistic systems (if so desired).

Another key difference between Churchman, Ulrich and myself is that, unlike these other two writers, I do not accept the principle of universalisation. In other words, I do not agree that it is possible for a moral judgement to be consistently applicable to all people in all circumstances. Even the most 'obvious' moral judgements, like the proscription of one human being killing another, is not easy to generalise to all situations. What about war for example? Or what about a situation where an adult kills another who is about to take the life of a child? I suggest that, while the judgement 'killing is wrong' will be the subject of consensus in the vast majority of contexts, there are nevertheless boundaries to the application of it that can be revealed through the exploration of cases like the ones I mentioned above. Hence, in law, there are legitimate defences against the charge of murder (such as self defence, the absence of an intention to kill, and diminished responsibility). Therefore, I argue that to make universalisation a core principle of a methodology risks making morality uncritical: one could easily end up with simplistic moral pronouncements in which people are discouraged from considering the limitations of their judgements. When universalised morality of this kind becomes solidified in plans, and ultimately in institutions, all kinds of anomalies and potential injustices may arise. This is particularly noticeable in pluralistic societies where lifestyles may differ greatly.[85]

In some ways, presuming that I have correctly understood Ulrich's position on universalisation, I am drawn to conclude that Ulrich did not take the idea of boundary critique far enough. As I see it, if value and boundary judgements are intimately linked, then it is contradictory to claim that any moral idea can be universally applied. It can only be applied in relation to what is taken as pertinent to the analysis being

---

[85] A good example is the UK immigration system which allows people to bring a genuine marriage partner into the country. This legislation is supposedly based on a universal human right for people not to be separated from those they love. However, more than this has been generalised: the person has to be a different gender, and only one spouse is allowed. Thus, the immigration system ends up discriminating against gay men and lesbians who might wish to bring in a partner of the same sex, and Muslims (amongst others) who may wish to marry more than one person. These forms of discrimination could be avoided if planners were willing to accept that their own morality might be bounded, and need not be imposed on others with different moral values. At the very least, these moral issues could become the subject of public argumentation, resulting in informed decisions about whether the morality of the majority should be imposed on minorities.

conducted at the time. In principle, if other factors are considered, they may suddenly reveal the need to limit application. As I see it, to discuss the boundaries of the application of moral ideas without recourse to universalisation—in other words, to deliberately discuss the *selective* application of moral ideas—does not imply moral relativism (a state where all value systems are looked upon as equally acceptable, regardless of the consequences for others). I am simply arguing for sensitivity to the limits of moral generalisation, and recognition that, because it is invariably necessary for planners to extend their morality into the design of social systems which have effects on others, the nature and size of this extension should be the subject of reflection and justification—as far as possible in meaningful dialogue with the affected (or their representatives if the affected cannot speak for themselves). On occasion, disagreement over the extension of values may not be resolved (for instance, if a small minority insist that they have the right to steal from others), and in such circumstances limits to the freedoms of this minority will need to be imposed—but in my experience such situations are minimised when there are genuine attempts to address the needs of all parties.[86]

7.4 Elaborating the Theory of Boundary Critique

When I first set out to write this book, one of my aims was to extend the theory of boundary critique to incorporate a new model, which is perhaps a more usual representation of the relationship between two boundary judgements. When stakeholder groups with different purposes and some common concerns come together, it makes intuitive sense to represent their boundary judgements as overlapping circles rather than as one circle contained within the other. An example is a health service and a social services department[87] which have different remits in the

[86] An example is a participative planning exercise I engaged in, which looked at the need for reform of the mental health and criminal justice systems in the UK (Cohen and Midgley, 1994; Chapter 16 of this book). The people with mental health problems who participated in the planning, all of whom had been detained against their will in the recent past, were quite willing to acknowledge that there are situations (when the safety of others is put at risk) where it is right to use forced detention. When I started this planning work, I anticipated resistance to the whole idea of forced detention, but I found that nobody took such a stand—and neither did anybody want detention to be enforced in all circumstances. Everybody involved was keen to explore the detailed circumstances in which detention could be seen as legitimate.

[87] A social services department, in the UK local government system, is a department which meets many non-financial needs (e.g., child protection, social support for women at risk of violence, services for disabled people, etc.). It has overlapping concerns with health,

FIGURE 7.4: Overlapping concerns of two stakeholder groups

UK welfare system, but which have overlapping concerns in some areas (e.g., providing aids and adaptations for disabled people). Figure 7.4 shows the area of overlap which can become a focus for either collaboration or conflict, or indeed a mixture of the two.

However, since making my decision to extend the theory in this way, Yolles (1999a, 2000) has started work in this area too and has introduced his own model based on the same idea. As some of Yolles's theory and language is different from mine, I present a hybrid version in Figure 7.5 that uses my own terminology so as to maintain consistency with Figure 7.3.[88] Then, following some explanation of this model (and before concluding this chapter), I discuss a difference between Yolles and myself on our understanding of ethics. I believe that it is important to make this difference explicit because, in my opinion, the two views have very different implications for how we should be developing the methodology of systemic intervention.

7.4.1 Modelling the Overlapping Concerns of Stakeholder Groups

In Figure 7.5 (on the next page) you will see that the ethics of the two stakeholder groups with overlapping concerns come into conflict, but there is no marginal area (representing people and/or issues lying outside one boundary and inside the other) that becomes sacred or profane. Rather, in the eyes of each stakeholder group, it is the other

---

housing and other departments both within and outside of local government.

[88] Yolles has used a new language of cybernetics, synthesising complexity theory (Nicolis and Prigogine, 1989; Cohen and Stewart, 1994) and viable system theory (Beer, 1975, 1979, 1981; Schwarz, 1994, 1995) to explain the functioning of systems of this type. This is quite complex, and I recommend the reader consult Yolles (1999a) for further information. I will not review it in detail here because, from the point of view of process philosophy, it provides just *one way* of understanding the operation of systems of marginalisation (it is just one kind of second-order analysis) which, in my view, should not be used entirely to the exclusion of other possibilities.

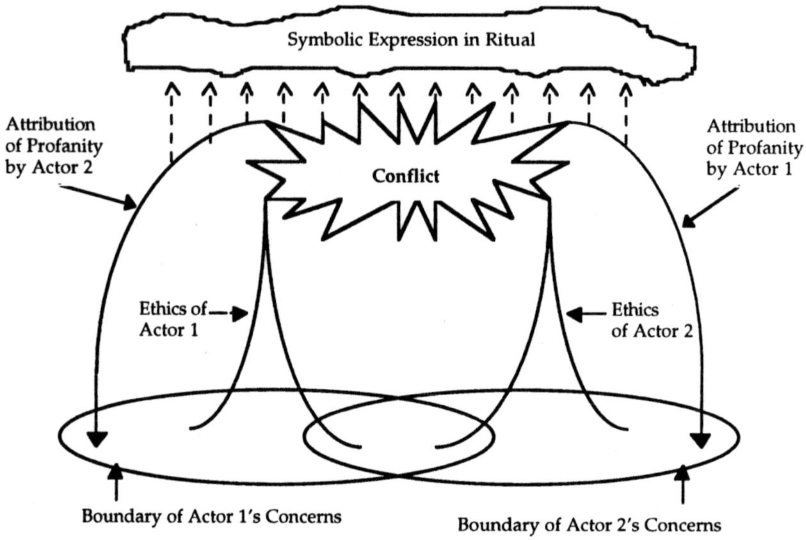

FIGURE 7.5: Model of overlapping stakeholder concerns, with attributions of profanity

group and its concerns that become sacred or profane.

When one group sees the other as sacred, and there is no element of profanity in the picture, it is unlikely that anyone will view the conflict as a major problem because there is an incentive for the group which sees the other as sacred to strive for co-operation—even to the extent of moderating their own values and ethics to come into line with what are perceived as sacred expectations. Therefore, while acknowledging that it *is* possible for one stakeholder group to view another as entirely sacred, I will put this situation to one side and concentrate attention on what happens when stakeholder groups regard one another as profane. This is when participants are most likely to experience their situation as problematic.

Having said this, I should nevertheless note that it is possible for a mix of attributions of sacredness and profanity to obtain: stakeholder groups may initially see each other as sacred if they have a shared interest in dealing with a particular concern, but their differences may nevertheless impede co-operation and ultimately lead to each group regarding the other as profane. If they are locked together by their need or desire to co-operate, but can't actually trust each other, then this can make for a very difficult and stressful situation which the parties find almost impossible to resolve. This is just the kind of situation that appears to exist in many instances of problematic multi-agency working in the public sector (Fargason *et al*, 1994; Kintrea, 1996; Yerbury, 1997; Watson, 1997; and Midgley *et al*, 1997).

An attribution of profanity is most likely to obtain when each group wishes to treat the overlapping area differently because of its connection with a separate set of concerns. As a result, each comes to see the other's concerns as profane and their own as sacred. Yolles (2000) stresses that power relations and the subjective perspectives of actors are vitally important in determining how processes of marginalisation will pan out in any local situation. Power relations come into play when one set of stakeholders is able to pursue its purposes through action in a manner that constrains the actions of the other set of stakeholders. The subjective rationalisations of the actors are important because this sort of power play will only become problematic if one or other set of stakeholders perceives it as such.

In a power struggle, the dominant group imposes controls on any activities relating to the area of overlap that support pursuit of the concerns of the other stakeholders. These controls may be active, in the sense of setting agendas, constructing rules, withholding information or even using force to prevent the other group from getting its way. Alternatively, controls can be passive: a dominant group may simply ignore the voices of other interested parties. In such a situation, the subjugated group may try to sabotage pursuit of the dominant group's purposes, and in response the dominant group is likely to get more active in imposing controls on the actions of the subjugated group. As the power struggle unfolds, rituals form. These may or may not be directly functional, but they symbolically express the pattern of dominance, subjugation and attributed profanity.

Yolles (2000) illustrates this with reference to a long-running strike on the Liverpool Docks (on the West coast of England). Details are not provided here because the issues are exceedingly complex, and truncating them in the interests of brevity will inevitably introduce distortions. However, I recommend reading Yolles's paper as the strike provides an illuminating case study of conflict, power relations and the attribution of profanity.

Before closing this section, I should emphasise that Figures 7.5 is not designed to *replace* Figure 7.3, and the analysis associated with it, but to supplement it.

7.4.2 A Key Difference between Yolles and Myself

Now, an important difference between my own thinking and that of Yolles (1999a) should be highlighted. This is that Yolles talks about *cognitive purposes* rather than ethics. Ethics, in Yolles's theory, are part of the worldviews guiding both the boundary judgements and purposes pursued by agents. In contrast with Yolles, I prefer to be more explicit about ethics, which I define as *values in purposeful action.*

As Churchman (1979) says, it is common for professional interveners to think about how to pursue their purposes, but less common for them to consider the values that inform them. This is because of the 'instrumental' rationality (reasoning geared to fulfilling pre-set goals) that dominates most social affairs: 'practical' rationality (reasoning geared to moral reflection and communication) tends to be marginalised (Habermas, 1972; Ulrich, 1983). Talking in terms of purposes in relation to boundary judgements without being explicit about values therefore risks slippage back into a purely instrumental rationality where there is reflection upon the means of achieving ends, but the ends themselves (the purposes) are taken for granted. This split between means and ends, which is widely used, cannot be sustained once we realise that means as well as ends have moral consequences (Ulrich, 1983; Flood and Ulrich, 1990): for example, Hitler's means of eliminating the Jewish people was at least as morally reprehensible as his stated end of purifying the Aryan race. The advantage of using the word 'ethics' in relation to boundary judgements is that it ties values to the purposive actions of agents. Thereby, we do not have to choose between being explicit about values (Midgley, 1992b) or purposes (Yolles, 1999a), but can fuse the two by talking about *ethics*.

7.5 Conclusion

We can now summarise the main contributions of Churchman, Ulrich, myself and Yolles to our understanding of the process of making critical boundary judgements. First, Churchman introduced the fundamental idea that the boundaries of analysis are crucial in determining how improvement will be defined during a systems intervention, and hence what actions will be taken. He also argued that pushing out the boundaries to make a systems intervention more inclusive may well involve sweeping in new stakeholders. While I agree with the questions raised by Ulrich over Churchman's almost exclusive concentration on the professional intervener, who becomes the sole focus for ethical decision making in an intervention, we do not have to lose Churchman's insights when we widen the boundary of who or what should be considered a moral agent: while there may be many kinds of agent (individuals, families, groups, organisations, societies, etc.), one kind of agent is indeed the professional intervener (acting, for example, as a consultant, employee, political activist or action researcher). Indeed, I recognise that many readers of this book will have professional experience of intervention (or be studying intervention methodologies).

After Churchman had advanced his new understanding of systems, which was very influential during the 1960s and 1970s (especially in the systems and operational research communities), Ulrich built on this by pointing out the need to rationally justify the setting of boundaries. He suggested, following Habermas, that rationality is dialogical. Therefore, if boundaries are to be established rationally, they should be defined in dialogue by all those involved in and affected by the intervention. The philosophical assumption underlying this idea is that, because systems judgements are framed in language, 'dialogue communities' should be seen as the origin of moral judgements, not autonomous individuals. The burden of moral responsibility is therefore shifted from the shoulders of the individual (especially the professional intervener) to the participative group. However, just as Ulrich questioned Churchman's focus on the individual, I have suggested that *any* agent (whether an individual, dialogue community, organisation, etc.), in interaction with the knowledge generating system of which it is a part[89], can be seen as morally responsible. We can therefore use Ulrich's understanding of dialogue communities, but do not have to see it as the only way of understanding systems.

Another important contribution from Ulrich was his method of Critical Systems Heuristics, which offers twelve questions that can be used by citizens to reflect on boundary and value judgements in social situations—both boundaries that are currently being used by planners and managers, and boundaries that citizen groups think *ought* to be used. I have found this to be a very useful tool for facilitating critical reflection in group work (see Chapters 14, 16 and 17 of this book), and it remains one of the few well-tried methods explicitly designed to support people in reflecting on boundaries. I anticipate that this will be one of Ulrich's lasting contributions.

Just as Ulrich built on the work of Churchman (criticising and reconstructing some aspects), I have built on the work of Ulrich. Specifically, I conducted an examination of the systemic forces that work to stabilise conflictual situations, and produced a model of marginalisation processes (Figure 7.3) that can be used to inform critical reflection during interventions. In writing about Figure 7.3, I demonstrated the need to be aware of how some stakeholders and issues

---

[89] As we saw in Chapter 4, these knowledge generating systems only become visible in the event of second-order reflection to define their boundaries. In principle, the whole Universe can be seen as a knowledge generating system, but in practice the boundaries are likely to be viewed more narrowly.

may be stigmatised by systemic processes, resulting in their potential marginalisation during interventions. I also argued that it will sometimes be necessary to challenge a consensus on boundaries by seeking the involvement of people who might not be defined as directly affected or involved, but who may nevertheless have an important perspective to bring to bear on the boundaries of the intervention. This understanding of marginalisation has since been added to by both Yolles and I (Figure 7.5)—although, as I argued earlier, there are some differences of opinion between us over terminology and the importance of being explicit about ethics.

In the next chapter, I will briefly review how the theory of boundaries enables theoretical pluralism, which in turn provides a basis for *methodological* pluralism—and I will argue in Chapter 9 that the latter is essential if systemic intervention is to be flexible and responsive in practice.

8

Theoretical Pluralism

In Chapters 1, 4 and 6, I argued that process philosophy (as I have described it) allows theoretical pluralism. This is because different theories assume different boundaries of analysis. If it is justifiable to use any one of a number of boundaries, then it becomes equally justifiable to draw upon the full variety of theories that support, and are supported by, the various boundaries. This is the case when making first-order analyses (about 'the world') or second-order analyses about agents within knowledge generating systems giving rise to first-order analyses (including, for some purposes, the self, however it is bounded in a particular instance).

Before moving on (in Chapter 9) to discuss how methodological pluralism is based on this theoretical pluralism, I first want to discuss four important implications of the above paragraph: the first is the impossibility of accepting the idea, at least in any absolute sense, that knowledge is cumulative (building into a more and more accurate picture of reality). Following on from this is a second implication: I argue that theories should be seen as more or less useful in terms of whatever purposes of intervention are being pursued. The third implication of accepting theoretical pluralism is that there is always an agent (who is embedded in a wider knowledge generating system) making choices amongst a plurality of options. However, to be consistent with the proposed pluralistic stance, we should accept some theoretical pluralism about what it actually means for an agent to 'choose'. The fourth and final implication follows on from the observation that there is always an agent making choices when theoretical pluralism is embraced: this kind of pluralism does not imply absolute relativism. It is still possible to talk in terms of principles or standards for choice—albeit standards that are locally and temporarily relevant rather than universal. Let me start by explaining the first of the implications of theoretical pluralism: that it takes us beyond the cumulative view of knowledge.

8.1 Beyond the Cumulative View of Knowledge

In order to accept theoretical pluralism, we must (like Kuhn, 1962) give up the common assumption made in traditional scientific circles (e.g., by Popper, 1959) that knowledge is cumulative: in other words, that scientists are developing a 'body of knowledge' that is moving inexorably closer and closer to the 'truth' about reality. We must give up this assumption for three reasons. First, if we accept the systems idea that everything is ultimately interconnected, then no theoretical knowledge, however well elaborated, can accurately reflect reality (at a very basic level, setting aside all the ideas about interpretation explored in Chapter 6, theories assume boundaries which cut interconnections). Second, as Fodor (1974) argues, if different forms of language are used to produce theories that have relevance at different 'hierarchical' levels (e.g., theories about eco-systems, organisms, organs, cells, etc.), then the idea of a single body of knowledge will always be problematic.[90] Third, as Popper (1972) realised, it is not even possible to know whether the development and testing of a given theory is taking us closer to, or further away from, 'reality'. This does not mean giving up talk of reality altogether, but (as discussed in Chapter 4) it means realising that discussion of a material world assumes one particular kind of boundary judgement—just as talk of consciousness, linguistic systems, etc. (dealt with as phenomena without an explicit material root) assumes different boundary judgements.[91]

Having said this, it *is* actually possible to say that a body of knowledge is growing, and to defend this knowledge as useful and/or true—*but without assuming that it is useful or true in any absolute sense.*

---

[90] Of course, the whole project of general systems theory (e.g., von Bertalanffy, 1968) is to evolve just one language of relevance to all the hierarchical levels. However, as Flood and Robinson (1990) note, this creates a focus on the similarities between levels, so the differences (which are also important) tend to get neglected. Therefore, Fodor's (1974) argument that different languages are necessary for understanding different hierarchical levels raises just as many questions for general systems theory as it does for the view that science is progressively constructing a single, more and more complete picture of reality.

[91] Here I part company with both Kant (1787) and Wittgenstein (1953) who suggest that we should stay silent about the 'real world' because, if it exists, it is beyond our knowledge: we can only have knowledge of consciousness (Kant) or language (Wittgenstein). Although I agree with them (and the vast majority of philosophers of science writing in the latter half of the 20th Century) when they say that we cannot have unmediated access to a 'real world', I nevertheless believe it is acceptable to talk about it as real—but in the knowledge that this is just one way of speaking about phenomena, implying the use of one kind of boundary judgement. The capacity for movement between boundary judgements and theories, some of which might be about material reality and others about cognitive or linguistic systems, prevents the absolutism associated with naive realist positions from creeping in (refer back to Chapter 4 for a more detailed argument).

What is true today may be myth or falsehood tomorrow, and in another context on the very same day, that body of knowledge may simply be irrelevant: for example, we may believe that modern physics is saying something meaningful about the origins of the Universe (the production of a theory of astrophysics is an intervention into the knowledge shared by a community of interested parties, most of whom will be physicists), but this body of knowledge, even if accepted as true, cannot help us plan the distribution of food and medical supplies to injured people after an earthquake. In such a situation, a very different body of knowledge is required. Therefore, knowledge is relative to the purposes of agents (especially individuals and communities of scientists and practitioners)—who are, of course, under the influence of wider systems which can be bounded in any number of ways.

8.2 Theory in Action

This last point, that knowledge is relative to the purposes of agents, indicates that theory serves social roles. Theory can never be 'pure'—nothing other than a reflection of reality, as some scientists would claim—but is always meaningful in terms of the actions or interventions pursued by agents. This explains why, from a pluralist perspective, it is perfectly possible for people to work with a theory even though they know that it excludes certain factors from analysis. A good example is Luhmann's (1986) theory of ecological communication which talks about linguistic systems, and consciously excludes from analysis the human biological systems which give rise to them. Luhmann does not deny that biological systems exist, but *the analysis of them is not relevant to his purpose* (which is explaining the extraordinary imperviousness of modern capitalist societies to discourses of radical change).

Another example can be found in the discipline of physics. Many physicists seem quite happy to use both Newton's theory of gravity and Einstein's theory of relativity, even though the latter is more comprehensive. The rationale for continuing to use both, rather than wholly replacing Newton's work with Einstein's, is that Newtonian mechanics is still satisfactory for a limited domain of application (Bohm, 1980). If Newton's theory is *adequate for certain purposes*, then it is perfectly legitimate to use it.

Even though this example comes from the heart of traditional science, some scientists (e.g., those following in the footsteps of Popper, 1972, and Bhaskar, 1986) may nevertheless protest that theories *do* reflect reality more or less imperfectly, even if we cannot know the exact nature of their imperfections. This takes us back to the point made in Chapter 4, and picked up again in the previous section (and footnote 91), that it is still possible to accept theoretical pluralism and talk about reality—but with a recognition that talking about reality implies a particular kind of boundary judgement that focuses attention on the material world. Importantly, however, we should also acknowledge that using this kind of boundary *only has meaning in terms of the purposes of scientists' interventions*: scientists who talk about theory as a reflection of reality are intervening in scientific (and other) discourses about matters that they anticipate will be of common interest to the participants in that discourse.

Despite this argument, variations of which have been advanced previously (e.g., Dewey, 1946; Lewin, 1948), there is a tendency amongst some authors (like Seidman, 1988) to *oppose* the practice of intervention (which assumes that theories are more or less useful) to that of observation (which assumes that theories reflect reality). In contrast, I argue that using methods of observation, and developing theories which say something about reality, is just as interventionary as engaging in those practices we usually associate with the term 'intervention' (see the discussion of intervention and observation in Chapter 6).

8.3 Choice between Theories

Another important implication of theoretical pluralism follows on from the above two points: there is always an agent (either seen as a lone decision-maker or as embedded within wider systems) making choices amongst a plurality of options. Exactly what it means for an agent to exercise choice is an interesting question: 'choice' looks very different depending which theoretical perspective and associated boundary judgement is being used to understand it. In other words, it is possible to accept theoretical pluralism with regard to the meaning of choice between theoretical options! Let me give a few examples.

Prioritising a boundary around a single human being gives the impression of autonomous individual choice, or free will. This is one boundary used by Merleau-Ponty (1962) when he says that individuals are either totally constrained or totally free—there is no middle ground. When they are constrained, their actions are determined by larger systems (the individual boundary is deprioritised), but when they are free they are truly autonomous agents (the individual boundary comes to the fore). Similarly, Maturana and Varela's (1992) theory of autopoiesis (reviewed in Chapter 3) prioritises a boundary around the individual organism, and hence sees all choices as determined from within that organism (with the environment, including language, acting as a potential trigger, not a cause).[92] While Maturana and Varela talk about the "braiding" of rationality and emotion, suggesting that choice is a property of both, Kelly (1955) sees it as a purely rational business—but in a similar manner to these other authors, Kelly still attributes choice to the autonomous individual. Likewise, Rawls's (1971) liberal theory of justice views the individual as "somehow pre-social", and hence capable of semi-autonomous moral thought—but note that Rawls qualifies the word "autonomous" with the prefix "semi" because he recognises that individuals make choices in social settings where the limits to freedom are circumscribed and/or negotiated.

In contrast with these 'individualist' theories of choice, many family therapists have seen the choices made by individuals as both determining the nature of, and as being determined by, the 'family system' (see, for example, Watzlawick *et al*, 1968; and Selvini-Palazzoli *et al*, 1978). In this view of choice, there are two levels of agency: the agency of the individuals and the agency of the family, both of which are systemically linked. In one sense the family system determines the 'choices' available to its members, but the members can nevertheless shift the family dynamics through strategic interventions (but in families exhibiting 'pathological' behaviour, they may need support from a family therapist to achieve this).

Another author who sees choice as a property of wider systemic connections, stretching beyond the individual, is Bateson (1972). In particular, he talks about how individuals are embedded in circular information pathways, so 'choices' need to be seen in the context of

---

[92] It should be noted that Maturana and Varela (1992) have a view of boundaries that is similar in some senses to that proposed within general systems theory (von Bertalanffy, 1968): the boundary is created and maintained by the autopoietic activity of the organism. Maturana and Varela may therefore object to me saying that the boundary judgement is made by them: they may well prefer to see boundaries as natural functions of autopoiesis. However, I have two answers to this. First I suggest that, when producing the theory of autopoiesis, Maturana and Varela had to make a judgement that the level of the individual organism is an appropriate level to focus upon—and they had to make this judgement *prior* to the accumulation of evidence in support of their theory (otherwise they would not have known what evidence to look for). Second, given that they claim an anti-realist (or, in my terms, idealist) ontological stance (see Chapter 3), it would be contradictory if they were to refuse to acknowledge their own roles in constructing the 'reality' of the theory of autopoiesis. Of course, I am not arguing against the use of boundary judgements distinguishing physical entities (like organisms)—that would contradict my stance on both process philosophy (Chapter 4) and theoretical pluralism (this chapter). Rather, I am trying to counter a possible objection to my argument that the theory of autopoiesis involves an *a priori* boundary judgement that defines the individual organism as the most appropriate focus for analysis.

these. Also, these circular information pathways can be viewed as parts of larger systems with their own self-producing capacity, meaning that choice at the individual level is, in the main, explicable by the requirements of systems that include both human and non-human elements. Here, the boundaries being prioritised are those identifying information loops and socio-ecological systems: the boundary identifying the individual is very much secondary.

Finally, there are two sets of authors who operate with the boundary of the Universe, but who have produced theories with radically different implications for understanding the nature of choice. First, there are those who have an entirely mechanistic view of the Universe, in which free choice is a meaningless concept. An example is Skinner (1971) whose theory of behaviourism views all 'choices' made by individuals as reactions to environmental stimuli based upon previous learning—and the nature of this learning can be described in terms of simple, deterministic, biological processes. Thus, Skinner's Universe is like the one described by Descartes (1642, 1644): basically a clockwork toy, but without the realm of the spirit intervening through the exercise of human will. Similarly, Maze (1983) embraces an "unqualified determinism" and claims that uni-directional causality is impossible: a person who 'chooses' to walk around a chair has not caused this to happen—it is equally possible to say that the chair is the cause of the person's actions. Thus, free choice is illusory: there are only interactions between organisms, between organisms and inanimate objects, and between inanimate objects, in which no one organism or object can be considered prime. In the view of Maze (1983), the Universe is the sum total of these deterministic interactions.

The other set of authors who operate with the boundary of the Universe, but whose work has radically different implications for understanding the nature of choice, are those developing and/or applying the theory of dissipative structures (e.g., Glansdorff and Prigogine, 1971; Nicolis and Prigogine, 1977; Prigogine and Stengers, 1984; Prigogine, 1989; Straussfogel and Becker, 1996; Capra, 1996; Allen, 1998). The basic idea of the theory of dissipative structures is that, while the Universe as a whole exhibits ever-increasing entropy (movement from an ordered state towards disorder), there are nevertheless entities within it—*neg*entropic open systems—that maintain their order, at least for a while. Prigogine and his colleagues call these negentropic systems "dissipative structures" because they are islands of structure (order) within the chaotic Universe, yet they dissipate relatively quickly. Both a human being and a candle flame are dissipative structures (when a human being 'dissipates', s/he dies), raising some fascinating questions about the similarities and differences between 'living' and 'non-living' phenomena (Prigogine, 1989). Interestingly, organisms are seen by some authors as having a key role in the production of entropy: a temporary manifestation of order (an organism) which acts as an open system to transform matter from a well-ordered to a less well-ordered state is an efficient entropy generator (Swenson, 1991, 1992).

Capra (1996) notes that the theories of autopoiesis (Maturana and Varela, 1992) and dissipative structures (Prigogine, 1989) both make a fundamentally different paradigmatic assumption to those theories which assume a mechanistic Universe (e.g., Skinner's, 1971, behaviourism). The theory of dissipative structures, for example, suggests that open systems exist in a state far from equilibrium, and often encounter "bifurcation points" where they must go in one direction or another. This 'choice' is driven by the teleology (goal-directedness) of the open system, which is buffeted by a complex environment, but which ultimately reacts to this environment according to the capacities given in its internal structure (which, in an organism, is a result of genetic predisposition and previous learning). Therefore choice, understood in this very specific manner, is a property of open systems which continually emerge and die within the Universe. Behaviourism, on the other hand, does not accept the possibility of teleology at all, but assumes "unqualified determinism" (Maze, 1983), and therefore also denies any kind of choice, however defined.

While both these sets of writers operate with a boundary demarcating the Universe as a whole, there is one significant exclusion from the behaviourists' boundary that makes sense in terms of the stated purposes of behaviourist theory—and it also explains the extreme divergence in the conclusions of the behaviourists and those arguing for the theory of dissipative structures. The behaviourists are primarily interested in explaining human and animal behaviour, and in the process draw upon the notion of the mechanical Universe, but they are not concerned with the *origins* of this so-called mechanism. In contrast, the theory of dissipative structures is particularly focused on the origins of order and disorder, and it proposes that order manifests itself continually within the Universe in the form of open systems—so therefore the Universe is not mechanical at all.[93]

[93] I find that a good metaphor for understanding the view of the Universe suggested by the theories of dissipative structures and autopoiesis is a glass of carbonated water. Bubbles appear in the water as if by magic, float to the surface and then burst. In a similar manner, organisms and other open systems are born into the Universe, move around, and die. Of course, the bubbles don't actually appear by magic: microscopic investigation shows that bubbles form on rough patches in the glass, suggesting that the conditions have to be right for a bubble to appear. Likewise, conditions in the surrounding environment have to be right for an open system to be born. Therefore, it is not simply a case of order emerging spontaneously out of chaos. Indeed, amongst many of the more sophisticated organisms on

This actually suggests that the two theoretical positions can be differentiated by the purposes of the agents who use them, related to their boundaries of inclusion. For the purpose of examining the relatively simple kinds of organism/object interactions studied by the behaviourists within laboratory settings, it is sufficient to look at the interactions of already-existing organisms and note that they can be explained with reference to the theory of behaviourism. This can then be generalised to the boundary of the Universe: if the concept of teleology is not needed to explain the interactions of already-existing organisms, then the act of generalisation will indeed give rise to a vision of a mechanistic Universe. Alternatively, generalisation may move from the general to the specific: if a Cartesian, mechanistic Universe is assumed, then interactions at a local level would have to be explained without recourse to teleology. However, for the purpose of explaining the unfolding of the Universe over time, and interactions between order and disorder (the purpose of those with an interest in dissipative structures), the birth and death of open systems as self-producing (rather than mechanically interacting) entities becomes important. This is an example of how different ways of seeing become possible depending on the purposes and boundaries adopted by agents (and the knowledge generating systems of which they are part).

So we see that there are many ways to conceptualise choice: as decision-making by autonomous agents; as decision-making by agents embedded in wider systems; as determined by the information loops a person is a part of; as a consequence of autopoiesis; as an activity of far-from-equilibrium systems faced with bifurcation points; or as wholly determined by the environment (in the latter case 'choice' is an illusion of consciousness with no corollary in reality). Just as theoretical pluralism can be welcomed in first-order inquiries (about 'the world'), where consideration needs to be given to the usefulness of theories for specific circumstances, it can also be employed in second-order reflections about both the nature of the agent engaging in pluralist practice and what it means for the agent to choose between theories.

8.4 Standards and Principles for Choice

Following on from the observation that there is always an agent making choices when theoretical pluralism is embraced (however 'choice' is defined in any particular circumstance), it remains for me to

this planet, the conditions for the birth of a new organism are created within the body of the organism itself.

show that theoretical pluralism does not imply absolute relativism—a complete lack of standards or principles to inform choice. If the latter were the case, 'choice' under conditions of theoretical pluralism might amount to little more than the whim of an agent, regardless of the effects on others.

Certainly it is the case that, from my point of view, there can be no *universal* standards for choice between theories. Even frameworks that have been developed to highlight the assumptions of various perspectives (e.g., Burrell and Morgan's, 1979, influential framework of sociological paradigms) cannot make any satisfactory claim to objectivity (and therefore universality). To take Burrell and Morgan's (1979) classification of sociological paradigms as an example, we find that it has been widely challenged: Willmott (1993), for instance, has claimed that the sociological perspectives don't all fit neatly where Burrell and Morgan claim. Indeed, whenever anyone tries to step up a meta-level and classify a plurality of theories, it transpires that their classification reflects just another theoretical stance. Hence, Gregory's (1992) strong claim that all meta-theories are suspect: they pretend to rise above all other theories, but are inevitably theoretical themselves. They therefore exist at the *same level* as the theories they try to classify.

If there are no universal standards for choice, where does this leave us? In my view, it does not leave us in a state of absolute relativism. Rather, it puts us in a position where we not only have to acknowledge that agents choose theories according to their purposes, but we also have to recognise that agents (individuals, teams, groups, organisations, etc.) create standards of their own that develop and change over time—and how the origins of these changes are seen will depend on how the agent is viewed (e.g., how the knowledge generating system influencing the agent is bounded during second-order inquiries). However, regardless of what means of viewing the agent is used, it is unlikely that the creation of standards for choice will be seen as random: it will always have an identified origin in the knowledge generating system of which the agent is a part (or be an emergent property of the rationality of the agent him or herself if s/he is viewed as an autonomous individual).

One particularly interesting theory of the origins of standards has been proposed by Foucault (1984b,c), although this is by no means the only possible view that can be taken: Foucault argues that knowledge, power (defined as 'action upon the actions of others') and the identity of the agent are woven tightly together, so it is most likely that choice between theories (forms of knowledge) will be shaped by power relations (which in turn find expression in, and are influenced by, the identities of agents). Thus, the plurality of theories visible to agents; the ways in which agents construe their purposes; and the ways in which they understand their 'choices' may all be influenced (but *not* determined[94]) by the operation of power. I would therefore expect norms for choice to be apparent, even if it were universally accepted that there are no legitimate universal standards! In the view of Foucault (1984a-c), the fact that norms and standards for choice emerge from relationships between power, knowledge and identity means that it falls to 'subjects' (in my terminology, agents) to reflect on the desirability (or otherwise) of these norms, and to act strategically to support or challenge them as appropriate.

The creation of standards does not lose meaning with the death of universalism, and will rarely be seen as random. Which theories will be seen as useful for what purposes will depend on the agent's relationships—including power relationships if we use Foucault's (1984b,c) understanding—with the wider systems in which s/he is embedded. Therefore, *locally* relevant standards for choice (as opposed to universal standards) can always be defined, and the construction of these can be subject to critique through second-order reflections on the nature of the knowledge generating systems in which the relevant agents are embedded.

8.5 Conclusion

In this chapter, I have explored some of the implications of theoretical pluralism: that knowledge cannot be seen as cumulative (building into a more and more accurate picture of reality) in any absolute sense; that theories should be seen as more or less useful in terms of the purposes of intervention being pursued; that there is always an agent making choices (variously defined) amongst a plurality of

[94] In Foucault's earlier writings, he tended to talk in terms of knowledge and power interacting to construct identities. However, in 1984a-c, it became apparent that he had adjusted his position to acknowledge that 'subjects' (agents) can have an impact on power-knowledge networks through strategic action. Therefore, Foucault's position cannot be described as determinist. For a very clear review of Foucault's changing ideas, see Darier (1999).

options; and that we should talk in terms of locally relevant, rather than universal, standards for choice. In the next chapter, I move from theoretical pluralism to propose *methodological* pluralism. I argue that methodological pluralism is vital if we are to build a flexible and responsive intervention practice.

9

Methodological Pluralism

Having established the link between process philosophy, boundary critique and theoretical pluralism, we can now focus in more detail on *methodological* pluralism. In Chapter 5, I defined methodology as "the set of theoretical ideas that justifies the use of a particular method or methods" (p.103). As methodology is essentially theoretical, it should be clear that, if it is possible to have theoretical pluralism (see the last chapter), then we should be able to have *methodological* pluralism too: we can accept a plurality of theories flowing into methodology, and hence a wide variety of methods may be seen as legitimate.

This chapter is about why methodological pluralism is valuable. Following a discussion of why we should welcome this pluralism, I briefly introduce the variety of methods and methodologies available (not in any detail, because the literature is so vast, but indicating general types). Consequently, the reader will begin to see a small part of the diversity that provides a resource for a systemic and pluralist intervention practice. Some previous work on methodological pluralism will be reviewed as part of this, in preparation for a more detailed look in Chapter 10 at the issues surrounding choice between methods, and the practice of mixing methods drawn from different paradigms.

9.1 Two Levels of Pluralism

Earlier in this book (Chapter 5), I differentiated the terms 'method' and 'methodology'. While 'methodology' refers to the theory that justifies the use of particular methods, a 'method' is a set of techniques operated in a sequence (or sometimes iteratively) to achieve a given purpose. When I talk of methodological pluralism, I mean embracing the possibility of engagement at two levels: at the level of *methodology*, where we can respect others' methodological ideas, thereby allowing their insights to inform our own methodology (either

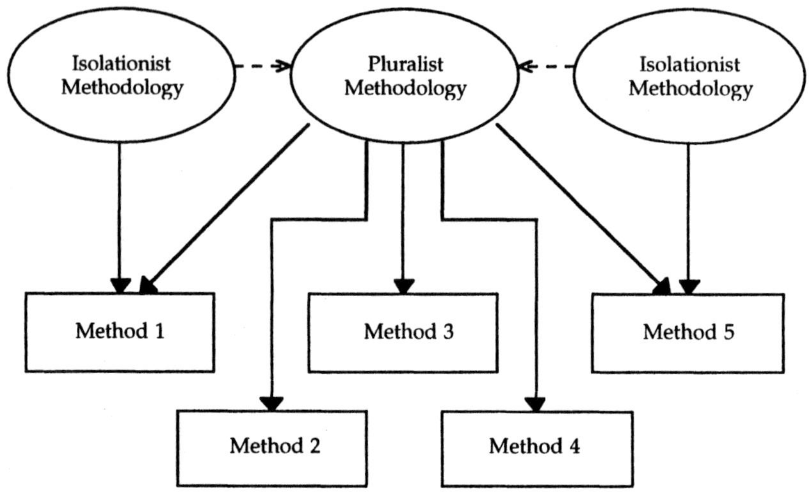

FIGURE 9.1: Relationship of a pluralist methodology to isolationist methodologies and a variety of methods

temporarily, during a particular intervention, or on a longer-term basis as continual reference points); and also at the level of *method*, where we can use a wide range of methods in support of particular purposes. Figure 9.1 illustrates these two levels.

In Figure 9.1, we see that different methodologies (in the ellipses) are linked with different methods (in the rectangular boxes) by arrows which represent the ascription of validity and/or legitimacy. In two cases (the left and right ellipses) this is a one-to-one linkage, with a particular methodology indicating the validity and/or legitimacy of just one method (or a narrow range of methods). When a methodology is proscriptive in this way, refusing validity or legitimacy to the majority of methods, it can be called 'isolationist' (Jackson, 1987a). Most methodologies produced during the 20th Century, whether for scientific study or intervention, are isolationist: they prescribe a 'one best way' of doing things. In contrast, a pluralist (using the methodology represented by the central ellipse in Figure 9.1) can use the full range of available methods, but they are seen through the theoretical lens of his or her own methodology, and are made meaningful in local situations by the way they meet (or fail to meet) the purposes of the agent(s) engaged in intervention (and of course these purposes may be evolved through the intervention itself).[95] The pluralist defines his or her methodology in

[95] For an agent engaging in intervention using systems methodology, his or her purposes are likely to be informed by the methodology itself, as well as his or her life experiences. For other agents involved in the intervention who may have less or no knowledge of systems

contrast with others (most of which will be isolationist). Although pluralist methodologies are distinguished from isolationist ones, the pluralist will still be able to learn from the latter—hence the dotted arrows going between the pluralist methodology and the isolationist methodologies in Figure 9.1.[96] The idea of developing a pluralist methodology through reflection on the strengths and weaknesses of other methodologies (including isolationist ones) will be expanded upon in Chapter 11, where an ideal model of interventionist learning is presented.

As there are different rationales for pluralism at the levels of methodology and method, they are dealt with separately below.

9.2 The Value of Learning from Other Methodologies

The essential value of being aware of, and learning from, a variety of methodological positions comes from the knowledge that no one theory, or set of theories—whether or not they have been codified into a methodology—can ever be comprehensive (Morgan, 1986; Francescato, 1992; Romm, 1996). Therefore, it is bound to be the case that others will have different insights to ourselves. While we may disagree with some of their assumptions and want to challenge them (e.g., the assumption made by some traditional scientists, such as Popper, 1959, that only observational methods are valid), other insights and assumptions will be valuable to us—indeed, valuable enough for us to want to use them in a methodology of our own.

The purpose of learning from other methodologies is therefore that reflections on the similarities with, and differences from, one's own ideas can enable the continued evolution of one's own methodology in a manner that enhances the conceptual basis with which interventions are planned (Gregory, 1992; Romm, 1996). The key to this learning is to welcome the insights of others without taking on any idea to the exclusion of all others, thereby losing the possibility of pluralism.

---

methodology, life experiences are more likely to dominate. As the systems methodology described in this book requires the intervener to open up to the perspectives of others involved in the intervention (Churchman, 1970; Ulrich, 1983; Midgley, 1992b; Chapter 7 of this book), the possible actions suggested by the methodology come to be seen as a resource for all the agents involved and affected, including the methodologist, to fulfil their purposes. Of course, when a systems methodology is used that has the exploration of boundary judgements at its heart, purposes are not pre-defined but can be developed through the process of intervention.

[96] Because of the proscriptive nature of isolationist methodologies, it is less likely that learning will go in the other direction, from a pluralist to an isolationist methodology, but such learning is not impossible.

Therefore, to say that (for example) use of the scientific method can be valuable at times should not lead to the conclusion that only this one kind of method is valid. Those aspects of scientific methodology that promote a worldview which invalidates other methods need to be opened to challenge, but the scientific method (and indeed any other method which may have originally been derived from a proscriptive methodology) can still be seen through the 'lens' of a methodology which welcomes pluralism, such as the one proposed in this book. Of course, this raises the thorny issue of the nature of learning across paradigm boundaries, and I will look more closely at different authors' views about cross-paradigm learning in Chapters 10 and 11.

9.3 *The Value of a Plurality of Methods*

The value of pluralism at the level of methods comes from observations of what happens if only a very narrow set of methods are used—indeed, it has been known for some people to specialise in the use of just one. With an armoury of just a couple of methods, three significant interlinked problems arise:

First, the intervener is likely to be unresponsive to diverse understandings of issues arising out of the different perspectives of those affected by them. If the way of seeing the issue isn't supported by the chosen method, then dissatisfaction is likely to result. If this dissatisfaction is experienced by those who are marginal to decision making (e.g., by an environmental group trying to influence business decisions, or by employment rights campaigners trying to have an effect on the formation of government economic policy), then the most likely outcome will be decisions which further marginalise stakeholders with a different point of view from the decision makers. An example, described in more detail in Chapter 2, is Spash's (1997) observation that methods of cost-benefit analysis which involve stakeholders in making trade-offs between economic and environmental values automatically marginalise environmentalists who cannot accept the trade-off mentality embedded in the methods. In the longer term, this kind of marginalisation can lead to significant social problems (not to mention environmental ones in the above example) including, in the more extreme cases, civil unrest and violence.[97]

---

[97] Pinzón and Midgley (2000) argue for boundary critique and methodological pluralism in the context of conflict resolution in Colombia because it is only by respecting the issues as they are seen from all sides that it becomes possible to start meaningful negotiations. The idea of trying to resolve conflict in the Colombian context without resorting to the use of a wide variety of methods for use in different local situations seems to me to be an impossibility.

In contrast, if dissatisfaction is experienced by the decision makers themselves, and the chosen method only supports the understanding of a marginalised group without opening up a dialogue with decision makers, then it is likely that the intervention will simply be ignored by the latter. After all, it is rare for somebody to listen to others who are not prepared to listen to them in return, and those who are already in control of decision making are simply able to turn their backs in such a situation (Gregory, 1992). The only scenario in which decision makers are likely to listen to marginalised groups without expecting the same courtesy in return is when their openness to the views of others is being evaluated by a 'higher' authority (e.g., when health professionals are obliged to listen to patient views because this is prescribed in quality procedures).[98]

The second significant problem with using a narrow range of methods is that, as intervention proceeds, the issues of relevance to those involved in, and affected by, the intervention may change as people's understandings develop. So what may have started out as, say, an issue of organisational restructuring to improve efficiency and effectiveness, might eventually come to be seen as an issue of fostering participative decision making to improve workforce satisfaction and commitment. If the intervener can only handle methods for restructuring, then s/he will be unable to deal with this change in focus (unless the new issue can be 'forced' into the mould of the old one). This is actually an example from an intervention I worked on in partnership with a Ph.D. student, Chia Hui Ho. We were asked by Tainan City Council in Taiwan to support their restructuring efforts, but we very quickly found that issues of democratisation came to the fore. Had we not been able to show how these could be dealt with alongside the restructuring by using different methods from those originally envisaged by the leader of the Council, the restructuring would almost certainly have been thwarted by employee sabotage (Ho, 1997).

---

[98] The issue of what may motivate decision makers to listen to marginalised groups is a difficult one. Other methodologies, such as Ulrich's (1983) Critical Systems Heuristics which (in many respects) is not naive with regard to power issues, have been criticised for failing to consider this in an adequate manner (Jackson, 1985c; Ivanov, 1991; Flood and Jackson, 1991b,c; Mingers, 1992b; Romm, 1995a; Midgley, 1997c). While Gregory (1992) suggests that a guarantee of being listened to in return is the only thing that can motivate decision makers to listen to marginalised stakeholders, Foucault (1980) and Vega (1999) both argue that dialogue tends to result from relations of force: decision makers will only listen to others if they would lose out from not doing so. It could be that, when decision makers engage in dialogue with marginalised groups, they do so, not because they fear direct retribution from those groups, but because they fear a more general loss of public trust. However, they can get away with ignoring marginalised groups if the latter are the ones who refuse to engage in dialogue (there is a general moral expectation in Western culture that dialogue will be the first recourse in cases of conflict).

The third problem associated with the use of a narrow range of methods is that interveners may see all issues through the methodological 'lens' that these methods have traditionally been associated with. There is an old saying that rings true in this context: if the only tool you have is a hammer, then everything looks like a nail. Therefore, it is not just that the use of a narrow range of methods makes the intervener unresponsive to a diversity of interpretations of issues, and unable to react appropriately to changing agendas, but also *s/he is likely to be unaware of this fact*. This lack of insight, which comes from an unwillingness to explore the possibility that there may be other ways of seeing and doing, is a substantial obstacle to practising systemic intervention.

9.4 A History of Methodologies and Methods

To make clear the wide diversity of approaches available to the pluralistic systems intervener, I will use most of the rest of this chapter to produce a very brief history of the development of intervention methodologies and methods. In principle, it would be possible for me to review the whole range of scientific disciplines and intervention practices, but this would be a huge task filling many encyclopaedic volumes. The variety is that great! Even if I were to confine the review to the methodologies and methods emerging from just one research community, the task would consume several chapters (see, for example, Jackson, 1991, for a useful review of management systems methodologies that takes almost half his book). I have therefore chosen to paint a more general picture, highlighting key paradigm changes in the 20th Century. It will then be the reader's task to follow up particular developments of specific interest to him or her.

However, I will start with an acknowledgement that this is *my* history of methodologies and methods—inevitably limited by the scope of my own reading, which has primarily been in the areas of management systems and operational research (plus, to a lesser extent, scientific methodology, management, psychology, family therapy and action research). I have placed a particular emphasis on management systems, partly because of the importance to this book of systems ideas, and partly because I have had most practical experience in this area. Also, I have tended to give more detail of later work than earlier (particularly ideas emerging in the last thirty years), again because these have been the primary influence on my own practice. I invite the reader, who will undoubtedly have his or her own preferred emphasis (which may be quite different from mine) to think about how the methodologies and methods I have reviewed might be complemented by others that I have omitted.

One further caveat about what it means to present a 'history' of ideas is also necessary before I can enter into the discussion proper:

9.5 What is History?

There are many different approaches to history, making quite different assumptions about what a 'history' is. Flood and Gregory (1988) identify four paradigms of historical inquiry. First, there is the *linear* view that history is a series of 'facts' about what has happened over a period of time. This assumes that it is possible to have an objective, or quasi-objective, representation of events. In a way, it is a 'common sense' view of history that many non-historians will take for granted.

Second, there is the *structuralist* view that the trajectory of events is pre-determined, or guided, by deep socio-structural relationships. An example is the classical Marxist argument that the end of capitalism is as inevitable as the end of feudalism (which capitalism replaced). According to Marx and Engels (1888), economic recession is an intrinsic feature of capitalism. However, some recessions are more severe than others, and a particularly severe recession (or series of recessions) will leave so many people in poverty that they will realise that their interests do not lie in rescuing the current system, but in creating a revolution and instituting a new system of common ownership—socialism. Socialism is said to be a system where laws are necessary to ensure common ownership and prevent a slip back into capitalism. The idea, however, is that socialism is only a transitional phase which again is pre-destined to be superseded: once all the people come to realise the benefits of common ownership, the need for laws to enforce it will recede. The end result will be a communist society in which people implicitly understand how to live in non-exploitative relationships with one another—there will be no need for laws regulating economics to enforce this. Essentially, this is a view that suggests there are 'deep' economic and social forces at work pre-determining the direction of history.

Another very interesting structuralist view of history is presented by Berry and Kim (1994), building on the work of Strauss and Howe (1991). Berry and Kim argue that interactions between economic and population cycles over a 200 year period in the USA have produced a variety of predictable 'zeitgeists' (spirits of the time) which have largely determined the political agendas pursued during that period of history—including the timing of wars and other major events.

The third view of history is that histories are written by the 'winners' of conflicts (Carr, 1961). Flood and Gregory (1988) call this an *interpretive* position, indicating that history is inter-subjectively constructed, not objective.[99] History is the result of power struggles in which the *desired* history comes to be solidified as fact, and competing interpretations are eradicated. For example, it is likely that the history of Hitler's rise to power in Nazi Germany would have been portrayed rather differently had Germany won the Second World War.

Fourth, there is the *interpretive-analytic*[100] view of Foucault (e.g., 1980, 1984a-c), that history is constructed through an interplay between knowledges (especially, in recent times, by the knowledges developed through the disciplines); power relations; and the identities of subjects engaging with these power-knowledge relations. There is some similarity with Carr's (1961) perspective, in the sense that history is an outcome of power struggles, but for Carr power is 'owned' by victors and exercised over victims. In contrast, for Foucault (1980), power is not 'owned' by anybody. It is closely associated with the development of forms of knowledge which people use to order their social relationships. What appears on the surface to be one person exercising power over another is actually the *end result* of a process of knowledge formation in which certain social practices come to be legitimated (see Chapter 4 for a more detailed explanation).

In this chapter, I shall be using a variant of the conventional, linear view of history. It is a 'variant' in two respects. First, I do not make the assumption that this history is 'true' in any absolute sense: there will be many omissions, and I recognise that my history could have been constructed in a variety of different ways. Therefore, in Flood and Gregory's (1988) terms, an element of interpretivism creeps in—or, in my own terms, I recognise the inevitability of using boundaries in constructing a historical perspective. Second, I see the value of the interpretive-analytic view. Indeed, Valero-Silva (1998) has produced a useful history of management systems that emphasises the role of power in shaping our knowledge of methodology and methods: in most cases, methods have become popular when they have 'fitted' the changes in worldview being promoted within industry. Therefore,

---

99 However, as Carr's (1961) view is that history is constructed as an end result of social conflict, there is an argument for saying that he has produced a non-deterministic, structuralist theory rather than an interpretive one.

100 The term "interpretive-analytics" was invented by Dreyfus and Rabinow (1982) to describe Foucault's theoretical stance.

methods enhancing competitiveness and profitability have gained wide-spread currency, while those which challenge the prevailing capitalist worldview tend to remain buried in the literature. While I accept that (inevitably) my own history will be a history of methodologies that have 'survived or thrived' in a world of power relations, I nevertheless intend to cover a set of ideas that have been applied beyond the domain of industrial relations as well as within it (e.g., in family therapy). Therefore, some of the effects of industrial power should be mitigated. Remembering that my purpose in constructing this history is to show that a *diverse resource* of methodologies and methods is out there, I suggest that this variant on a linear overview will be sufficient for readers to begin to start sampling the literature and, through their own on-going research programmes, begin to uncover some of the many gems that this overview will not have revealed.

Finally, it should be noted that, in constructing my history, I have tried to show how ideas from a number of traditions have intermingled over the years, and how some of the variety of methodologies and methods can be accounted for by understanding that previously existing ideas from seemingly disparate sources (such as quantitative applied science, the human relations tradition and psychoanalysis) have been synthesised with each other and with systems theory to produce quite distinct methodologies and methods.

9.6 Scientific Methods

I will start the review with scientific methods, which have been refined over several hundred years. Although these were first developed to enhance the control of observation and *prevent* intervention by the researcher into the functioning of that which was being researched, it is nevertheless the case that many writers have argued that science should be harnessed in the service of the 'social good'. The Pragmatists (discussed in Chapter 5) certainly took this line, as did early action researchers (e.g., Lewin, 1948). Of course, both these sets of authors challenged the idea of independent observation, but many others interested in the 'applied sciences' have not done so. For instance, it is still largely the case that scientists testing the efficacy of medicines wish to preserve a 'non-interventionary' stance: while the results of their experiments may have a deliberate biological and social effect (preventing or curing disease), their methodology excludes this element of intervention from their understanding of science.

I have therefore included scientific methods in this review, not with their original methodological intent in mind, but with the idea that they *may be explicitly used for intervention* in the development of scientific knowledge and social practice (see Chapter 6 for the case for seeing 'observational' methods as interventionary). Thus, interveners using scientific methods cannot be excused from considering the social consequences of their experimentation: while these consequences may not always be predictable, the fact that there is indeed an intention to intervene (at the very least into the development of knowledge) means that, if they wish their research to be considered systemic, they must (at minimum) explore its possible effects as a part of the research effort. This means considering boundaries for analysis, in terms of both issues to be studied and people to be involved in debate. It also means using appropriate quantitative and/or qualitative methods to secure an enhanced understanding of the likely systemic effects of their scientific work, and consequently to enable informed decision making about if and how it should be undertaken.

For many people schooled in the traditional sciences this may be viewed as a draconian restriction on their practice, but (as I see it) it is an inevitable consequence of wishing to take account of the systemic effects of an intervention (rather than refusing to even think about, let alone take responsibility for, how knowledge might be used). Far from limiting scientific practice, I suggest that engagement with those who may be affected by the scientific endeavour will give rise to many more (scientific and non-scientific) paths for inquiry, with clear benefits for others, than are currently being pursued (see also the ESRC Global Environmental Change Programme, 1999).[101]

Of course, there is an argument that this is not feasible because it will require many scientists to learn a whole new set of skills that are beyond their current knowledge and abilities (Brocklesby, 1997). Part of my answer, expanded upon in Chapter 11, is that the consequences of *not* following such a path are potentially very serious (witness the starvation caused by introducing crops engineered by scientists in the West into fragile third-world eco-systems[102]). Also, there are plenty of

[101] In 1999, the Economic and Social Research Council (ESRC), a State-funded research body with substantial influence on the formation of mainstream social sciences policy and practice in the UK, issued a special briefing which used the public concern over genetically modified foods as a case study to argue for a major change in scientific research, broadly in line with what I have proposed in the paragraph this footnote is linked with (ESRC Global Environmental Change Programme, 1999). If this change in thinking (previously, the ESRC tended to favour neo-positivist approaches) also occurs in other countries, then I suspect that we are well on the way to a substantial paradigm shift in mainstream applied science.

[102] See George (1976) for a shocking indictment of the effects of scientific research supposedly undertaken in the name of third-world agricultural improvement. Essentially,

examples of scientists who have made this leap already,[103] and if a scientist *really* feels unable to engage with new methods, there is always the possibility of constructing multi-disciplinary teams to do so.

There is, of course, a large volume of literature on experimental methods (and see Chapter 6 for another discussion of them). For those without a scientific background, I have found the following to be good introductory texts: Festinger and Katz (1954), Campbell and Stanley (1966), Wright *et al* (1970) and Campbell and Cook (1976).[104] For a basic introduction to statistics, I have found Robson (1973) invaluable: this presents the main concepts and a variety of fairly simple methods that are particularly accessible to those without a background in advanced mathematics (all mathematical operations are described in words as well as figures). In my view, mastering Robson's methods will provide any reader who is concerned about their ability to handle mathematics with an excellent foundation for moving on to tackle other texts with less discursive explanation (e.g., Siegel, 1956; Plutchick, 1968; McGuigan, 1968; Heermann and Braskamp, 1970; Hays, 1974; Ferguson, 1981).[105]

---

scientific research conducted in laboratory conditions has led to the development of 'better' crops without taking into account local knowledge of the eco-systems in which they are to be introduced. George also highlights the business interests that are served by this kind of research at the expense of meeting the immediate needs of subsistence farmers for whom lower-tech solutions to their problems may be more appropriate. In the context of agricultural development in Mexico, Rose (1988) suggests that "science is only as good as the political and economic system in which it operates" (p.14).

[103] Most of the people I have worked with, and know in the academic community, started their careers as disciplinary scientists. I also started out in this way as a psychologist, running my own research business for three years before finally deciding that I wanted to concentrate on developing the theory and practice of systemic intervention. In my judgement, for most people the stimulus for moving in this direction is the realisation that the traditional scientific methodologies propounded by many disciplinary scientists exclude the social effects of scientific activities from analysis. Once the realisation dawns that this is the case, and the judgement is made that it is not acceptable, the search is on to find an alternative paradigm of inquiry. Most often, this leads into reflections on the need for stakeholder participation in scientific decision making. See Flood (1990) for another example of reflections on personal and professional development following a similar trajectory.

[104] These were the books that I used when I studied the foundations of experimental psychology in 1979/80, but there will no doubt be many other texts around (including more recent ones) of value to people wanting an introduction to scientific methods.

[105] Again these were the texts that I used when studying psychology, but there will undoubtedly be more recent ones of value to those wanting to learn a variety of statistical methods.

9.7 Applied Science

Over the years, a wide-ranging literature focusing on intervention rather than observation has been produced. At the end of the 19th Century, this was of two types. The first type focused on the needs of individuals with a variety of what we now call 'mental health problems': thus, psychoanalysis and its later variants were developed. The second type of intervention methodology can be thought of (at least as it was first conceived) as 'applied science': using the techniques of experiment, observation and quantification for a social purpose. I will deal with the latter first, as it has been the most widely practised approach to intervention in a variety of disciplines throughout the 20th Century (only in the latter half of the 20th Century did the paradigm of 'applied science' begin to give way to alternatives in mainstream debates).

Examples of 'applied science' are many and varied. They include, for example, medical experimentation (e.g., Brodie *et al*, 1994) and the development of methods of economic modelling (e.g., McCormick *et al*, 1974). These both have a long history and are still central to the disciplines of medicine and economics, but will not be a focus of this review.

9.7.1 Scientific Management

The applied sciences also include 'scientific management' (Taylor, 1947), first proposed at the beginning of the 20th Century, which advocates the application of scientific methods (of a kind) to the business of improving industrial efficiency and effectiveness. Taylor (1947) conducted a variety of industrial efficiency studies. His method (at its very basic) is as follows. Taking a particular task, the scientist works out the most efficient way of doing it (using the minimum human time and energy). Once the task is standardised, s/he monitors a variety of workers undertaking it in order to figure out a reasonable expectation for the time it should take a worker to complete *n* repetitions of that task. Targets are then set for worker performance, and incentives offered to those who better the target set for them. Scientific management, as advocated by Taylor, also involved moving away from traditional manufacturing methods where "craftsmen" (*sic*) made products from start to finish. In contrast, with scientific management, the scientist's role is to decompose the whole manufacturing process into component parts that are measurable, and that workers can specialise in.

Scientific management saw the birth of the modern manager: tasks of financial and strategic management were seen by Taylor as separable from shop-floor work. Scientific management was highly successful in terms of improving productivity, especially in large manufacturing corporations such as Ford. It was due to scientific management, or 'Taylorism' as it became known, that Ford was able to reduce the price of cars to the point where they became affordable to many ordinary American citizens. Thus, at the beginning of the 20th Century, scientific management enabled the extension of mass production and consumption throughout Western societies. Indeed, this kind of activity was intensified after the First World War when politicians and industrialists were faced with the task of rebuilding industry. Scientific management was the fuel in the engine of America's post-war boom.

9.7.2 The Human Relations Movement

Also under the umbrella of 'applied science', but in stark contrast to Taylorism (which viewed workers as part of the 'organisational machine'), industrial psychologists began to take an interest in the individual differences between workers. Myers (1920, 1923, 1926), for instance, argued vociferously against the use of efficiency measures which assume that there is only 'one best way' for workers to do a job:

> "....mental and bodily differences between workers are such that it is impossible to train, or to expect, each worker to perform the same operations in identically the same way....it may also be harmful to the worker because it tends to discourage initiative" (Myers, 1926, p.27).

Myers's alternative approach was to seek to remove problems encountered by individual workers, thereby enhancing their productivity (see also Farmer and Eyre, 1922, for a practical example). This approach was later picked up by Mayo (1949), who conducted a whole series of scientific experiments on changing working conditions to enhance social relationships between employees. To over-simplify somewhat, his conclusion was that worker satisfaction results from good social relationships, and satisfied workers are more productive than unsatisfied or alienated workers. Mayo can be seen as one of the founders of the 'human relations' movement, which (unlike scientific management) took account of the subjectivity of the worker, and which still has strong advocates today.

The contrast between the proponents of scientific management (e.g., Taylor, 1947) and human relations (e.g., Myers, 1926, and Mayo, 1949) is particularly acute on the issue of worker initiative. Taylor believed that efficiency is improved by *removing* initiative from workers and placing it in the hands of management, while Myers and Mayo argued the very opposite: that output is enhanced when workers are largely trusted to make their own decisions. It is significant that both Taylor and Myers were using an applied scientific approach because, of course, both were able to provide evidence that their interventions were successful. Indeed, it has been argued that the fashion in management research and practice has swung backwards and forwards from scientific management to human relations many times during the 20th Century, with neither side being able to win the argument (Valero-Silva, 1998).

9.7.3 Operational Research

Despite the birth of the human relations approach, and perhaps because scientific evidence could still be provided in support of Taylorism, those methodologies and methods that objectified human beings and viewed them as mechanical parts of larger systems did not die out. For instance, the Second World War saw the birth of a new practice of intervention: OR (which stands for operation*s* research in the USA, and operation*al* research in Europe). Scientists from a wide variety of disciplinary backgrounds came together during the war to support the planning of military operations. They applied mathematical modelling techniques to find optimal solutions to complex problems. A typical example is provided by Churchman (1987), who describes how he applied mathematics to the problem of determining an optimal quality control procedure in the manufacture of ammunition. Many histories of the birth of OR have been written, but a brief one that I have found particularly useful is Trefethen (1954).[106] For introductions to the mathematical methods of OR see, for example, Jennings and Wattan (1994) and Targett (1996)—these being just two of the many relevant textbooks available.

After the war, most OR practitioners went into industry. In the UK, they had a large part to play in the development of the newly-nationalised industries—most famously, coal and steel (see, for example, Jones, 1992). However, over the years, many writers found

[106] Interestingly, Valero-Silva (1998) argues that the practice of the early operational researchers was not too far removed from scientific management: they both assumed the validity of using mathematical techniques to optimise efficiency and effectiveness. Valero-Silva suggests that it was only possible to establish operational research as having a separate identity because of the closeness of those war-time researchers to government. They had the *power* to establish their own identity as a research community rather than say they were following in the footsteps of scientific management.

that the meaning of the term OR was not immediately transparent to those in industry, and some decided to adopt a new label: 'management science'. Today, those calling themselves management scientists live alongside operational researchers: there is a general recognition that both communities have very similar, if not identical, interests (and indeed, in the mid-1990s, the two professional societies representing American operations researchers and management scientists merged into one larger unit).

Although OR and management science (OR/MS) certainly emerged out of the applied science movement, it should be noted that there is now some debate over the claim that they are 'scientific' (despite the fact that their practitioners commonly use quantitative methods with a scientific origin). This is because the focus on intervention within OR/MS challenges the usual scientific convention that intervention is to be avoided. Also, properly controlled experiments are difficult to conduct in the field, so a great deal of OR/MS knowledge is based on case study evidence (Lathrop, 1959; Barish, 1963; White, 1970; Dando *et al*, 1977; Raitt, 1979; Malin, 1981; Rosenhead, 1986; Keys, 1989, 1998).

9.7.4 Action Research and Action Learning

Interestingly, while significant work was being put into the development of OR, action research was also gaining popularity—especially amongst psychologists working in industry. Lewin (1947, 1948) was the key author in those early days of action research, and his work brought together methods drawn from quantitative applied science with some of the insights from the human relations tradition (allowing him to propose, in his 1948 work, a new approach to conflict resolution in organisational settings). Some of Lewin's ideas were explored in Chapter 6 (including the influence of systems theory that is particularly evident in his 1952 writings), so I will not give further details here. Suffice it to say that Lewin's methods were widely and successfully practised in a variety of organisations, although they have not escaped criticism either (see later in this chapter).

Another prominent thinker influenced by the human relations tradition in the 1940s and 1950s was Reg Revens (see Revens, 1982, 1983, for an overview of his approach), who worked with the newly-nationalised coal industry in the UK. Rather than talk about action *research,* however, Revens preferred the term 'action *learning*'. Although Lewin and Revens worked separately (and using different labels), many authors have placed them within the same broad action research movement (e.g., Flood and Romm, 1996a; Frank, 1997). Revens was particularly interested in how seemingly intractable operational and production problems could be resolved through a process of learning, catalysed through an on-going programme of issue-focused meetings between those with responsibility for problem solving. Importantly, Revens viewed the problem solver as *part of* the problematic situation rather than as an independent observer, and therefore a key question for people to ask both themselves and others is, to what extent am I part of the problem, and to what extent are you? Frank (1997), writing about action learning, says that for individuals to begin to see themselves as part of a problem situation is an essential first step in *learning about learning*, enabling people to work in teams to improve both their relationships and their problem solving capacities. In some ways this participative approach can be seen as pre-figuring many of the changes that swept through the systems, OR and action research communities in the 1970s and 1980s (see later in this chapter for details).

9.8 Psychoanalysis

Alongside the development of the applied sciences in the late 19th Century, the theory and practice of psychoanalysis was being formed. While many people practised the art of intervening in the individual psyche to alleviate emotional distress, arguably the most famous therapist was Sigmund Freud—largely due to the substantial body of theory he developed over his lifetime. See the volume of Freud's works edited by his daughter, Anna Freud, for an overview (Freud, 1986).

While psychoanalysis differs markedly from applied science (it is qualitative rather than quantitative, and develops knowledge through a theory-practice cycle rather than the use of experimental methods), Freud nevertheless inherited one crucial assumption from the scientific endeavour: subject/object dualism. Freud took for granted that psychoanalytic theory is a body of knowledge to be used by the therapist to interpret the words of the patient: in this sense, the patient is an object of study rather than a participant in dialogue.

Since Freud's time, other major works have been produced by the likes of Carl Jung (1946)[107] and Melanie Klein (selected writings edited by Mitchell, 1986). However, it is only in the latter half of the 20th Century that some of the assumptions embedded in the therapeutic relationship (in particular, the patient-as-object being interpreted by the therapist-as-subject) have been revised. Now, many psychoanalytic therapists see themselves as co-constructing the therapeutic reality with their clients, and they are prepared to share

[107] Also see the volume of Jung's writings edited by Storr (1983).

theoretical interpretations rather than keep them hidden from view. Alongside this development in psychoanalytic methodology, the theoretical base has also been transformed by a greater focus on language and the social construction of realities (e.g., Samuels, 1993). For an introductory look specifically at the *methods* of the psychoanalytic encounter, see McLoughlin (1995).[108]

9.9 The First Wave of Systems Thinking

Although there has historically been considerable antagonism between proponents of the applied scientific and psychoanalytic paradigms, particularly within the discipline of psychology where their object of study is the same (the human animal),[109] there have nevertheless been some attempts to bring ideas from both camps together to produce new positions. This was particularly noticeable in the middle of the 20th Century, although it is open to question how conscious writers of that era were about the origins of their ideas in applied science and psychoanalysis (most likely, some were conscious of what they were doing, while others were not). What is particularly noticeable about several of these new positions is their use of the systems ideas that were gaining popularity at the time—particularly the open systems theory of von Bertalanffy (1950) and Bateson's work on pathological communication (several of Bateson's highly influential papers on this subject, written between 1942 and 1971, are reprinted in Bateson, 1972).

---

[108] McLoughlin's (1995) text is a very clearly written introduction to the methods of "psychodynamic counselling", but (like the earlier Freudian and neo-Freudian forms of psychoanalysis) tends to assume subject/object dualism. This is evident in the words McLoughlin uses to describe the therapeutic relationship: e.g., "As you hold your stance and observe your client you will see him behaving within the container you have provided" (p.25). These 'behaviours' are viewed as (often unconscious) communications by the client, but are there to be interpreted without changing the "container", or boundaries of intervention. If the therapeutic reality were co-constructed, then it would be possible for the container to change in response to the client's communications. McLoughlin acknowledges that the *possibility* of changing the container is there, but most of his prescriptions suggest that it is the therapist's role to set and maintain the boundaries.

[109] In my own under-graduate study of psychology, which was heavily dominated by neo-positivism (in particular, the work of Popper, 1959), we were (as far as I can remember) only given one lecture on Freud—and the focus of this was on why we should set him aside. The reason, of course, was that Freud did not use scientific methods!

9.9.1 *Socio-Technical Systems Thinking*

One striking instance of a synthesis of ideas from different paradigms is Socio-Technical Systems Thinking (e.g., Trist and Bamforth, 1951; Trist *et al*, 1963; Emery and Trist, 1965; Emery and Thorsrud, 1969, 1976), which emerged from the work of the Tavistock Institute of Human Relations in London. Arguably, this can be seen as bringing together four traditions: human relations (the applied science movement which recognises the importance of subjectivity within the workplace); psychodynamics (understandings of group behaviour branching off from psychoanalytic theory); action research (at that time conceived by Lewin, 1948, as a union of human relations and quantitative applied science); and the theory of open systems (developed from the work of von Bertalanffy, 1950, amongst others). The Socio-Technical Systems Approach can be seen as one of the most enduring products of the human relations movement, but it is also a milestone for those with an interest in systemic intervention.

There are two core ideas in Socio-Technical Systems Thinking. First, there is the concept of semi-autonomous work groups: Emery, Trist and their colleagues argued that the most effective and satisfying way of organising work is in groups which take collective responsibility for completing tasks. The second core idea (which explains the 'socio-technical' label) is that organisations have both social and technical components which need to be addressed during intervention—but not by separating them and trying to optimise them independently. Separating the technical and social is a recipe for disaster, because it could lead to two sets of solutions that do not fit together. Rather, organisations should be viewed as systems in which the social and technical interact: if both are dealt with simultaneously then, while it may at first appear that sub-optimal solutions are being devised for each of the social and technical aspects, the outcome will be optimal for the organisation as a whole—which is what really counts in terms of improving worker satisfaction and hence performance.

Socio-Technical Systems Thinking was widely applied in the 1950s and 1960s, one of the best known initiatives being the Norwegian Industrial Development Project (Emery and Thorsrud, 1969, 1976; Bolweg, 1976)—an intervention conducted in partnership with the Norwegian Labour Organisation and the Norwegian Employers Federation (with the backing of the Norwegian Government), introducing semi-autonomous work groups throughout industry. A similar nation-wide intervention was also conducted in Sweden (Gustavsen and Engelstad, 1986; Gustavsen, 1992), but this had a slightly different theoretical rationale (incorporating some insights from Habermas's, 1984a,b, theory of communicative action, reviewed in Chapters 2 and 4).

9.9.2 *Systemic Family Therapy*

While the socio-technical systems thinkers brought together ideas from human relations, psychoanalysis and quantitative applied science under the banner of systems thinking, and worked on interventions in organisations, another set of practitioners found a very different domain for systemic intervention: the family. In the 1950s, Bateson mounted a sustained critique of the psychoanalytic tradition, which saw mental health problems (most notably schizophrenia) as pathologies of the individual. In contrast, Bateson (e.g., 1960) pointed to family dynamics and the participation of the person with a mental illness in pathological *conversations*. This theoretical work in systems and cybernetics had a profound influence on several groups of mental health practitioners who began to engage in systemic family therapy (e.g., Weakland and Jackson, 1958; Jackson, 1960; Haley, 1962, 1963; Watzlawick *et al*, 1968; Minuchin, 1974; Selvini-Palazzoli *et al*, 1974, 1978, 1980). Later, Laing and Esterson (1964) also picked up on Bateson's insight, and their work became a foundation stone of the 'anti-psychiatry' movement.

Although, at the time, family therapy was viewed as a departure from psychoanalysis, in retrospect we can see certain similarities between the two traditions that indicate (as with Socio-Technical Systems Thinking) the use of systems theory to create a new synthesis rather than a total abandonment of the older ideas. In particular, both psychoanalysis and family systems theory (at least as it was conceived in those early days) viewed the experience and behaviour of the individual as being caused by wider forces: psychoanalysis proposed the existence of the unconscious, and family therapy looked to family conversational patterns for the 'larger system' exerting control over the individual. As such, both traditions objectified their subjects: they interpreted their actions using hidden theory, refusing to engage in any mutual construction of therapeutic reality. Indeed, some family therapists went so far as to place 'experts' behind one-way mirrors, observing the therapist-family interaction, to enable greater 'objectivity' in diagnosis and the recommendation of prescriptions for change (see Watzlawick *et al*, 1968, and Roy-Chowdhury, 1997, for useful reviews of the methods of early family therapy). There is also a strand of applied science in this work, in that people were quite prepared to experiment on families in order to gain generalisable knowledge (Haley, 1962), although the kind of experimentation that is possible in fieldwork with families differs somewhat from the mainstream scientific practice of controlled experiment and observation.

Before moving on, it should be noted that the sophisticated set of theories and methods developed by systemic family therapists have also been successfully applied in domains other than the family, such as in consultancy with organisations (Campbell *et al*, 1994).

9.9.3 Systemic Operational Research

While both Socio-Technical Systems Thinking and systemic family therapy emphasised the human dimension over the scientific (although without entirely losing the latter), OR (which emphasised the scientific) also came to be influenced by systems thinking at approximately the same time. Several different practices, centred around mathematical modelling techniques, were developed and applied in the late 1950s and early 1960s. Perhaps the four best known of these are *System Dynamics* (e.g., Forrester, 1961), a method for quantitatively modelling complex feedback processes and considering the impact of changes to system relationships; *Systems Engineering* (e.g., Hall, 1962; Jenkins, 1969), an approach which focuses on the design of whole organisational systems, using quantitative methods, to meet given purposes in an optimal manner; *Systems Analysis* (e.g., Quade and Boucher, 1968; Optner, 1973; Quade *et al*, 1978; Miser and Quade, 1985, 1988), a method for assessing costs, effectiveness and risk given multiple scenarios; and *Viable System Modelling* (e.g., Beer, 1959, 1966, 1981), which facilitates the diagnosis of organisational problems through comparisons between a real organisation and an ideal model derived from cybernetic and systems theories.

Each of the above approaches has been widely applied with substantial success (although not without criticism either—see later in this chapter). System Dynamics, for example, was used by Meadows *et al* (1972, 1992) to inform *The Limits to Growth* and *Beyond the Limits to Growth*: seminal texts based on models of global human-environment interaction, suggesting that humankind will face serious ecological problems in the middle of the 21st Century if we continue on a path of unsustainable development.

Systems Engineering has been applied so widely in China that a whole discipline, with its own Academy of Science, has been named after it.

Likewise, Systems Analysis was adopted by the US Government in the 1960s to inform policy development, and has since been used in a wide variety of settings. It is also the principle approach associated with the International Institute for Applied Systems Analysis (IIASA), a research group supported by twelve nations which produces guidance for the management of global problems (Miser and Quade, 1985, 1988).

Lastly, like the other approaches, Viable System Modelling has been extensively applied in organisations world-wide, but arguably its most famous application was to the whole economy of Chile (Beer, 1981). Chapter 14 of this book includes an example of my own application of Viable System Modelling.

9.10 The Second Wave of Systems Thinking

The first wave of systems thinking (incorporating insights from both the quantitative and human relations branches of applied science, amongst other traditions) gained great popularity in the 1950s and 1960s. However, in the late 1960s (and even more in the 1970s and early 1980s), significant questions began to be asked, both about the philosophical assumptions embodied in the first wave, and the consequences of its practical application.

9.10.1 Criticisms of the First Wave

Some approaches in the first wave of systems thinking (particularly those reviewed above under the title 'Systemic Operational Research') were criticised for regarding models as representations of reality rather than as aids for the development of inter-subjective understanding (see, for example, Churchman, 1970; Checkland, 1981; Espejo and Harnden, 1989; de Geus, 1994). As the interveners who used these approaches believed that they had unique insights into the nature of complex systems, granted by the use of their systems methods, they set themselves up (and were regarded by others) as *experts* in systems thinking. Thus, they fell into the trap of making recommendations for change without properly involving those who would be affected by, or would have to implement, that change. The result could often be recommendations that were regarded as unacceptable by stakeholders, and were therefore not implemented, or were resisted if implementation was attempted (Rosenhead, 1989a).

These approaches were also criticised for viewing human beings as objects that could be manipulated as parts of larger systems, instead of individuals with their own goals which may or may not harmonise with wider organisational priorities (Checkland, 1981; Lleras, 1995). In consequence, several authors pointed out that the first wave of systems approaches, which emphasised quantitative applied science, failed to see the value of bringing the subjective insights of stakeholders into activities of planning and decision making (e.g., Churchman, 1970; Ackoff, 1981; Checkland, 1981; Eden *et al*, 1983).[110] Finally, it has been argued that most of these systems approaches assume that the goal of the person or organisation commissioning a systems project is unproblematic, when it is actually common to find that goals are unclear or there are multiple viewpoints on which goal it is most appropriate to pursue (Checkland, 1981; Jackson and Keys, 1984). In such circumstances, to take a cynical view, it is relatively easy for the commissioner to subvert application of the systems approaches from the quantitative applied science tradition: unless the intervener has a strong sense of ethics, and some understanding of participatory practice, use of these approaches will tend to support only the views of the person providing the money, allowing the opinions of others to be ignored (Lilienfeld, 1978; Jackson, 1991).

Likewise, the early systems approaches to family therapy were criticised for maintaining the culture of expertise: the tendency of therapists to view families as objects open to theoretical interpretation stopped any possibility of families and therapists working together to co-construct therapeutic realities (Andersen, 1987; Cecchin, 1987; Hoffman, 1988; Anderson and Goolishian, 1988). Thus, only a limited range of possible ways of seeing the family were allowed to surface in therapy, which imposed corresponding limits on the changes that could be pursued (Rosenblatt, 1994).

The only first wave systems approach to largely evade the criticism of objectivism was Socio-Technical Systems Thinking. In my view, it managed to do so because of the emphasis on semi-autonomous work groups (embodying the human relations view of human beings, which values subjective perspectives). However, Socio-Technical Systems Thinking was criticised on different grounds: it inherited the emphasis of von Bertalanffy's (1950) open systems theory on organisms (and organisations) adapting to survive in ever-changing environments. Thus, the goals of the organisation are not generally open to critique (Jackson, 1991)—except, I would argue, when there is a threat of annihilation and a radical change is required. Jackson also points out that, although workers are able to participate in semi-autonomous work groups, there is still an assumption in Socio-Technical Systems Thinking that the manager needs to be an expert systems practitioner:

> "In....sociotechnical theory managers act paternalistically, for the good of all, by using their expert knowledge to adjust the organization in ways that will ensure its survival. Sociotechnical theory even gets the workers to control themselves, relieving

---

[110] However, Socio-Technical Systems Thinking was largely exempt from this criticism.

managers of one onerous chore, by convincing employees that they are getting a form of genuine control over their working lives" (Jackson, 1991, p.69).

All these criticisms led to a significant paradigm shift in systems theory, and simultaneously in the application of this theory to intervention. A second wave of systems thinking was born. In this new wave, 'systems' were no longer seen as real world entities, but as constructs to aid understanding. The emphasis was on dialogue, mutual appreciation and the inter-subjective construction of realities. Arguably, the authors who are best known for generating this paradigm shift in management systems are Churchman (1979), Ackoff (1981) and Checkland (1981)—the first two working in the USA, and the third in the UK, although many more authors than these actually contributed to the change. I have already reviewed Churchman's fundamental contribution to rethinking the systems idea (Chapters 3 and 7), in which boundary judgements came to be seen as conceptual (and ethical) rather than real-world entities, so here I will focus on the work of two of his colleagues, Mason and Mitroff (1981), who converted some of Churchman's thinking into a directly applicable method. I will also say something about the methodologies produced by Ackoff and Checkland, and will briefly touch upon second wave developments in System Dynamics that have revolutionised the way in which that methodology and its methods are seen. These four examples should be sufficient to make clear what the second wave of systems thinking was all about. I will then broaden the discussion beyond management systems to look at the impact of the second wave on systemic family therapy. Finally, I will argue that the same paradigm shift that has occurred in management systems and family therapy has also taken place in OR and action research.

9.10.2 Some Management Systems Methods from the Second Wave

Mason and Mitroff (1981)[111] were particularly influenced by Churchman's (1979) position that rational argument means pursuing a "dialectical process": seeking out the most articulate "enemies" of our ideas and debating with them. In this way the assumptional boundaries of our ideas can be tested. The result may be the victory of one or other position, or a new synthesis that is stronger than either of its contributory parts. This is the idea that informed the development of

[111] This 1981 reference is out of print at the time of writing. If it is impossible to obtain, I recommend consulting Mason (1969), Mitroff and Emshoff (1979), Mitroff *et al* (1979) and/or Flood and Jackson (1991b), although these give less detail.

their method, Strategic Assumption Surfacing and Testing (SAST). To reduce it to its bare essentials, the method can be described as having four stages: *group formation* (gathering all those involved in, and affected by, a situation and splitting them into small groups according to their views on key issues); *assumption surfacing* (identifying the preferred strategy or position that each group is adopting, then revealing and quantifying the assumptions upon which it is based); *dialectical debate* (presenting the case for each position and discussing them all in a single, large group); and *synthesis* (achieving an accommodation amongst participants to find a practical way forward).

Mason and Mitroff's method has been well tested in practice, and their 1981 book reports many applications to address significant social policy issues and business planning problems.

Another 'second wave' method is Interactive Planning (Ackoff, 1981). The purpose of this is to harness the knowledge and creative abilities of everybody in an organisation to produce a plan of the ideal future that the organisation can work towards. The plan may take some time to implement, perhaps many years, but it offers a feasible set of targets for the long term. A key idea is that the plan should be wide enough and creative enough to "dissolve" any disagreements between participants. The transformation it proposes should result in the commitment of all concerned. The method itself can be represented in the form of three stages: *establishing planning boards* (every role in the organisation should be represented in planning, with participation being as wide-spread as possible); *generating desired properties of the organisation's products and/or activities* (this is 'ends planning', conducted under conditions of minimum constraint, with only technological feasibility, viability and adaptability limiting proposals); and *producing the plan itself* ('means planning', where all sections of the organisation agree on how to move forward).

Interactive Planning has been applied with great success in a wide variety of corporations, both large and small. It has also been used to provide long-term blueprints for the redesign of cities, including Paris in France (here, the number of people participating by questionnaire ran into the millions) (Ackoff, 1981). I have also used aspects of Ackoff's work in some of my own interventions (see Chapters 14, 16 and 17).

The third approach that provides an example of second wave systems thinking is Soft Systems Methodology (SSM) (Checkland, 1981; Checkland and Scholes, 1990). This encourages participants in intervention to generate issues through on-going explorations of their perceptions, allowing people to model desirable future human activity. Given the necessary commitment from individuals involved in and affected by possible changes, these models of future human activity can be used as a basis for guiding actual human activity in the world. However, to ensure that the models will indeed be useful, it is necessary for participants to relate them back to their perceptions of their current situation. In this way, possibilities for change are tested for feasibility.

The methods of SSM, which are usually operationalised in a workshop format, can be summarised as follows: (i) consider the problem situation in an unstructured form; (ii) produce a "rich picture" (a visual representation of the situation people currently find themselves in); (iii) identify possible "relevant systems" that might be designed to improve the situation, and harmonise understandings of these by exploring key questions[112] (these questions are prompted by the use of a mnemonic, which is explained in Chapter 15 of this book); (iv) produce a "conceptual model" for each relevant system (a 'map' of the human activities that need to be undertaken if the system is to become operational); (v) refer back to the rich picture to check the feasibility of the ideas; (vi) produce an action plan; and (vii) proceed to implementation. Of course, participants need to move backwards and forwards between these activities, harmonising the outputs from each one with the others—the activities should not be implemented mechanically in a linear sequence. More details about the methods associated with SSM are provided in Chapter 15.

Of all the second wave systems methodologies, Checkland's is arguably the one that has been applied most widely, largely because of the Masters degree at Lancaster University that ran until the mid-1990s. This supported large numbers of students in applying Checkland's methods, and many of these students went into academia and passed on their knowledge to the next generation. If I was to pick one single application to recommend, however, it would be Checkland's work in supporting new developments in the UK National Health Service (Checkland, 1997).[113]

It should be clear from the descriptions of these three approaches that they are considerably different to those advanced by the first wave systems methodologists. If the intervener can be regarded as an expert at all, his or her expertise is in *facilitation* rather than in systems modelling. While there is some quantification involved in SAST, this is merely a device to clarify the subjective perspectives of participants, who can then test out their assumptions in debate. In all

[112] What will the system do? What worldview lies behind the desire to establish it? Who will be affected and in what ways? Who will operationalise it? Who can stop it from working? What factors will have to be taken as given by the system?

[113] Also see my own application of SSM in Chapter 15 of this book, in which nineteen organisations came together to plan the development of a post-disaster counselling service.

these approaches, there is no pretence to identifying an *objective* 'one best way': rather, it is proposed that the 'best way' emerges from intersubjective, rational argumentation.

Because of the emphasis on participation, it is arguable that these second wave systems thinkers have embraced the human relations position that people are most productive when they *join together* to produce solutions, rather than having solutions given to them by experts. However, a residual element of quantitative applied science rationality is still present: all the second wave authors accept that they occasionally encounter defined disagreements that are amenable to resolution through scientific experimentation, and in these cases quantitative applied science has a legitimate place.

Before moving away from management systems to discuss family therapy, I should also touch on second wave developments in System Dynamics. While System Dynamics (e.g., Forrester, 1961) was originally a method for quantitatively modelling complex feedback processes, in the 1980s a revolution in both methodology and methods took place. This was based on the insight that a System Dynamics model can be seen as a device for aiding communication between stakeholders on complex issues, and does not necessarily have to be viewed as a reflection of reality. Essentially, this represents a leap from a synthesis of systems and quantitative applied scientific thinking to a synthesis of systems and human relations.

Once this leap had been taken, some authors began to argue that quantification is less important than the quality of the relationships that are expressed in the model, and this quality can only be assessed in terms of the needs of the participants in debate. When engaging in quantitative analysis, System Dynamics practitioners were used to using a computer to build the model and test scenarios. However, in accepting qualitative modelling techniques, some writers abandoned the computer, saying that a flip-chart provides a more user-friendly interface for communication. Perhaps more than any other methodology, System Dynamics demonstrates the paradigm shift that took place in the movement from the first to the second wave of system thinking: the core idea of modelling complex relationships is still there, but its meaning and practical application are thoroughly transformed. For a book of edited readings on second wave System Dynamics, see Morecroft and Sternman (1994).[114]

[114] A similar paradigm shift has also taken place in the use of Beer's (1959, 1966, 1981) Viable System Modelling. See Espejo and Harnden (1989) for some edited readings.

9.10.3 Dialogical Family Therapy

Having demonstrated the impact of the second wave of systems thinking on management systems methodologies and methods, it remains for me to show what effect it had on systemic family therapy. Alongside critiques of the first wave of expert-driven approaches, some authors began to look for new theories that could inform an alternative practice (e.g., Dell, 1982a,b; Keeney, 1982, 1983; Watzlawick, 1984; and Andersen, 1987). They seized upon Maturana's theory of autopoiesis (Maturana, 1988a,b; Maturana and Varela, 1992)—see Chapter 3 for a review. The most important aspect of this theory to family therapists was not Maturana's focus on the self-producing individual organism, but the idea that people participate in a variety of 'rational domains' expressed in language. This is Maturana's 'conversation theory', which suggests that individuals are able to shift from one rational domain to another via a change of emotional state. Therefore, the focus of the new family therapy was on the therapist facilitating families in moving from old rational domains, embodying destructive understandings of their relationships, to new ones which allow the family members to see their relationships in a different light. The therapist also has a role in supporting family members in making new connections with their emotions to enable the shift to fresh rational domains.

In recent years, Maturana's theory has come in for some criticism (e.g., from Perelberg and Miller, 1990; Goldner, 1991; and Mingers, 1997b), but the focus on dialogical practice has continued unabated—mostly informed by Gergen's theory of constructionism which emphasises the role of the therapist in co-constructing new family 'stories' that people can tell about their relationships (McNamee and Gergen, 1992).

9.11 Parallel Movements in OR and Participative Action Research

At the same time as the second wave of systems thinking was being developed in the 1970s and 1980s, a new wave of operational research (OR) was also being launched (see Rosenhead, 1989a, for an excellent edited book of examples), and a parallel movement in action research was also emerging (see, for example, Schein, 1969; Argyris and Schön, 1974; Reason, 1988b; Whyte, 1991a). Each of these are discussed below, starting with action research.

9.11.1 *Participative Action Research*[115]

During the 1970s and 1980s there was considerable dissatisfaction amongst action researchers about some of the assumptions made by Lewin (1947, 1948)—in particular, that action research should be seen as an 'applied science', in the sense that it involves expert-led experimentation in the social domain (albeit modified to be more responsive to local agendas than traditional applied scientific approaches). A new generation of writers were more concerned with the benefits of fostering participation in workplaces and local communities, and this meant either dispensing with the 'applied science' label (Reason, 1988b), or seeking to democratise science (Whyte, 1991a). There was also a renewed interest in learning as a group process, involving both individual and collective reflection.

It is arguably the case that this second generation of action researchers placed more emphasis on methodological principles of participation and reflection than the design and communication of actual methods (at least in comparison with their peers in the systems and operational research communities)—but this is a generalisation, and some work on methods can be found (e.g., in Reason, 1994). Below, I provide some references to three second generation action research methodologies that I have found particularly interesting, but I should be clear that this is by no means a comprehensive listing of the available literature.[116]

First, there is Action Science (Argyris and Schön, 1974, 1985; Schön, 1983). This focuses attention on the attitude of the individual to relationships with others, particularly in situations where the individual is being confronted with new ideas, bad news, or different points of view. Argyris and Schön argue that people often react defensively: they seek to control the interaction in a manner that allows them to 'filter out', dismiss or interpret away the unwelcome information, thereby preventing their own assumptions from being challenged. Indeed, Argyris and Schön claim that some people's whole

[115] I have given this section the general title "Participative Action Research", but this should not be confused with the specific methodology of Participat*ory* Action Research which will be reviewed later.

[116] We can also see action learning (reviewed earlier) as part of this group. Although this was originally devised in the 1940s and 1950s, its proponents continued to develop it in the latter half of the 20th Century, and a renewed interest was shown in it when the second generation of action researchers came along. Revens (1982) was its original author, but others have since become involved in its development (see, for example, McGill and Beaty, 1992; Mumford, 1997). The focus of action learning, as we saw earlier, is on the establishment of reflection and action cycles, encouraging problem-focused learning amongst individuals and groups.

management styles are dominated by their defensive routines, and when this happens across an organisation, the organisation cannot take full advantage of the knowledge of its members. Unfortunately, people are rarely aware of their defences at a conscious level, and they often espouse participation and openness whilst simultaneously (and unconsciously) undermining any possibilities for participation and openness in their own dealings with others. Therefore, a gap appears between their *espoused theories* and their *theories in use* (Schön, 1983). Schön argues that people can rarely identify and address their own defensive routines unaided, and the job of the action researcher is to facilitate movement to greater self-awareness and new ways of working with others, often by introducing a 'detached observer' into the organisation to provide feedback on interactions. In its focus on the role of the individual in relation to others in an organisation, Action Science can clearly be seen to be influenced by both the psychoanalytic and human relations traditions.

Second, there is Participatory Action Research, which was developed over a period of many years, even though some of the key writings were only published recently (e.g., Whyte, 1991a): for example, one of Whyte's best known case studies (his work in the Xerox corporation) describes an intervention undertaken in the early 1980s to help reverse Xerox's ten year decline in fortunes (Whyte *et al*, 1991; Pace and Argona, 1991). The emphasis in Participatory Action Research is on professional researchers *working with* organisational members (including "low-ranking" people) to collaboratively evolve locally relevant knowledge. A variety of methods may be drawn upon (including some traditional social science methods, such as surveys), but a key principle is that organisational members are involved from the very beginning in the design and execution of the research, and this participation should continue to the very end with the production of conclusions and the implementation of recommendations. It is precisely this participatory aspect that makes Participatory Action Research different from the earlier work of Lewin (1948). Whyte (1991b) acknowledges the link between his own methodology and earlier work in Socio-Technical Systems Thinking, and indeed one of the chapters in Whyte's (1991a) edited book (Elden and Levin, 1991) discusses Scandinavian variations on Participatory Action Research—the Norwegian and Swedish Governments having implemented Socio-Technical Systems Thinking in industries nation-wide (see earlier).

Finally, I will introduce Co-operative Inquiry (Reason, 1988b, 1994; Reason and Heron, 1995; Heron, 1996). Unlike most action research approaches (and also most systems and operational research methodologies), Co-operative Inquiry is not dependent on the presence of a facilitator coming in from 'outside'. Participants with a common interest work together in cycles of collaborative inquiry. Although this 'common interest' is often a task that needs to be undertaken, Co-operative Inquiry is not task-orientated (in the sense of neglecting to explore human relationships). On the contrary, these are an explicit focus, and many emotional issues can surface which need to be dealt with by the group. As people engage in collaborative, task-orientated learning, they also learn about the learning process itself, which involves each individual gaining some awareness of his or her own role(s) in the group dynamics. Each cycle of collaborative inquiry involves (i) explorations of individual and group purposes and ideas; (ii) engagement in agreed actions, with the processes and outcomes being discussed and interpretations recorded; (iii) deepening of the inquiry, exposing previously hidden assumptions and patterns of learned behaviour to group and individual analysis, enabling new ways of seeing; and (iv) consideration of the learning experience in relation to the original purposes of the collaborative inquiry, and the formulation of new purposes for a further cycle of learning (if this is considered appropriate).

9.11.2 Problem Structuring Methods in OR

As with the second wave of systems thinking and the second generation of action research methodologies, the focus of the new wave of OR was on the facilitation of debate rather than the use of expert-driven modelling techniques. At the time, there was substantial communication between the OR and management systems communities: indeed, all four second wave systems approaches reviewed earlier (Strategic Assumption Surfacing and Testing, Interactive Planning, Soft Systems Methodology and the dialogical version of System Dynamics) were seen as contributions to both the systems and OR literatures.[117] Nevertheless, several OR methodologies were produced that were not explicitly based on systems thinking, and three examples of these are reviewed briefly below.

The first example is Cognitive Mapping (Eden, 1988), which is based on Kelly's (1955) personal construct theory (discussed in Chapter

[117] Ackoff (1979) turned his back on the OR community, believing it would never accept any methodology or method into the mainstream other than those based on quantitative applied scientific thinking (the first wave of OR being focused almost exclusively on expert-driven, mathematical modelling approaches). Therefore, it is unlikely that he would appreciate the idea of Interactive Planning being labelled as an OR approach. Nevertheless, despite Ackoff's reservations, I and several others have had applications using Interactive Planning published in OR journals (e.g., Midgley *et al*, 1998), suggesting that the OR community *has* opened its doors somewhat, even though first wave ideas still dominate.

2), which suggests that the individual human being has a unique view of the world that is actively constructed through decision making. This has been applied to a wide variety of problem situations where individuals are required to make a choice between discrete alternative options. Cognitive Mapping involves a facilitator working with an individual to explore the subjectively perceived variables influencing the decision. These are then mapped onto paper, showing the choice that has to be made; the goals that it is hoped will be realised through making the choice; and the factors (and their interrelationships) to be considered in reaching a decision. Essentially, a cognitive map is a decision maker's personal record of the process of exploring options. Visualising the issues simply makes it easier to see the whole picture (from the decision maker's point of view). Cognitive Mapping can be seen as a synthesis of preoccupations from psychoanalysis (Kelly and Eden can be seen as inheritors of Freud's interest in what drives individual behaviour) and its applied-scientific cousin, the human relations tradition.

While Cognitive Mapping is aimed at supporting *individuals*, Strategic Options Development and Analysis (SODA) (e.g., Eden, 1989) applies a similar logic to *group* decision-making. For an example of the application of SODA in an employment agency with a specific emphasis on ethical practice, see Eden and Simpson (1989).

Another widely-applied problem structuring method developed during the 1970s and 1980s in the OR community is Strategic Choice (Friend and Hickling, 1987; Friend, 1989). Strategic Choice was born following reflection on interventions with a variety of public and private sector decision makers operating in environments where inter-organisational collaboration was essential to successful service delivery. It is usually practised in a participative workshop format. The problematic situation is examined in terms of three areas of uncertainty: uncertainty associated with the *working environment* (which can be dealt with using methods from the quantitative applied sciences); *values* (which can be dealt with through 'political' exploration); and *related decision fields* (requiring analysis of relationships between the decision-in-hand and other strategic and organisational priorities, often involving multi-agency working). By examining all three types of uncertainty, priorities will emerge and the appropriate path(s) for analysis can be chosen. Friend and colleagues also talk about planners and managers moving between four "modes" of decision making: *shaping* (formulating problems and issues in appropriate ways); *designing* (identifying options); *comparing* (exploring the consequences of taking the different options); and *choosing* (taking action and planning the management of future decision making processes). A variety of methods and techniques are provided to support each of these modes of decision making: see Friend (1989) for a useful introduction to these methods, and Hickling (1989) for an example of the application of Strategic Choice (it was used to formulate national policy in the Netherlands for the storage, handling, transport and use of Liquefied Petroleum Gas).

While Cognitive Mapping, SODA and Strategic Choice are some of the most widely used problem structuring methods developed in the OR community, they are by no means the only ones. Others include Robustness Analysis (Rosenhead, 1989b,c); Metagame Analysis (Howard, 1989a,b); and Hypergame Modelling (Bennett, Cropper and Huxham, 1989; Bennett, Huxham and Cropper, 1989).

What all these problem structuring approaches have in common is a focus on the facilitation of the participatory exploration of ideas, in the tradition of the human relations movement (indeed John Friend, the principle author of Strategic Choice, worked at the Tavistock Institute of Human Relations for many years). It is also common for these approaches to grant a restricted place to the methods from the quantitative applied science tradition (e.g., in Strategic Choice these methods are used to address just one of the three types of uncertainty). Even though the authors associated with these problem structuring approaches rarely include an explicit discussion of systems thinking in their writings, it is nevertheless easy to see its influence: a recurring theme is the exploration of relationships between phenomena in the clarification of decision options (Keys, 1991). Therefore, I believe it is safe to conclude that exactly the same paradigm shift took place in OR as we saw in the systems and action research movements, and there was a substantial cross-fertilisation of ideas between the systems and OR communities. The result has been a wealth of new methods and methodologies. If we put these alongside the methods from the pure and applied sciences, psychoanalysis, scientific management, the human relations tradition, operational research, action research, and the first and second waves of systems thinking (in both management systems and family therapy), we have a truly substantial resource to draw upon for systemic intervention.

9.12 The Third Wave of Systems Thinking

Finally, I wish to discuss a third wave of systems thinking which is clearly evident in the management systems community; shows some signs of emerging in family therapy and action research; and has its corollary in the OR community too. This section takes us from the 1980s to the present day. While *some* new methods are associated with the third wave, the primary focus has been, firstly, on discussing the limitations of the earlier approaches (especially their purported naiveté with regard to handling power relations); and secondly, on using the great variety of methods in a pluralist intervention practice. Indeed, the book you are now reading can be seen as a contribution to third wave thinking. Over the coming pages I will again start with a discussion of management systems before going on to look at systemic family therapy, action research and OR.

9.12.1 Critiques of the Second Wave of Systems Thinking

In the late 1970s and early 1980s several critiques of second wave systems thinking were launched, primarily on the grounds that the participative methodologies that characterised this wave did not account sufficiently for power relationships within interventions, and/or conflicts built into the structure of society (e.g., Thomas and Lockett, 1979; Mingers, 1980, 1984; Jackson, 1982).

The criticism of the lack of attention paid to power relations during interventions stems from the observation made by most of the above authors that, if (in an industrial context) you bring managers and shop-floor workers together, the latter may feel unable to speak openly for fear of threatening their employment if they say something that is unpopular with management. Thus, the second wave methods are likely to reinforce the vision of change being promoted by the holders of authority.

However, there was also a set of criticisms focused on conflicts built into the structure of society. One such criticism came from an explicitly Marxist position (Thomas and Lockett, 1979): the authors tried to draw out similarities between the Marxist agenda and second wave systems thinking, and commented on the absence of a theory of society in the latter. From a Marxist perspective, it is a definite problem that managers and workers can reach collaborative agreements through the use of systems methods without necessarily changing the basic relationship of employer/employee: it suggests that systems methods are being used to facilitate a false consciousness amongst the workers that it is acceptable for an employer to profit from the labour of employees.

Other criticisms came from a Habermasian perspective. Mingers (1980, 1984) and Jackson (1982) suggested that the focus of second wave systems thinking on participation is right, but a theory of emancipation (of a non-Marxist variety) is needed to enable 'second wave' methods to be harnessed in the service of real social change. Mingers and Jackson argued that second wave systems methodologies are 'regulative': that is, intervention usually has such a local focus that wider unjust political and economic relationships are taken as the environment of the system of concern, and therefore participants are encouraged to adapt their organisations to these relationships instead of challenging them. Both authors drew upon Habermas's (1972) theory of knowledge-constitutive interests, which suggests that human beings have three inherent interests: a *technical* interest in predicting and controlling our natural and social environments; a *practical* interest in achieving mutual understanding; and an *emancipatory* interest in freeing ourselves from power relationships and the false ideologies they give rise to (more details of this theory are provided in Chapter 10). Mingers and Jackson argued that second wave systems thinking is useful for facilitating the practical interest in mutual understanding—but without explicit consideration of power relations and their ideological effects, second wave methods could give rise to "distorted communication" (Habermas, 1972) in which mutual understanding comes to be based even more firmly on false ideological premises.

Very soon after these criticisms began to surface in the literature, the second wave systems thinkers came under attack from a new direction. People became increasingly concerned that the academic systems and operational research communities were being torn apart by a paradigmatic war between first and second wave thinkers—and yet both forms of thinking are necessary to deal with different kinds of problem (see Dando and Bennett, 1981, for an interesting analysis of this paradigmatic war). In 1984, Jackson and Keys published what was to become a highly influential paper, arguing that the first and second waves should be regarded as complementary rather than in competition with one another. This paper became one of two key foundation stones upon which the new, third wave of systems thinking (advocating methodological pluralism) was built (see below).

9.12.2 The Birth of Critical Systems Thinking

By the end of the 1980s, the third wave of systems thinking had begun to take shape and came to be called *Critical Systems Thinking* (CST). CST was built upon two foundation stones: Jackson and Keys's (1984) argument for methodological pluralism (mentioned above) and Ulrich's (1983, 1988, 1994) social theory and systems methodology, Critical Systems Heuristics (CSH).

I reviewed CSH in Chapter 7. However, to refresh our memories, we should note that the central idea is the need to be critical of the value and boundary judgements made by planners. Those involved in and affected by planning are encouraged to reach agreement on the key assumptions upon which planning ought to be based. When dialogue is avoided by planners, those affected by their plans have the right to make a 'polemical' case against the planners in order to embarrass the latter into starting discussions. It should be noted that CSH was not launched as a criticism of second wave systems thinking: while Ulrich drew upon the prior ideas of Churchman (e.g., 1970, 1979) about system boundaries, CSH was primarily formed out of reflections on mainstream philosophy of science (e.g., Popper, 1959) and the limitations of critical philosophy (particularly Habermas, 1976): it is really a synthesis of Habermas's theory of communicative action (including the idea that rationality is dialogical, so a rational plan is one that is based on free and fair discussion between planners and those affected by their plans) and Churchman's theory of system boundaries. Churchman's concepts are used to 'pragmatise' the idea of dialogue, making it clear that the boundaries of both participation and subject matter need to be determined as part of discussions. Ulrich's (1983) work was launched fully-formed into the systems community, and had a gradual but ultimately substantial influence on third wave thinking.

The other key argument contributing to the birth of Critical Systems Thinking (CST)—the idea that systems practitioners should embrace methodological pluralism (Jackson and Keys, 1984)—took several more years to fully evolve. Jackson (1987b) noted that none of the first and second waves of systems thinking dealt with issues of power and coercion in an acceptable manner. He identified Ulrich's (1983) Critical Systems Heuristics as the only methodology that was adequate in this regard. He therefore proposed that the three different types of systems thinking are useful to deal with three different types of problem: first wave systems thinking is useful when there is agreement on the nature of the problem situation; second wave thinking is useful when there is non-coercive disagreement between key players; and Critical Systems Heuristics is useful in situations characterised by coercion. This is actually an over-simplification of Jackson's argument, but it will be covered in more depth in the next chapter. Also, Jackson (1985a, 1988) suggested that Habermas's (1972) theory of knowledge-constitutive interests can be used to underpin methodological pluralism: all three kinds of systems thinking are necessary for us to adequately address the three human interests (in prediction and control; mutual understanding; and emancipation from restrictive power relations).

Other authors rapidly began to join the growing movement, contributing to the evolving understanding of methodological pluralism (e.g., Oliga, 1988; Midgley, 1988, 1989a,b, 1990a,b; Flood, 1989a,b, 1990; Gregory, 1989, 1990), until it finally became possible to define a coherent perspective called 'Critical Systems Thinking'.[118] This was consolidated in a book of readings edited by Flood and Jackson (1991a). It became clear that a third wave of systems thinking was very much with us: while the first wave took a quantitative, applied scientific line on systems (with the exception of Socio-Technical Systems Thinking), and the second wave stressed participation and human relations, the third wave emphasised the value of both—and shifted attention to how choice between the great variety of systems methods can be exercised in a critical and systemic manner.[119]

9.12.3 Revisioning Critical Systems Thinking

However, even before Flood and Jackson (1991a) produced their book of edited readings on CST, there were considerable tensions in the CST movement. I suggest that these primarily centred on three problems,[120] and resolving them led to a significant revisioning of CST. It could conceivably be argued that the revised version constitutes a *fourth* wave of systems thinking, as the revision has substantial implications. However, in this book I prefer to describe it as part of the third wave for two reasons: first, the name CST has survived the revisioning process; and second, the interest in boundary critique and methodological pluralism remains (even though the meaning of these terms has evolved). Let me explain the three problems and their resolution.

First, by saying that Ulrich's (1983) Critical Systems Heuristics is only useful for dealing with situations characterised by coercion, Flood and Jackson (1991b) effectively confined boundary critique to a minority of situations. This begs the question of how coercion gets to be defined in the first place (Ulrich, 1990, 1993; Midgley, 1996c): if there is no scope for boundary critique, coercion may be hidden by whoever is presenting a

---

[118] It is not easy to pin down the origin of this label, but a story has circulated that the name was first suggested by David Schecter: it expresses the interest in Habermasian critical theory and its combination with systems thinking, but is differentiated by the word 'thinking' from Ulrich's (1983) Critical Systems Heuristics. Schecter did not publish his own ideas until 1991, but Jackson, Flood and I all began using the label CST simultaneously in 1988.

[119] In the above review, although I have outlined the main events in the formation of CST, I have provided very little information about the vast amount of work on methodological pluralism that was conducted in the late 1980s and early 1990s. For some more detailed reviews of this early CST literature, see Midgley (1992d), Munlo (1997) and Ho (1997).

[120] Other problems were identified, for example by Midgley (1989b, 1990a, 1992a,c), Gregory (1990, 1992), Tsoukas (1992, 1993b) and Wooliston (1992). However, in my view, most of these have been less significant in terms of revisioning CST: they were resolved 'along the way' without needing to be an explicit focus of research.

description of the problem situation. It is undoubtedly the case that more is needed by way of boundary critique than Critical Systems Heuristics alone (Midgley, 1996c), but if people are serious about taking account of power relations, then boundary critique cannot be subordinated to a form of methodological pluralism that limits its application based on superficial diagnoses of problem situations. Inevitably, the answer to this problem is to accept that boundary critique needs to be practised up-front in all interventions (Midgley, 1996c, 1997a,c; Mingers, 1997a; Munlo, 1997; Flood, 1999a; Han, 2000). This does not contradict the wish to embrace methodological pluralism, because boundary critique is never enough on its own: it is always necessary for agents to find practical means to deal with problematic issues following boundary critique, and the use of a plurality of methods may be helpful in this regard (interspersed, if possible, with periodic checking on boundary questions).

The second problem is that Flood and Jackson's (1991a) use of Habermas's theory of knowledge-constitutive interests *and* Ulrich's (1983) use of Habermas's theory of communicative action both emphasise the universalisation of morality. This means that, when people make moral claims, they should first be sure that the principle they are using can be universally applied. So, for example, if killing another human being is wrong, it should be argued that it is wrong in every case and in all circumstances, otherwise the claim should not be made. I argued in Chapter 7 that it is rare to be able to identify a moral principle that can be regarded as genuinely universal, so insisting on universality risks making morality uncritical: either people end up making simplistic moral pronouncements without any consideration of their limitations, or they are discouraged from making moral judgements at all. The answer to this problem is simply not to insist on universality in the first place (Taket and White, 1993; Flood and Romm, 1996a; Vega, 1999). In my view, this makes perfect sense in terms of the theory of boundary critique outlined in Chapter 7: all moral judgements are spatially and temporally located. Even if it is argued that a moral principle should be applied very widely indeed (as is the case, for example, in some discourses on human rights), *absolute* universality should not be assumed. Morality should always be seen as a matter of political assertion, supported by argument concerning the benefits and limitations of its application.

The third and final problem is the evident paucity of systems methods in the face of coercion, which I argued in 1997 had not yet been sufficiently well-addressed by critical systems thinkers (Midgley, 1997c). The failings of first and second wave systems approaches in this regard are well documented (e.g., Mingers, 1980, 1984; Jackson, 1982), but it seems to me that the 'solution' to the problem of coercion proposed by Flood and Jackson (1991a,b)—use of Ulrich's (1983) Critical Systems Heuristics—is equally problematic. Contrary to the writings of Jackson (1987b), Oliga (1988), Flood and Jackson (1991b) and Gregory (1997), it is my contention that Critical Systems Heuristics (CSH) is not capable of dealing with coercion because *coercive situations are generally characterised by the closure of debate*. Either those with authority simply refuse to talk to other people; they use their position to subdue or get rid of people who challenge them; or else they have 'reasons' why everything that is being said during debate misses the point. As Willmott (1989), Ivanov (1991), Romm (1995b) and Midgley (1997c) have all observed, in any of these circumstances the use of CSH to guide debate becomes redundant.

Of course, Ulrich (1983) does not ignore the ability of coercive agents to close off debate. He says that, when debate is obstructed, CSH can still be used to support the "polemical employment of boundary judgements". This means that those affected by coercion can expose the coercive agent's ideology (expressed in the boundary judgements that s/he takes for granted) by offering a counter argument (a polemic) that does not even attempt to refute the 'facts' advanced by the coercive forces. The affected may simply assert what ought to happen from their own point of view. The CSH questions can be used to guide the construction of this point of view, which will of necessity involve making different boundary judgements to those made by the agents of coercion. The result, if dialogue is not forthcoming, is the embarrassment of those practising coercion, as their arguments are revealed as just another stakeholder viewpoint, with its own ideology and no superior claim to objectivity.

Let us leave aside for now Ivanov's (1991) counter-claim that the polemical employment of boundary judgements will not necessarily result in the embarrassment of those practising coercion. If we assume that engendering embarrassment *is* possible, then this suggests that *there is another participant to whom coercive forces must yield*. Coercive agents need only be careful about their image if there is a 'higher authority' to whom they must defer. Debate is therefore not genuinely closed: there is an arbitrator with whom both coercive forces and those affected by their activities continue to talk.

This point is borne out by examination of the hypothetical example provided by Ulrich (1983) of one particular polemical employment of boundary judgements. Ulrich describes a situation in which local residents construct a polemic to challenge 'experts' who are proposing to site a nuclear power plant in their community. However, the *context* of the construction of this polemic is a planning inquiry where adjudication will take place on the rights and wrongs of the proposal. I suggest that, for the polemical employment of boundary judgements to have any effect, some forum for debate must still be in operation where coercive agents have to be mindful of how their activities are perceived by others. We therefore see that the successful use of CSH (whether in direct dialogue or by supporting the polemical employment of boundary judgements) *does* depend on the possibility of debate taking place.[121]

In short, CSH can only ever be of use when communication is possible, either face to face or via an arbitrator. Therefore, rather than seeing CSH as dealing with coercion (which in my view is characterised by closure of debate), I argue that it is more appropriate to see it as a method of *value clarification*. Indeed, I suggest that there are two 'modes' in which CSH can be used. Mode One involves value clarification within a stakeholder group. Communication with other stakeholders then happens via an arbitrator (usually with the hope that more flexible, face to face communication may be possible in the future). For some practical examples of Mode One CSH, see Cohen and Midgley (1994); Midgley *et al* (1997); and Chapters 14, 16 and 17 in this book. Mode Two, in contrast, comes into operation when stakeholders can work together to generate answers to the CSH questions that transcend the narrowly defined interests of any one group. In this mode, CSH is used to generate an accommodation between stakeholders through the mutual exploration and clarification of values that, to borrow a term from Ackoff (1981), 'dissolves' conflict. It is also possible to combine the two modes so that stakeholder groups first clarify their own values using CSH, and then seek to transcend them in debate with others (for a practical example of this, see Gregory *et al*, 1994).

Once we have reconsidered the role of CSH in this manner, we are again left in a position of having no methods to deal with coercion (involving closure of debate)—which was the situation that, in part, brought the third wave of systems thinking into being. The solution, as I

[121] There is also a deeper problem. Even when there is a willingness to engage in debate, coercion of a kind may still be introduced into a situation *by the use of CSH itself*. Some people find it very difficult to engage effectively in the kind of rational argumentation that is central to the practice of CSH. This will obviously be a major problem when people with severe learning disabilities are involved. Methodologists such as Jahoda *et al* (1989) and Whittaker *et al* (1991) have attempted to find ways in which interveners can elicit the views of people with learning disabilities, but even with help most are unable to engage in debate at the level of abstract complexity needed to participate usefully in CSH. Even amongst those without a learning disability there are many people who will find participation in CSH problematic. Anyone who lacks confidence in their own ideas, or who is self-conscious and self-effacing, may have difficulty representing an argument persistently or persuasively. This affects enough people for the issue to be taken seriously in thinking about CSH. Hence, if CSH is used with participants who find rational argumentation difficult, it may itself be perceived as coercive.

see it (Midgley, 1997c), is to recognise that systemic intervention can take a variety of forms: as well as activities of information production and debate (which are the usual purposes pursued through first and second wave systems methods), there is a need for the explicit inclusion within systemic intervention of *political action and campaigning*. The idea is that, when it is perceived that debate is blocked and coercive forces have control, changes can be fought for within the wider system to free up the situation. Systems methods (of various kinds) may, of course, be useful in support of this wider political action. Dealing with coercion is therefore not so much a matter of which method to use, but *what are the appropriate boundaries for analysis and engagement*: when coercion is experienced, this suggests the need to widen the boundaries. Indeed, Flood and Jackson's (1991a,b) earlier version of CST, which kept CSH in reserve for handling coercion, was criticised for tending to take organisational boundaries for granted (Midgley, 1996c): intervention was usually for organisational change alone. As I argued in Chapter 7, and underline here, a truly *critical* systems thinking must prioritise boundary critique. The reframing of coercive situations can rarely take place without it.

As with the birth of the first version of CST, the revisioning process I have described was again consolidated with the production of a book of key readings, this time edited by Flood and Romm (1996b). I particularly recommend this book to readers wishing to find out more about what CST has to offer. Also, other management systems thinkers outside of the CST community have explored similar issues, suggesting that the third wave of systems thinking is still gaining momentum (see, for example, Francescato, 1992; Müller-Merbach, 1994; Gu and Zhu, 1995, 2000; de Raadt, 1997; Linstone, 1999; Zhu, 2000).

9.12.4 *The Third Wave of Family Therapy*

Earlier I said that, while the third wave is clearly evident in management systems, it only 'shows some signs of emerging' in systemic family therapy and action research. Family therapy is discussed in this section, and action research in the next.

The issue of power has been just as important for family therapists as for management systems practitioners. The critique of both first and second wave thinking in family therapy has tended to come from feminist writers (e.g., James and McIntyre, 1983; Goldner, 1985, 1988, 1991; Taggert, 1985; MacKinnon and Miller, 1987; Goldner *et al*, 1990; Ussher, 1994). The crux of their argument is that systems and cybernetic approaches have been naive with respect to conceptualising power. Bateson (1979), for example, argues that power is a "mythical abstraction" (p.223): if all actions take place within circles of causation, then power cannot be located in the hands of one party—it is always a property of a relationship. In the family, no one individual has the power to create problems or make things better: the relationships between the family members have to be considered as a whole. In the eyes of feminist writers, this kind of assertion gives rise to the suspicion that Bateson believes that physical or sexual abuse is as much a responsibility of the abused as of the abuser. Of course, this neglects the fact that a child, for example, has far less ability to stop the abuse than the parent who is doing the abusing. In other words, power can, with some justification, be seen as something that is exercised by one party over another. Even Maturana (1988a,b), whose work has been used extensively by second wave family therapists, is largely silent on the issue of power: in joint-authored work he has made similar pronouncements to Bateson (Mendez *et al*, 1988; Krull *et al*, 1989), but in his sole-authored work he does not set out any clear position of his own (Mingers, 1997b; Vélez, 1999).

Similarly, the idea of methodological pluralism has surfaced in family therapy (Rosenblatt, 1994). The basis of Rosenblatt's argument is that the family can be viewed in many different ways through the use of a variety of metaphors, and each metaphor may suggest a different approach to intervention. Interestingly, the same idea was proposed in the practice of management systems by Flood and Jackson (1991b) and Flood (1995a): Rosenblatt, Flood and Jackson were all influenced by the earlier work of Morgan (1986) on metaphor.

The key difference between third wave systemic family therapy and management systems is that, as far as I am aware, in family therapy nobody has yet synthesised the critical idea expressed by the feminist writers with the practice of methodological pluralism. It was this synthesis that consolidated the third wave in management systems.

9.13 Power Issues and Action Research

As in management systems and family therapy, there has been considerable interest amongst action researchers in how to handle power relations. Whereas the second generation writers in action research mentioned earlier (e.g., Schein, 1969; Argyris and Schön, 1974; Reason, 1988b; Whyte, 1991a) tend to focus on the promotion of open communication and learning within group and organisational settings, other writers pay more attention to the wider political and economic systems that interventions take place within. Two of the best known writers who have such an interest are Fals Borda and Rahman (1991), and much of their practice is with the poorest communities in developing countries. They advocate the promotion of political awareness as part of the empowerment process, and emphasise community self-reliance in the face of local, national and international political and economic pressures. Rahman (1991) puts it like this:

> "....PAR [Participatory Action Research] is a philosophy and style of work with the people to promote people's empowerment for changing their immediate environment—social and physical—in their favor. In situations characterised by sharp class exploitation and oppression at the micro level, as observed in many countries (particularly in Asia and Latin America), this usually involves some form of class confrontation, which is often combined with collective socioeconomic initiatives to improve the short-run livelihood of the people. In situations where micro-level class exploitation is not so sharp, as in a number of African countries, people's collective action takes the form more of socio-economic initiatives. These often confront or assert *vis-à-vis* those state bureaucracies and technocracies that seek to impose their ideas of "development" (modernisation)—ideas which typically are alien to the people's way of life and culture and are also often destructive of the physical environment.... Additionally.... [people often become involved in] negotiating with or challenging the relevant state organs for better service in areas where they are supposed to serve" (Rahman, 1991, p.16).

Other writers (e.g., Levin, 1994, and Reynolds, 1998) adopt an explicitly Habermasian model of action research, where the emphasis is placed on dialogue between stakeholders, encompassing challenges to intelligibility, truth claims, moral positions and the sincerity of speakers. Not only is this seen as the best means of enabling collective learning, but the promotion of this challenging form of dialogue is viewed as part of the reconstitution of civil society—an antidote to the increasing dominance of instrumental rationality (where social *ends* are not open for negotiation, and only the *means* to reach pre-determined ends can be discussed). According to Habermas (1984a,b), the increasing dominance of instrumental rationality has been brought about by the extension of capitalist economic relations into many aspects of our lives that were previously under non-economic community control (see Chapter 4 for a more detailed explanation).

I also want to mention the Critically Reflexive Action Research methodology proposed by Weil (1998a). This has many of the same concerns as the later work in Critical Systems Thinking: in particular, the need to link individual and group learning into explicit reflection on "intended/un-intended social/systemic outputs/ outcomes" (Weil, 1998a, p.45). In other words, the starting point (the organisational context or group task) is not taken for granted: participants in action research focus some of their inquiries on their role(s) in the wider system. Thereby, the ends they set out to pursue are considered, not just the means of achieving already-given ends.

In recent years there has been a growing dialogue between action researchers and critical systems thinkers, with many writers beginning to draw upon ideas from across the board. See Levin (1994); Flood and Romm (1996a); Wilby (1996a,b, 1997); Ulrich (1996a); and Reynolds (1998) for some conscious efforts to bridge the gap between these two communities.[122] However, it's my impression that, while the action research community is very much getting to grips with power issues, there is (so far) less interest in methodological pluralism. If my impression is correct, this might be because there is more of a focus on methodology than methods within the action research community. It is certainly the case, having talked with a number of action researchers with an interest in power issues, that many actually take it for granted that methodological pluralism is useful—as long as the overarching principles of participation and critically reflective inquiry are not compromised.

9.14 Multimethodology

Having discussed the beginnings of a 'third wave' in family therapy and action research, I now want to move on to OR. Earlier, we saw that there was a significant amount of communication between second wave systems thinkers and operational researchers developing problem structuring methods. During the 1980s and 1990s, when the third wave of systems thinking began to break, this communication intensified, and a substantial number of third wave authors actually wrote for both communities. Unsurprisingly, therefore, we now find very similar third wave ideas in the domains of operational research and management science (OR/MS) as we do in systems thinking. In OR/MS, however, the term 'methodological pluralism' has largely been replaced by the word 'multimethodology'. The key text which consolidated the multimethodology movement was a book edited by Mingers and Gill (1997), which explicitly acknowledges the origins of many of these ideas in, amongst other movements, Critical Systems Thinking. Indeed, while there are a group of authors who have chosen not to use critical ideas (e.g., Bennett *et al*, 1997; Bentham, 1997; Gill,

[122] It should also be noted that the journal *Systems Practice*, which featured a great deal of work on Critical Systems Thinking from 1988-1997, changed its name to *Systemic Practice and Action Research* in January 1998 specifically to provide a forum in which the systems and action research communities could share ideas.

1997; Leonard, 1997; Ormerod, 1997; Schwaninger, 1997), others are arguing for the same exploration/synthesis of critical thinking and methodological pluralism as we find in CST (e.g., Flood and Romm, 1997; Jackson, 1997; Midgley, 1997a; Mingers, 1997a,c; Spaul, 1997).

9.15 A Key Implication of Third Wave Thinking

In terms of the three discourses (quantitative applied science, human relations and psychoanalysis) that seem to have ebbed and flowed throughout 20th century developments in the various research communities concerned with intervention, it appears that we have reached a point where the value of all three are being recognised from within the one perspective. Indeed, this is seen most graphically in the works of Midgley (1992a) and Mingers (1997a) who have both drawn upon Habermas's (1976, 1984a,b) theory of 'three worlds' to support methodological pluralism. In brief (see Chapter 4 for more details of the theory), this is the idea that, in any sentence intended for communication, the speaker implicitly claims that what s/he is saying is intelligible, true, appropriate (right), and that s/he is being sincere. Intelligibility is simply a precondition for effective communication, but the other three claims refer to three 'worlds': the objective natural world; the normative social world; and the subjective internal world of the individual. Hearers may challenge any of the three claims made about the three worlds.

It is striking to observe that each of the three discourses that have flowed through the debates about intervention in the 20th century also seem to prioritise investigation into just one of these worlds: the quantitative applied science discourse primarily refers to the objective natural world; the human relations discourse (which originated as a form of applied science, but has shifted its ground over the years) primarily refers to the normative social world[123]; and the discourse of psychoanalysis primarily refers to the subjective internal world of the individual.[124] The theories of methodological pluralism proposed by

[123] The human relations tradition is concerned with subjectivity too, but the exploration of subjectivity tends to be undertaken in relation to what will motivate the individual to participate in normatively constructed tasks. Thus, the emphasis of human relations work is mostly on structuring activities using mechanisms where people can participate in producing (within limits) the normative direction of an enterprise. The exploration of subjectivity is therefore subordinated to normative exploration. See Midgley (1992a) for a discussion of this point in relation to second wave systems thinking.

[124] I use the word 'primarily' advisedly. None of the discourses is *exclusively* concerned with just one 'world', but tends to *prioritise* investigation into one.

Midgley and Mingers therefore allow these three discourses, which have previously competed for attention (and have been synthesised over the years in various different ways that prioritise one discourse over the others) to fully co-exist for the first time as aspects of the one discourse on methodological pluralism.

In my own 1992a work, I agreed with Habermas (1976) that the theory of 'three worlds' reflects the structure of language, which is universal for all human beings (I was somewhat equivocal about this, but ultimately mounted no real challenge to it). However, in the light of the above analysis, I wish to distance myself from this aspect of universalism that I inherited from Habermas. Having researched the history of these three discourses (quantitative applied science, human relations and psychoanalysis), I now believe that what *superficially* appears to be an inherent property of language is actually a reflection of the history of Western intellectual thought. I suspect that Habermas picked on truth, appropriateness (rightness) and sincerity as the three 'fundamental' claims, not because they are *really* more fundamental than any others, but because they *seem* more fundamental due to the way they link into the three discourses that have dominated our thinking for a number of generations. Perhaps there are no fundamental claims in language after all—just non-fundamental claims that reflect the discourses that have historically had importance for particular research communities.

9.16 Conclusion

In this chapter, I have argued for pluralism at the methodological level, in the sense of respecting the fact that others may have useful insights that we may learn from in constructing our own methodological ideas. I have also argued for pluralism at the level of methods, meaning that we can draw upon methods originally produced within other methodologies and reinterpret them through our own methodology. This means that, if we are using a systems methodology, even methods developed outside systems paradigms can be used as part of systemic intervention.

Having argued for pluralism at both these levels, I went on to outline an explicitly partial history of the development of intervention methodologies and methods during the 20th Century, focusing in particular on management systems (with a lesser emphasis on family therapy, operational research and action research, amongst others). I argued that the same three discourses—quantitative applied science, human relations and psychoanalysis—have been drawn upon again and again during the 20th Century. They have been synthesised together, and with systems theory, in a multitude of different ways. The result is a plethora of methods and methodologies which represent a substantial resource for the systems intervener wishing to practice methodological pluralism. Certainly, there are so many methods that it is impossible for any one intervener to be competent in the use of them all. However, as I shall argue in Chapter 11, comprehensive coverage is not the point—the point is to engage in a continuous process of learning and reflection, building new skills over time.

In this chapter, I skated over some of the complexities of the debates surrounding methodological pluralism: in particular, how it is possible to justify drawing upon methods from a variety of paradigms when these paradigms make fundamentally different and supposedly irreconcilable assumptions about the nature of the world (ontology) and our knowledge of it (epistemology). This will be discussed in Chapter 11. Also, I have not yet explained how, practically speaking, methods may be chosen from the vast array available and mixed to the best effect. This is the subject of the next chapter.

10

Mixing Methods

Having explained why methodological pluralism is valuable, and having shown just a little of the great variety of methodologies and methods that are available for the systemic interventionist to learn from and draw upon (Chapter 9), I can now present a strategy for selecting and mixing methods in practice. This was a strategy that I began developing in the late 1980s (Midgley, 1988, 1989a, 1990a; Midgley and Floyd, 1988, 1990) when Critical Systems Thinking (CST) was first coalescing into an identifiable perspective, and then I altered it somewhat in the mid-1990s when CST was revisioned (Midgley, 1997b) (see towards the end of Chapter 9 for a discussion of the revisioning of CST).

In the text below, I contrast my approach, which I call the *creative design of methods*, with an earlier strand of CST research centred around the development of a framework, the *System of Systems Methodologies* (Jackson and Keys, 1984; Jackson, 1987b, 1990; Flood, 1990; Flood and Jackson, 1991b), which seeks to align systems methodologies with their most appropriate contexts of application. The System of Systems Methodologies has received a great deal of attention in both the management systems and operational research literatures. As part of my presentation of the creative design of methods, I will explain why I (along with most other critical systems thinkers) chose to abandon research into the System of Systems Methodologies after recognising its initial promise.

Before presenting and contrasting the two approaches, I should mention that the creative design of methods provides one, but by no means the only, alternative strategy for choosing and mixing methods during intervention. For other strategies see, for example, Gregory (1992, 1996a,b), Flood and Romm (1995a, 1996a), Mingers (1997a), White and Taket (1997) and Taket and White (2000). I have referenced these texts in preference to the many other strategies that have been discussed in the literature because each of them are explicit about philosophy, methodology and practice. In my view, it is essential for the credibility of any argument for methodological pluralism that its proponent(s) take a position on philosophy and practice as well as methodology: this is because of the challenges to the philosophical coherence and practical applicability of methodological pluralism that have been mounted by critics (see Chapter 11), requiring well worked out counter-arguments. Simple discussions of methodology that take the philosophical and/or practical issues for granted are not sufficient, especially in the face of a scientific orthodoxy which does not accept methodological pluralism (see Chapter 2).

I should also clarify some terminology before entering the argument proper. When writing about the System of Systems Methodologies, Jackson and Keys (1984) talk about aligning *methodologies* with appropriate problem contexts. In contrast, I talk about selecting, designing and mixing *methods*. This reflects a difference between our philosophical stances. The System of Systems Methodologies is regarded as "meta-paradigmatic" (Flood, 1990), allowing choice between various methodologies and their associated paradigmatic assumptions, while the creative design of methods is viewed as embodying its own paradigmatic assumptions (it does not claim to sit above and beyond all other paradigms). Therefore, in the creative design of methods, *methods* are drawn from other methodologies and interpreted through the intervener's own methodology. My own view is that it is not possible to take a "meta-paradigmatic" position (see Chapter 11 for a more detailed argument).

So, let me start by introducing the System of Systems Methodologies before moving on to present the creative design of methods.

10.1 The System of Systems Methodologies

As mentioned in Chapter 9, the first writers to consider the issue of methodological pluralism in the management systems community were Jackson and Keys (1984).[125] They were concerned to show that different systems methodologies have different strengths and weaknesses, making them suitable for application in different circumstances. Their approach was quite straight-forward: they simply developed a grid with four boxes, representing four different types of perceived problem context, and then aligned different systems methodologies with each of

[125] Mingers (1997c) reviews some earlier contributions from the operational research literature, but arguably these have been less influential.

Relationships between Participants

System		*Unitary*	*Pluralist*	*Coercive*
	Simple	Simple-Unitary: key issues are easily appreciated, and general agreement is perceived between those defined as involved and/or affected	Simple-Pluralist: key issues are easily appreciated, but disagreement is perceived between those defined as involved and/or affected	Simple-Coercive: key issues are easily appreciated, but suppressed disagreements are perceived between those defined as involved and/or affected
	Complex	Complex-Unitary: key issues are difficult to appreciate, but general agreement is perceived between those defined as involved and/or affected	Complex-Pluralist: key issues are difficult to appreciate, and disagreement is perceived between those defined as involved and/or affected	Complex-Coercive: key issues are difficult to appreciate, and suppressed disagreements are perceived between those defined as involved and/or affected

***FIGURE 10.1**: The System of Systems Methodologies*

them. These four boxes were later expanded to six by Jackson (1987b), and the authors called the resulting grid of contexts the System of Systems Methodologies. This has been described in the literature using a number of different terminologies. In producing my own description over the coming pages I have chosen to adopt the terminology of Flood and Jackson (1991b), which is now the most widely used.

The grid defining the six contexts of application in the System of Systems Methodologies has two axes, and is presented here in Figure 10.1. One axis is labelled *Relationships between Participants* (referring to perceptions of the relationships between people in the problematic situation being addressed) and the other is labelled *System* (referring to perceptions of complexity).

Let us look at each axis in turn, starting with Relationships between Participants. The Relationships between Participants axis has three states: *unitary* (a perception of full agreement between participants on definitions of the problem situation), *pluralist* (a perception of disagreement between participants) and *coercive* (a perception of disagreement that is masked, or potential disagreement that is not being allowed to surface, due to power relationships between participants). The System axis has two states: *simple* (easy to understand) and *complex* (difficult to understand).

The six contexts in the System of Systems Methodologies are arrived at by cross-referencing the two axes, so these can be labelled simple-unitary, complex-unitary, simple-pluralist, complex-pluralist, simple-coercive and complex-coercive. Various systems methodologies have been aligned with these different contexts: see Jackson and Keys (1984), Jackson (1987b), Banathy (1987), Oliga (1988), Flood and Jackson (1991b) and Midgley (1992d, 1995a, 1996d) for details. In broad terms, when Jackson and Keys (1984) and subsequent authors conducted this alignment of methodologies with their ideal contexts of application, first wave (quantitative, modelling) systems approaches were said to be most appropriate for the unitary contexts; second wave (qualitative, participative) methodologies were regarded as best for pluralist situations; and third wave (confrontative, boundary-challenging) methodologies [Ulrich's (1983) Critical Systems Heuristics was the only such approach identified] were aimed at coercive contexts (see Chapter 9 for fuller descriptions of the first, second and third wave systems approaches).

These were not arbitrary alignments. First wave (quantitative, modelling) methodologies were said to be best suited to unitary contexts because formulating models in response to a set of questions will only be of relevance to those people who agree that this set of questions is the *right* set. If there is disagreement over what the basic issues are (i.e., the context is pluralist), then this will not be addressed by provision of a simple set of facts, or projections of future scenarios, that are orientated to answer questions that only some people regard as important. Similarly, if we are dealing with coercion, the 'improvements' introduced by using first wave methodologies will simply strengthen the hand of those who have control over what issues are addressed by allowing them to pursue their aims more effectively. When there is genuine agreement on the nature of the problem, however, then first wave methodologies can provide useful answers.

In contrast, second wave methodologies were aligned with pluralist contexts because, when there is open and non-coercive disagreement, debating techniques can be helpful in providing a basis for mutual understanding and decision making. However, when there is agreement on what the problems are (i.e., the context is unitary), then there are few differences between viewpoints to explore, so debate becomes redundant. Debating methodologies are equally unhelpful in coercive contexts because open disagreement is not easy to surface, and the intervention inevitably ends up supporting the dominant vision.

In contrast once again, Critical Systems Heuristics (CSH) was aligned with the coercive contexts because, when mutual understanding is difficult to achieve and a necessity for 'taking sides' arises, it can help in subjecting dominant visions to dialectical challenge (Jackson, 1987b; Flood and Jackson, 1991b). When there is agreement on the right course to pursue (i.e., the context is unitary), such dialectical challenge will usually be redundant. Also, if we try to use CSH when disagreement is open (i.e., the context is pluralist), then its challenging nature may well threaten the potential for mutual understanding that could make conflict easier to handle in other ways.[126]

We therefore see that each type of methodology in the System of Systems Methodologies has its strengths, but each also has significant weaknesses. This gives rise to the possibility for methodology *choice* following diagnosis of the problem context. It also provides a means to consider how to *mix* aspects of different methodologies. Jackson and Keys (1984) talk about this in the following terms:

> "Some problem contexts will, of course, not fit exactly into any one of the.... categories. Faced with such an intransigent problem context, the problem solver may still gain benefits from the analysis. It will be possible, using the analysis, to see how a particular methodology might be extended by making use of aspects of other approaches. For example, a problem solver who is armed with a soft-systems methodology appropriate for a [complex]-pluralist context may find it possible to 'harden up' his methodology for a problem context which has some [simple]-pluralist aspects. The resolution of conflict over objectives may be helped by the use of a quantitative approach to aid the decision makers in investigating the effects of their own preferred solutions relative to the solutions of others" (Jackson and Keys, 1984, p.484).

The authors also refer to the possibility of dynamism in the problem context, necessitating movement between methodologies (although it must be said that this is only mentioned in passing):

> "The emphasis is on the key variables in problem contexts which can, in changing their character, lead to qualitative changes in such contexts, affecting the problems therein and thereby demanding a significant re-orientation in problem-solving approach" (Jackson and Keys, 1984, p.474).

[126] However, I cannot agree with Flood and Jackson (1991b) that Critical Systems Heuristics (CSH) (the only boundary-challenging method identified in the System of Systems Methodologies) is appropriate for dealing with coercion. This is because CSH is dependent on debate taking place, and coercive situations are usually characterised by the *closure* of debate. Refer back to Chapter 9 (and to Midgley, 1997c) for further details of this argument.

Before moving on, one final point needs to be made. Jackson (1990) stresses that the System of Systems Methodologies is most expressly not a 'rule book' to be followed systematically. Indeed, he is highly critical of authors [e.g., Banathy (1984, 1987, 1988) and Keys (1988)] who treat it in this way. Rather, it should be regarded as an ideal of intervention practice that is useful for critical reflection on methodology design. To explain, ideals are theoretical constructs, and to be critically reflective is to question assumptions. By saying that we should be critically reflective about methodology design, Jackson is suggesting that there is a need to look carefully at the situations we are going into, trying not to take too much for granted. We also need to consider the possible consequences of the methods we might use, and design our approach accordingly. So, by saying that the System of Systems Methodologies is an ideal that can be used to guide critical reflection, he is saying that the theoretical insights it provides can offer *direction* to our thinking, but should not *determine* it. Jackson (1990) recognises that practical situations may require compromises with what we might like to do with methodologies in an ideal world, and interveners must think critically about how they should manage non-ideal situations.

10.1.1 Total Systems Intervention

In later work, Flood and Jackson (1991b) embedded the System of Systems Methodologies into a "meta-methodology" (a methodology for choosing other methodologies), thereby signalling more clearly how the framework can be used. The name of this meta-methodology is Total Systems Intervention (TSI), which is said to embrace three "phases" of intervention: *creativity, choice* and *implementation.*

The *creativity* phase is based on the idea that each of the main systems methodologies embraced by TSI embodies a particular "metaphor of organisation" (here Flood and Jackson, 1991b, follow Morgan, 1986). For instance, they suggest that System Dynamics implicitly assumes that organisations are like machines. Cybernetic methodologies, on the other hand, look at organisations as if they are neuro-cybernetic learning systems (brains). In contrast, most second wave (qualitative, participative) planning methodologies assume a culture or a coalition metaphor, and emancipatory (third wave) methodologies, such as Critical Systems Heuristics, view organisations as if they are prisons. Six metaphors are identified in all. According to Flood and Jackson (1991b), these metaphors are helpful because their use in debate can enhance creativity. Participants in intervention can use them to think in different ways about the issues with which they are concerned. For example, they may explore the possibility that their organisation is 'broken' (the machine metaphor), finds learning difficult (the brain metaphor), or is failing to grow (the organismic metaphor). The output of the creativity phase should be the identification of a key metaphor, or a set of key metaphors related together, that seems to be particularly apt in describing the problems facing the organisation.

The *choice* phase then follows. To choose an appropriate methodology, or set of methodologies, the metaphors generated during the creativity phase are used together with the System of Systems Methodologies (see Flood and Jackson, 1991b, for specific details). Having chosen a methodology or methodologies, TSI asks the practitioner to move to their *implementation*. The implementation of systems methodologies yields change proposals.

10.1.2 Philosophical Underpinnings

In the introduction to this chapter, I noted that critical systems thinkers are not only concerned with the 'practical' aspects of methodological pluralism, but also its philosophical underpinnings. The System of Systems Methodologies has been underpinned by an epistemological theory (a theory about the nature of knowledge) originally proposed by Habermas (1972). Habermas calls this the "theory of knowledge-constitutive interests". This was first discussed in the Critical Systems Thinking literature by Mingers (1980) and Jackson (1982), but Jackson (1985a) was the first to relate it to the System of Systems Methodologies. Of course, Habermas's work is immensely broad, and cannot be summarised adequately in a few paragraphs. However, Jackson (1985a) offers his own understanding of the theory of knowledge-constitutive interests which I have reproduced below:

> "According to Habermas there are two fundamental conditions underpinning the socio-cultural form of life of the human species—'work' and 'interaction'.
>
> 'Work' enables human beings to achieve goals and to bring about material well-being through social labour. The importance of work to the human species leads human beings to have what Habermas calls a 'technical interest' in the prediction and control of natural and social events. The importance of 'interaction' calls forth another 'interest', the 'practical interest'. Its concern is with securing and expanding the possibilities of mutual understanding among all those involved in the reproduction of social life. Disagreement among different groups can be just as much a threat to the reproduction of the socio-cultural form of life as a failure to predict and control natural and social affairs.
>
> While work and interaction have for Habermas.... pre-eminent anthropological status, the analysis of power and the way it is exercised is equally essential, Habermas argues, for the

> understanding of all past and present social arrangements. The exercise of power in the social process can prevent the open and free discussion necessary for the success of interaction. Human beings therefore also have an 'emancipatory interest' in freeing themselves from constraints imposed by power relations and in learning, through a process of genuine participatory democracy, involving discursive will-formation, to control their own destiny" (Jackson, 1985a, p.523).

While Jackson was the first to claim that this theory could be used to underpin the System of Systems Methodologies, we should note that there has been a difference of opinion between Jackson (1985a, 1991) and Flood (1990) concerning how this underpinning should be achieved. A review of their individual positions can be found in Midgley (1992d). To keep matters simple, I will concentrate on the position that has been discussed most widely in the literature—that proposed jointly by Flood and Jackson (1991b). In short, they suggest that

> "'hard' and cybernetic systems approaches [first wave methodologies] can support the technical interest, soft [second wave] methodologies the practical interest, and critical systems heuristics [a third wave methodology] can aid the emancipatory interest" (1991b, p.49).

To explain in more detail, first wave (quantitative, modelling) systems approaches are viewed as supporting one particular human interest—our technical interest in predicting and controlling our environment. In contrast, second wave (qualitative, participative) methodologies involve managing debate between people so that learning may be facilitated, ideas evaluated, and plans for action developed. In relation to the theory of knowledge-constitutive interests, these too are seen as supporting one interest—this time, our practical interest in achieving mutual understanding. Finally, Critical Systems Heuristics (CSH) is concerned with subjecting assumptions in planning to ethical critique. As we saw in Chapter 7, CSH asks both the intervener and participants in dialogue to address a number of questions concerning the issue of whose views should enter into the planning process, and how this should be achieved. According to Flood and Jackson (1991b), this can support the remaining human interest—our emancipatory interest in freeing ourselves from restrictive power relations.

10.1.3 Summary

In concluding this section, we see that the System of Systems Methodologies aligns systems approaches with contexts for use, and supports this alignment with an epistemological theory of universal human participation in work and interaction. It is the notion that work and interaction are fundamental to the human condition which gives rise to our interests in prediction and control, mutual understanding and freedom from oppressive power relations. Complementarity between paradigms is granted by this theory. While the main focus of the creators of the System of Systems Methodologies has been choice between whole methodologies [see Carter *et al* (1987) and Flood and Jackson (1991b) for some practical examples], they do acknowledge that contexts of intervention may appear sufficiently complex or dynamic to warrant allowing first wave methodologies to be influenced by second wave thinking, or second wave methodologies to take in first wave ideas. In these cases, they say that the System of Systems Methodologies can still provide guidance.

Before moving on to discuss the creative design of methods, I will provide some references to criticisms of the System of Systems Methodologies and its underlying philosophy which the interested reader might like to follow up. See, in particular, Gregory (1990, 1992), Midgley (1990a,b, 1992a,d, 1995a, 1996c), Mansell (1991), Mingers (1992b-d, 1993), Tsoukas (1992), Jones (1993), Dutt (1994) and Sutton (1995). However, in consulting these, it will be important to read several replies that have also been written in defence of the System of Systems Methodologies: see, for example, Jackson (1992, 1993a,b) and Schecter (1993).

Flood and Jackson's meta-methodology, TSI, has also been subject to critique (Taket, 1992; Gregory, 1992; Elstob, 1992; Ghosal, 1992; Tsoukas, 1993b; Green, 1993a; Brocklesby, 1994; Cummings, 1994; Flood, 1995a,b; and Midgley, 1996c), but again see the various replies to get a balanced picture (Jackson, 1993a; Green, 1993b; and Ho, 1994). Furthermore, it is worth noting that a second version of TSI has been produced by Flood (1995a,b), who chose to abandon the System of Systems Methodologies in light of the criticisms raised in the literature.

10.2 The Creative Design of Methods

So let us now look at the creative design of methods. I first developed this approach following reflection upon a particularly complex intervention (Midgley, 1988, 1989a, 1990a; Midgley and Floyd, 1988, 1990). I problematised the notion of simple methodology choice, arguing that *most* situations are perceived as sufficiently complex to warrant the use of a variety of methods, and there is often a need to develop new methods from scratch. Therefore, it is more useful to think in terms of the *design* of methods than simple choice between 'off-the-shelf' methodologies.

This line of research gave rise to the concept of "creative methodology design" (Midgley, 1990a), which was later changed to the *creative design of methods* (Midgley, 1997b) to avoid confusion between 'method' and 'methodology'. This involves understanding the situation in which an agent wishes to intervene in terms of a series of systemically interrelated questions, expressing the agent's purposes for intervention. Each purpose might need to be addressed using a different method, or part of a method. The purposes are not necessarily determined as a complete set in advance, but may evolve as events unfold and understandings of the situation develop. In this sense, it is important to acknowledge that interventions take place over *time*, and that different purposes may emerge at different 'moments' of inquiry, requiring the use of different methods (Midgley, 1992a). The concept of time is therefore as crucial to the creative design of methods as it is to process philosophy (see Chapter 4).[127]

Another particularly important idea is that the methods that are finally designed (or which emerge) are often different from the sum of their parts (Midgley, 1997b). It is not usually a matter of 'stitching' methods together in an additive fashion (although this can be done): a whole *system* (interrelated set) of purposes can be pursued through a *synergy* of different methods. An alternative way of expressing this idea is to think in terms of 'multi-layered' intervention (Weil, 1998a), where methods have to be responsive to different 'levels' of analysis. Therefore, if there is a need to engage in planning (to give a hypothetical example), the agent might need to consider the emotional dynamics of the planning team; the effects of work with that team on other stakeholders; and the nature of the planning task itself. Each of these 'levels' of analysis might be equally important, and when they are all responded to through the creative design of methods, the resulting synergy of methods is a new, more comprehensive whole.

Let us elaborate the hypothetical example of planning to show how questions expressing purposes might be asked. An agent might start with the question, how can I get wide-spread commitment to planning our future? The agent's knowledge of the literature might tell him or her that, if people are able to participate in creating their own direction during planning, they are much more likely to be committed to it (e.g., Ackoff, 1981). Thus, the search will be on for a method, or synergy of methods, that embodies the principle of participation. Then a related question might be, how can we make sure that marginalised

[127] Also see Wang (1995) for a discussion of interventions 'spiralling' through time.

groups participate? Depending on what the needs of these groups are, it might be appropriate to adapt the participative method or use a second method in association with it. Another question might be, how can we make sure that the plan is sufficiently detailed to guide short-term action as well as provide a long term vision? This question might spark a synergy of two or more participative planning methods to enable both a long-term and a shorter-term focus. Then the agent might ask, how should we deal with the fact that particular individuals often seem to monopolise discussions, thereby silencing others? The answer might be to consider the style of facilitation, ensuring that everybody is asked for an opinion. Alternatively, the planning methods might be altered to incorporate moments when individuals and/or small groups are asked to independently generate their own contributions before a synthesis of ideas is sought. Finally, the question might be asked, how can we design an organisational structure that can effectively implement our new plans? Again, the agent's knowledge of the literature might tell him or her that there are various 'off-the-shelf' organisational structures that may be appropriate, but s/he will remember that s/he needs to keep the intervention participative if wide-spread commitment is to be maintained. Therefore, s/he may adapt an expert-led method for organisational redesign to be used through a participative process. How the creative design of methods actually works during interventions should become clearer in Chapters 14-17, where several examples from my own practice are provided.

10.2.1 The Role of Intuition

Now, the process of identifying questions that express purposes, and designing appropriate methods to pursue these purposes, may not be as formal as some reports of intervention using this approach might suggest. For example, Midgley and Floyd (1990) list a variety of situations they faced when evaluating a computer training service for people with disabilities, together with the choices of methods these situations gave rise to. This kind of listing generates an impression of a meticulously pre-planned approach. However, at times, the intervener can be faced with the need to make an instant decision on what action to take (say, when a strongly expressed disagreement surfaces during a workshop), and in such a circumstance the question 'what should I do?' might not be consciously articulated at all. The intervener may need to draw upon his or her knowledge and experience and act intuitively. However, when this happens, it is usually possible to reflect back on the situation and identify the connection between what was happening and the actions taken. This connection then tends to be described erroneously in reports *as if* a question and/or purpose had been articulated.

In the past, I have been just as guilty as other authors of writing reports of practice that hide the use of intuition. Upon reflection, I would now prefer to see the use of intuition made more visible so that we can begin to destroy the illusion so often created of flawlessly pre-planned interventions. To aid reflection on choices of methods, I believe it is certainly useful to articulate those choices as questions, but in my view agents should be honest about when the questions have been articulated retrospectively, and they should declare this openly. If this suggestion is taken up, it could have several important effects. First, students of systemic intervention might feel less daunted by the prospect of practising if they are encouraged to value their own intuition as an important resource. Second, if the exploration of theory comes to be seen (amongst other things) as a means to enhance learning to improve the individual's intuitive resource for the future, then theory will be perceived as less divorced from practice than is currently the case for many interveners. Third, when people make mistakes based on erroneous intuitive judgement, they will be less likely to attempt to hide them with rational justifications. Everybody knows that mistakes can be made in the heat of the moment, and it is important to be able to acknowledge these and reflect upon them so as to identify possible alternative actions that could have been taken. In this way learning may take place, and future judgements (both deliberative and intuitive) may be made more successfully.

Clearly, the final method that is implemented in an intervention is a product of the choices made by the intervener, usually in interaction with others, but these choices may be the result of either conscious deliberation or intuitive reaction (or a mixture of both) depending on the circumstances. The term 'choice', in this context, therefore takes on a wider meaning than its usual definition as 'rational decision making between clearly expressed alternatives' (also see Chapter 8 for different ways to view choice).

10.2.2 The Pivotal Role of the Intervener

In previous work (Midgley, 1989a, 1990a), I have made it clear that, if an intervener is coming into a situation from outside,[128] it is important to develop questions and their associated purposes in

[128] An intervener may initially be 'internal' or 'external' to a situation. I deliberately talk about *agents* as interveners so as not to create the impression that an intervener is always an external consultant. Indeed, an agent may be a group considering its own development and the learning of its members (as in Reason and Heron's, 1995, Co-operative Inquiry).

dialogue with stakeholders, but that interveners should also take care to allow people time and space to surface issues confidentially (indeed, this can also be necessary even when the intervener is an 'insider' who thinks s/he knows the other stakeholders intimately—precisely because of his or her insider role, s/he may not be privy to all the relevant issues). Confidential space is needed to facilitate the identification of power issues that people might not be willing to discuss openly. Here, the intervener cannot avoid taking a lead in guiding the development of the intervention (Midgley, 1989a): unlike Ulrich (1990), who wishes to transfer responsibility for ethical decision-making wholly to participative stakeholder groups, I believe that the need to talk with people individually and confidentially places the intervener in a unique position of responsibility. He or she must manage the possible tensions between his or her own, and various stakeholders', different viewpoints.

Of course, the idea of managing these tensions, and possibly also conflicting purposes being pursued by different agents involved in the intervention process, once again raises the issue of the need for boundary critique (refer back to Chapter 7 for details). In previous writings (e.g., Midgley, 1997b), I have made it clear that the creative design of methods prioritises boundary questions because the selection and/or design of methods will be influenced by whatever boundaries are accepted during, or become dominant in, the intervention. Boundary critique is an active process, and there are many methods that can be used to practice it (Midgley *et al*, 1998; Chapters 7 and 14 of this book).

It is possible to identify two different types of question expressing purposes which guide the selection and/or design of methods (a third will be mentioned shortly):

- *Boundary questions*, leading to the design of methods for defining issues; and
- *Issue-related questions*, leading to the design of methods for addressing the issues already defined.

For an intervention to be conducted in a critical and systemic manner, wherever possible boundary questions should be explored first. However, just because boundary questions may be asked up-front, this doesn't absolve agents from reflecting on them again periodically during intervention to identify new issues in an on-going manner. For an example of how boundary questions can be raised throughout an intervention, see Chapter 14.

Although it might appear at first sight that boundary and issue-related questions are all that is needed during an intervention, Mingers (1997a) identifies a third category of question that is just as important:

- *Knowledge-related questions*, enabling explorations of "relations between agent(s) and intellectual resources" (p.421).

These questions probe the forms of knowledge that agents bring with them into interventions. They help agents identify appropriate knowledge resources from their previous experience, and also knowledge gaps which might need to be filled as part of the intervention. There have been many times in my own intervention practice when I have realised that I do not have the appropriate knowledge or expertise to undertake a particular task, so I have either had to conduct some research or invite someone else to join me in the intervention (for example, all the interventions described in Chapters 14 to 17 were undertaken collaboratively). Knowledge-related questioning is vital if we are to begin to see the development of systemic intervention methodology as a *learning process* for intervening agents (see Chapter 11 for details).[129]

10.3 Practising the Creative Design of Methods

Practically speaking, to get from a set of questions expressing one's purposes to a method (or synergy of methods) that will help realise these purposes, it is possible to draw upon one's intuitive knowledge and/or reflect on a variety of aspects of the armoury of methods available: their stated purposes; the methodological principles usually associated with them; the theories that have informed their development; the ideologies they assume; and the ways in which they have been used in past practice. Certainly, if an agent's knowledge bank of methods is to grow, this kind of theoretical reflection (partnered with experiential learning about uses of new methods in practice) is essential. Each of the above aspects is explained below:

---

[129] Like me, Mingers (1997a, p.421) identifies three categories of question: those exploring "relations between agent(s) and intellectual resources"; "relations between agent(s) and problem situations"; and "relations between problem situations and intellectual resources". The second category bears some comparison with my boundary questions, and the third category is similar to my issue-related questions. Mingers provides a list of useful questions for each of these categories which make for interesting reading. They give a good indication of the sorts of questions agents might need to ask during interventions.

10.3.1 *Purposes*

The purposes of the agent may or may not correspond exactly with the stated purposes of a method. However, even if there is an exact match, it is likely that a variety of methods will have been designed for the same purposes: for example, Checkland's (1981) Soft Systems Methodology; Ackoff's (1981) Interactive Planning; Mason and Mitroff's (1981) Strategic Assumption Surfacing and Testing; Friend and Hickling's (1987) Strategic Choice; and Eden's (1989) Strategic Options Development and Analysis are all useful methods for collaborative strategic planning. So the stated purposes of methods, while important, are not enough on their own to facilitate choice amongst methods.

10.3.2 *Principles*

Most methods were originally designed to support the operationalisation of particular methodologies, so one can gather further information by looking at the principles of those methodologies. For example, Checkland's (1981) Soft Systems Methodology (SSM) embodies the principle of participation, so it is likely that the methods of SSM will be particularly user-friendly as part of participatory practice. However, methods and principles *are* separable: for example, it is perfectly possible to use the methods from SSM in a non-participatory manner.[130]

The separability of methods and principles is actually of great benefit to agents wishing to enact the creative design of methods: if a method from one methodology does not do everything that one wishes, it may be enacted using the principle(s) from another methodology. A good example is Flood and Zambuni's (1990) use of the Viable System Model (VSM) (Beer, 1985) in an African tourism company: the authors, who were acting in a consultancy role, not only wanted to support a restructuring of the organisation (the purpose of the management in commissioning the intervention), making the VSM an appropriate

[130] I have seen many student projects where the student conducts an SSM by interviewing people in an organisation and then producing a rich picture, relevant systems, conceptual models, etc., on their own through an analysis of the interview data. At no stage do any of the interviewees get to talk with each other about their views. In such cases, the methods from SSM have been divorced from the principle of participation and have instead been enacted using the principle of expert-led design. In the context of a student project, the main purpose of the student is often to demonstrate knowledge of the SSM methods, and the results for the organisation are considered as being of secondary importance (if relevant at all). Of course, students who understand the importance of *methodology* as well as methods do not act like this: they are well aware that they are not actually practising Soft Systems *Methodology* if they do not respect the principle of participation.

choice (restructuring is one stated purpose of the VSM)—they also wanted to deal with issues of corruption, which the management were trying to ignore but which were important to other stakeholders. They therefore consciously enacted the VSM using the principle of emancipation from Critical Systems Heuristics (CSH) (Ulrich, 1983). Although no *formal* use of CSH was evident, key questions were introduced at appropriate moments, allowing the issue of corruption to be surfaced and dealt with. See Flood and Romm (1995a) and Midgley (1997b) for two different interpretations of this intervention.

10.3.3 Theories

The theories that originally informed the development of methods by their originators can also be a useful guide. For instance, faced with a need for participative strategic planning, and a choice between the five approaches given as examples earlier [Checkland's (1981) Soft Systems Methodology (SSM); Ackoff's (1981) Interactive Planning (IP); Mason and Mitroff's (1981) Strategic Assumption Surfacing and Testing (SAST); Friend and Hickling's (1987) Strategic Choice (SC); and Eden's (1989) Strategic Options Development and Analysis (SODA)], theoretical analysis can reveal some crucial assumptions.

SODA, for example, is based on Kelly's (1955) personal construct theory which assumes that human action is structured around choices between clearly defined options. In comparison with SODA, SC seems more open to accepting an initial uncertainty about options, but is nevertheless still influenced by the same body of theory: it therefore guides participants in debate towards a point where the necessary options can be specified. SAST, on the other hand, is based on a different theoretical understanding [Churchman's (1979) interpretation of Hegel's (1807) theory of dialectics], but with similar practical consequences. Churchman's understanding of dialectics proposes that critical thinking involves argumentation between two opposing ideas, leading to a synthesis. Therefore, SODA, SC and SAST all assume that it is possible and desirable to identify discrete options fairly early on in strategic planning, around which participative debate can be structured.

In contrast, both SSM and IP are based on evolutionary learning theories which suggest that a single coherent position can emerge from explorations of ideal (but feasible) scenarios, starting with the acknowledgement of a 'mess' of issues rather than clearly defined strategic alternatives. Indeed, it is not strictly necessary for participants in debate ever to identify clear alternative strategies and systematically evaluate them: just one option may be developed and tested through debate (although the exploration of multiple options is not precluded).

By understanding the theoretical commitments made by the authors of different methods, choice between those methods becomes possible. For instance, when clear options already exist and need to be evaluated, SAST or SODA might be preferable. When options are initially unclear, but it is considered important for more than one to be defined so that a systematic evaluation can be conducted, SC might be best. Alternatively, when there is a lack of clarity and the learning of participants is considered to be the main priority, then IP or SSM might be more useful.

Actually, even if there are clear options on the table, SSM and IP can be useful in certain circumstances—that is, if there is a suspicion that all the options reflect an overly narrow view of the situation. Therefore, choosing methods by reflecting on theory is not simply a matter of finding out which method(s) embody theoretical assumptions that reflect current priorities. There is also a need to subject these priorities to boundary critique. Thereby new purposes may be surfaced with important consequences for understanding the relevance of particular methods and their associated theories.

10.3.4 Ideologies

Very close to theory is ideology. Methods may make ideological assumptions: that is, assumptions with an identifiable political consequence.[131] A good example, explored in more detail in Chapter 2, is Spash's (1997) observation that methods of cost-benefit analysis assume a utilitarian rationality, where trade-offs between costs and benefits can be made with the ultimate aim of reaching a 'balance' that is recognised by stakeholders as providing the greatest good to the greatest number of people. However, utilitarianism is not politically neutral: it is one of the discourses that has informed the political philosophy of liberalism, and is generally viewed as unacceptable by environmentalists who prefer to take a deontological stance where there are some baseline commitments (for example to environmental standards) that, for the good of everyone in the longer term (beyond current stakeholders), should be regarded as non-negotiable. Clearly, to

[131] This is 'politics' with a lower case 'p', not a capital 'P'. In other words, ideology is not only about party politics, or the 'big' political debates, but is any set of assumptions that, when flowing into intervention, may result in an outcome that is viewed as political., or is seen as having political implications Of course, this means that just about any set of assumptions can be classed as ideological. In my view, whether a set of assumptions is 'ideological' or just 'theoretical' is a matter for argument in local contexts (including in academic debates).

choose a method of cost-benefit analysis is an ideological choice as well as a practical one, so ideological analysis (ideology critique) can be useful for revealing this fact.

Ho (1997) has developed a participative method, informed by Critical Systems Thinking (especially Gregory's, 1992, work on ideology critique), for interrogating the ideological assumptions of methodologies and methods as part of the creative design process. This involves examining the commonalities and tensions between the ideological assumptions made by candidate method(s)[132], stakeholders in the local situation, and the person or people who are being commissioned to conduct an intervention (whether 'insiders' or 'outsiders'). It is through this kind of ideology critique that the possible consequences of employing method(s) in a local situation can be considered. Ho also presents an interesting practical example, where he worked (as a Ph.D. student under my supervision) with Tainan City Council in Taiwan whose Leader wanted to design an intervention for restructuring the organisation. This makes clear the value of ideology critique to the creative design of methods: Ho supported stakeholders in revealing that the candidate method—Beer's (1985) Viable System Model (VSM)—would be met with resistance by employees who felt that a hidden agenda could be introduced into the intervention. The result was a redesign of the intervention so that the VSM could be operationalised participatively, allowing the employees to have a meaningful say in setting the agenda. Of course, there may not always be time to employ a participative approach such as the one described by Ho (1997), but even so ideology critique can still be valuable (Ho discusses the issue of tailoring the format of ideology critique to the constraints of the local situation).

Interestingly, while many methods keep their ideological assumptions hidden (indeed, it is probably the case that most of their creators are unaware of these assumptions), there are a handful of methods where the ideology being promoted is made explicit. Two good examples are Program Analysis of Service Systems (PASS) (Wolfensberger and Glenn, 1975) and Program Analysis of Service Systems' Implementation of Normalization Goals (PASSING) (Wolfensberger and Thomas, 1983). These are both methods which provide quantitative evaluations of the quality of service systems designed for 'disadvantaged' groups. They have been extensively used to evaluate services for people with learning disabilities, and have also been widely applied in services for other disabled and older people

[132] Ho (1997) calls methods "candidate methods" when they are being considered for use, but have not yet been chosen.

[see, for example, Williams (1995) who analyses data from over 400 PASS and PASSING evaluations]. The scores that are generated through the use of these methods show how the service system measures up against an ideal of service delivery derived from a particular ideology which is spelt out in detail by Wolfensberger and Thomas (1983), and is most commonly referred to as the ideology of *normalisation* (a term first coined by Nirje, 1960, and later adopted by Wolfensberger, 1972).[133]

The ideology of normalisation suggests that people with disabilities, and all 'disadvantaged' groups, should be able to live as 'normal' a life as possible. Wolfensberger argues that many groups in society are devalued, and the way services are provided to them often reinforces this devaluation by treating them in inappropriate, degrading or humiliating ways. In addition, services tend to isolate people with disabilities from the rest of the community, and present an image of them to others as either abnormal or deviant. PASS and PASSING both evaluate services by assessing how much they contribute to the devaluation of their clients. The ideal service is one which supports people in becoming valued members of their community, and which does not isolate them or mark them out as different in a negative way. Wolfensberger and Thomas (1983) describe this as "the use of culturally valued means in order to enable people to live culturally valued lives".

The ideology of normalisation has been highly controversial, but in my view Wolfensberger should be given credit for making it explicit: the result has been an extremely high-quality debate in the literature about the political and life-practical consequences of using PASS and PASSING, with many unforeseen side-effects being identified that might otherwise have remained invisible [see Burton (1983); Brown and Smith (1989, 1992); and Pilling and Watson (1995) for some particularly strong contributions to the debate].

If the intervener wishes to be conscious of the wider political consequences of systemic intervention, then this kind of ideological

---

[133] In some of the literature the ideology behind PASSING is described as *social role valorisation*. This rather less user-friendly term was introduced by Wolfensberger (1983) to counter a common misunderstanding of normalisation. The word 'normalisation' is often assumed to mean 'making people normal'. This smacks of social engineering, where people with disabilities are forced to conform to a stereotypical norm of the average citizen. As Wolfensberger makes clear, this is a misinterpretation, but its continual recurrence in the literature has made him abandon the term in favour of social role valorisation. However, like many other authors (e.g., Brown and Smith, 1992), I have chosen to stick with Nirje's (1960) original terminology which is already widely known and, in the UK at least, is showing no sign of being replaced.

analysis is essential (also see Midgley and Ochoa-Arias, 1999, for another argument in favour of politically aware systemic intervention).

10.3.5 Practical Results

Finally, observations of the past practical results of uses of different methods (from personal experience and/or from the literature) can be of great value in selecting and mixing the right methods for a particular intervention. Once one realises that the Viable System Model (VSM), for example, can be successfully imbued with an emancipatory principle during practice (this is outside the scope of the stated purposes, principles, theory and ideology of the VSM) (Flood and Romm, 1995a), then it becomes a possibility to use an emancipatory approach to it in future interventions. In my view, observations of practice are of greatest help when the intervener is faced with a choice between two methods (or sets of methods) with very similar stated purposes and principles. A good example is Interactive Planning (Ackoff, 1981) and Soft Systems Methodology (SSM) (Checkland, 1981; Checkland and Scholes, 1990) which are both approaches to planning that value creativity and participation for 'mess management'. When one looks at the methods and their practical results, however, the differences become obvious: in my experience (e.g., Cohen and Midgley, 1994; Midgley *et al*, 1997, 1998; Gregory and Midgley, 1994, 2000; Chapters 14-17 in this volume), Interactive Planning is much more likely to give rise to the generation of long-term plans, whereas SSM is particularly good for detailed medium-term planning. Indeed, aspects of the two approaches can usefully be synergised or harnessed together to give us the best of both worlds (I have used them together in my own intervention practice on several occasions).

10.3.6 Learning through Reflection on Methods

To a novice, the idea of assessing the purposes, principles, theory, ideology and past practice of a variety of methods may sound excessively complex. However, bear in mind that a great deal of learning about methods quickly becomes intuitive when one begins to practice. Indeed, describing the creative design of methods in words is a bit like the party game where one person tries to instruct another to drink a glass of water, and the person drinking the water is not allowed to do anything other than what s/he is told. The water invariably ends up everywhere except in the person's mouth. The problem is that the words are only really meaningful in relation to practical experience: in the absence of this experience, it may sound much more complex and time-consuming than it really is. See Chapter 11 for a model of interventionist learning about methods and methodology that complements the creative design of methods presented here, and Flood (1995a), Wilby (1996c) and Ho (1997) for other writings on the assessment of methods.

10.4 Writing Up Interventions

While activities of questioning are vital to the creative design of methods to ensure that agents' purposes (particularly the purposes of the professional intervener[134]) are not taken for granted, it would be overly cumbersome if write-ups of interventions went into great detail about the questions asked. When writing up interventions for publication, I (and others who have used the creative design of methods in practice) tend to articulate the *purposes* being expressed in the questions, rather than the questions themselves. Lists of questions are less reader-friendly than saying how particular purposes are arrived at in local situations; why they are important to the agent(s) concerned; and how they are pursued through the creative design of methods. See Chapters 14 to 17 for some examples.

10.5 Moving Away from the System of Systems Methodologies

Now, when the creative design of methods was first developed (Midgley, 1990a), it was conceived as a set of methodological ideas that could help improve the practical use of the System of Systems Methodologies—it was not originally intended to replace Jackson and Keys's (1984) framework.[135] However, in the later incarnation of the creative design of methods (Midgley, 1997b), the System of Systems Methodologies was abandoned. While the reasons for this abandonment were not made explicit in that work, they can be detailed now.

[134] See Chapter 13 for a discussion of professional identities and their consequences for understanding intervention.

[135] An improvement was thought to be necessary because the issue of the intervener's responsibility was not addressed in earlier descriptions of the use of the System of Systems Methodologies. The System of Systems Methodologies was also criticised for paying insufficient attention to the dynamism and complexity of most situations (Midgley, 1990a).

10.5.1 Limitations of the System of Systems Methodologies

First, as Gregory (1992) points out, the System of Systems Methodologies encourages people to accept only one interpretation of each methodology. Flood and Romm (1995a) have demonstrated that it is actually possible to use methods for a variety of purposes, some of which go beyond those they were originally designed for. By placing the methodologies in boxes, alternative interpretations and uses of methods which may be quite justifiable are discouraged. That this is indeed the case is demonstrated by the fact that there has only been one change in the structure of the System of Systems Methodologies (Jackson, 1987b), and only a handful of changes in the alignment of methodologies with the boxes (Banathy, 1987; Oliga, 1988; Flood and Jackson, 1991b; and Midgley, 1992d, 1995a, 1996d), in its sixteen year history. This is despite repeated calls for further critical reviews of both individual systems methodologies and the framework itself (e.g., in Flood and Jackson, 1991b).

Not only does the System of Systems Methodologies 'freeze' interpretations of methodologies in an unnecessarily restrictive manner, but it cannot easily take account of the methodological developments that occur when researchers learn from other perspectives (Gregory, 1992). A classic example is System Dynamics, which has always appeared in the System of Systems Methodologies in the 'simple-unitary' box. However, second wave research on System Dynamics has suggested that, rather than claiming the model represents reality (the traditional view), it is more useful to focus on the modelling *process* as a vehicle for the development of learning and social co-ordination (see, for example, de Geus, 1994; Lane, 2000; and the discussion of second wave systems thinking in Chapter 9 of this book). System Dynamics practitioners have therefore changed their understanding of 'system' to one that is much closer to that embraced by Churchman (1979), Checkland (1981) and Ackoff (1981), and presumably this new way of using it makes it *equally* applicable to unitary and pluralist contexts.

Also, as we saw in Chapter 9, the System of Systems Methodologies is problematical in the way that it confines the idea of making critical boundary judgements (in the form of Ulrich's, 1983, methodology of Critical Systems Heuristics) to simple-coercive contexts. This means that, in practice, critical reflection on, and discussion of, boundary judgements will only happen on an occasional basis. Of course, I am not suggesting that Critical Systems Heuristics gives us all we need to enact boundary critique (several criticisms can be raised against it, as a variety of authors have pointed out[136]), but the

[136] Jackson (1985c, 1991), Willmott (1989), Ivanov (1991), Flood and Jackson (1991b,c),

question remains, how is boundary critique to be enacted in situations where coercion is *not* identified?

Now, defenders of the System of Systems Methodologies may well reply by saying that boundary critique is redundant in situations where coercion has not been identified. However, the most immediate question that springs to mind is, how do we identify coercion?—and, related to this, whose views do we take into account? Answering these questions will involve the intervener and other interested parties in *making critical boundary judgements*. In other words, making up-front boundary judgements cannot be avoided in *any* intervention. Failure to realise the full implications of this will inevitably result in some of the most important boundary judgements—those which determine who the intervener will talk to and how the initial remit of the work will be defined—being made in an uncritical manner. Therefore, when generating questions for the creative design of methods, it is important to prioritise boundary questions. This is why, in the book you are now reading, a whole chapter (Chapter 7) is dedicated to boundary critique: methodological pluralism, in the form of the creative design of methods, cannot be operationalised in a critical manner without it.

10.5.2 Challenging the Philosophy behind the System of Systems Methodologies

It should be clear from the above that there were substantial practical and ethical reasons for moving away from the System of Systems Methodologies. However, a move away from Habermas's (1972) theory of knowledge-constitutive interests, used by Jackson (1985a) and others to underpin the System of Systems Methodologies, was also considered to be necessary. Two critiques were conducted: a critique of the use of this theory to underpin methodological pluralism (Midgley, 1989a,b), and a critique of the legitimacy of the theory of knowledge-constitutive interests itself (Midgley, 1992d, 1996c). Brief details of each of these critiques are given below.

First, Flood and Jackson (1991b) seem to see the theory of knowledge-constitutive interests as something that can take us above and beyond inter-paradigm debate. This is important because, if we claim that a pluralist theory is in some sense "meta-paradigmatic", as Flood (1990) does, we are heading for the same trap that Jackson and Carter (1991) identify in their critique of systems theories of the unification of science: creating a new Grand Truth that is beyond

Mingers (1992b), Romm (1994, 1995a,b), Brown (1996), Midgley (1997c) and Vega (1999).

question, and which seeks to invalidate any ideas that oppose it. Let me explain.

As agents trying to embrace diversity, we are inevitably selective. We cannot be aware of either the existence or the relevance of all other methodological positions. Nevertheless, we still aim towards the ideal of comprehensiveness in learning from others, and we allow whatever diversity we have appreciated to filter into our own methodological position (see Chapter 11 for further details). However, if we try to claim that our own position is meta-paradigmatic in relation to others, we are dismissing the possibility that the proponents of those other positions could legitimately disagree with it. Our own position will therefore no longer be open to change and further development.

Now, when it was first published, the above critique made me aware of the need to be explicit about the paradigmatic nature of my own work on pluralism, but was not the only spur to move towards a new position. While the theory of knowledge-constitutive interests may be internally coherent, I had doubts about its legitimacy (Midgley, 1992d). The problem is that it describes the relationship that human beings have with their social and non-human environment as one of 'prediction and control'. If this is used to inform the development of systemic intervention, it is likely to reinforce the humanist assumption that the natural world is a resource for human control and consumption. In my view, it is far better to view human beings as having an interest in preserving and/or building a sustainable, interactive relationship with their non-human environment. For further details of this argument, see Eckersley (1992) and Midgley (1992d).

It is also the case that Habermas (1972) subscribes to the rather dubious view that human society is in a process of social evolution. It is not clear whether Jackson and colleagues also believe this, but if they do, this would mean that the "emancipatory interest" which they say should be pursued during interventions could be tied in with the idea of humankind's supposed 'march of progress'. Given the problems of deciding what is or is not 'progress' (progress for some may be a setback for others, as Churchman, 1970, so eloquently argues), many authors (e.g., Rorty, 1989) have maintained that theories of social evolution have little credibility. I do not want to put words into the mouths of Jackson and his co-workers, as there is no positive indication that this is what they actually do believe, but there is certainly a need for further clarification here before we are able to accept their use of the theory of knowledge-constitutive interests (Midgley, 1996c).

As a result of these critiques, I decided to look for an alternative philosophical rationale for methodological pluralism. I have actually engaged in two periods of research on this: in 1992 I drew upon

Habermas's (1976, 1984a,b) theory of 'three worlds', but in the run-up to writing this book I abandoned this (see Chapters 4 and 9 for the reasons) and produced the work on process philosophy presented in Chapters 3 and 4. Therefore, the creative design of methods should be seen as part of a wider methodology which prioritises boundary critique during systemic intervention, and which draws upon process philosophy to make this methodology theoretically coherent.

10.6 Conclusion

In conclusion, in contrast with the System of Systems Methodologies (which tends to emphasise choice between 'off-the-shelf' methodologies), the creative design of methods involves the development of a dynamic set of interrelated questions, expressing purposes for intervention that evolve over time, each of which might need to be addressed using a different method, or part of a method. This is not simply a matter of 'stitching' methods together in an additive fashion: a synergy can be generated that allows a whole system of purposes to be addressed together. Also, it is important to note that, in generating the purposes, the need for critical thinking and debate about boundary judgements is crucial. To know which methods it might be appropriate to use in any particular situation, the agent(s) involved in an intervention may draw upon their intuitive resources and/or they may consider various methods' purposes, principles, associated theories, ideological assumptions and examples of past practice. This need not be as complex and time-consuming as it sounds, however, as the creative design of methods values interventionist *learning*: I do not assume that a great deal of theoretical work has to be done in advance of this learning. On the contrary, theoretical learning in the absence of practical experience is relatively empty. In the next chapter, I go into much more detail about what it means for interveners to engage in this kind of learning, and in Chapter 12 I again discuss the importance of practice for those with an academic interest in methodology.

11

Learning about Methodology and Methods

So far in Section Two of this book, I have argued for thinking about methodology in terms of systemic intervention; up-front boundary critique; theoretical and methodological pluralism; and the creative design of methods, allowing agents to draw upon, and mix, a wide variety of methods to realise their purposes. Adding all these elements together appears, at first sight, to present a daunting challenge to agents wishing to engage in systemic intervention—that is, if there is an expectation that agents enter intervention with a near-comprehensive knowledge base to draw upon. Also, doubts have been raised as to the cultural feasibility and intellectual credibility of methodological pluralism (e.g., Brocklesby, 1994, 1997; Mingers and Brocklesby, 1996): many of the different methods reviewed in Chapter 9, for example, were conceived in different paradigms, each of which makes fundamentally different assumptions about the nature of reality (ontology) and our knowledge of it (epistemology). How then may we mix methods without philosophical muddle, or without falling into the atheoretical eclecticism I criticised in Chapter 5? The present chapter is designed to raise these concerns and answer them by presenting a *model of learning* which interveners may use to develop their understandings of methodology and methods over time.

Here, I follow Mingers and Brocklesby (1996) in distinguishing between three types of challenge to the feasibility of methodological pluralism:

> "i) *philosophical*—paradigm incommensurability;
> ii) *cultural*—the extent to which organizational and academic cultures militate against multi-paradigm work; and
> iii) *psychological*—the problems of an individual agent moving easily from one paradigm to another" (Mingers and Brocklesby, 1996, p.111-112, my emphases).

Each of these problems will be discussed in turn, and then the arguments of various authors who have tried to address one or other of them will be reviewed. Finally, my own model of interventionist learning will be presented, and I will argue that acceptance of this model (which builds on the creative design of methods introduced in Chapter 10) addresses two of the three problems, and offers a new understanding of methodology which will be particularly valuable to agents wishing to engage in systemic intervention in a critical and pluralist manner. The third problem (the cultural one) will not be addressed by the model of learning—but then, I suggest that no methodology or model can create culture change except by demonstrating that it 'works' in ways which others value. It is the task of the whole book, not just the model presented in this chapter, to make this demonstration.

So let me start by clarifying the nature of the philosophical challenge.

11.1 The Philosophical Problem

At the level of philosophy we have to face the 'paradigm problem', which can be summarised as follows. All methodologies make different philosophical and theoretical assumptions—i.e., they are born in different paradigms—so if we wish to mix them, or bring them together in a framework, we have to justify this at the level of philosophy. Some authors (e.g., Burrell and Morgan, 1979; Jackson and Carter, 1991) claim that philosophical paradigms are irrevocably incommensurable. This might lead one to suppose that methodological pluralism is a non-starter. Others claim that rational analysis may bridge the paradigm gap, allowing for a 'unification' of paradigms (Reed, 1985; Han, 2000), or that communication across paradigm boundaries is possible even if unification is neither feasible nor desirable (Gregory, 1992; Willmott, 1993). Proponents of methodological pluralism claiming coherence must inevitably develop a position on the paradigm problem, otherwise they risk being accused of theoretically contradictory eclecticism.

11.2 The Cultural Problem

The 'cultural problem' has been described by Brocklesby and co-workers (e.g., Brocklesby, 1994; Brocklesby and Cummings, 1995; Mingers and Brocklesby, 1996; Mingers, 1997c) as follows:

"The question....is whether the existing cultural constitution of the management science community [and other communities concerned with intervention] will facilitate or act as a barrier against the widespread adoption of multimethodology as a research strategy. Obviously this depends on the size of the cultural gap between where we are now, and where—in relation to multimethodology—we would like to be. ....Fundamentally the problem is that very few of our colleagues are trained across two or more paradigms or work in groups where the sorts of multimethodology we have described are widely practised" (Mingers and Brocklesby, 1996, p.115).

11.3 The Psychological Problem

Authors identifying the psychological barrier to methodological pluralism (or multimethodology) include Brocklesby (1995, 1997) and Mingers and Brocklesby (1996). In his 1997 work, Brocklesby claims the following:

"It is one thing to say that there has been some degree of accommodation between the various.... paradigms because the combatants no longer completely ignore one another, or because it is now possible for "alternative" researchers to publish in dominant paradigm journals. But, for an individual agent, multimethodology demands a form of accommodation that is altogether more daunting. Reorienting educational programmes with the intention of creating a new breed of.... scientist who can routinely traverse the boundaries of the various paradigms is, itself, a difficult enough proposition, but transforming someone who has been thoroughly socialised in a single paradigm and has years of investment in a particular approach is an even more ambitious project" (Brocklesby, 1997, p.190).

Essentially, the problem is psychological 'resistance' to methodological pluralism. Brocklesby (1997) explains this resistance in terms of Maturana and Varela's (1992) theory of autopoiesis (reviewed in Chapter 3 of this book): individuals are inherently 'conservative', in that they will not move into a new 'rational domain' (paradigm) unless there is a very good reason (with an associated emotional commitment) to do so. However, in my view it is not necessary to view the issue through the theoretical 'lens' of autopoiesis to be aware that the phenomenon of resistance does indeed raise serious questions for the pursuit of methodological pluralism.

Mingers and Brocklesby (1996) also ask whether individuals have "cognitive predilections" (p.117) which predispose them to prefer one paradigm, and therefore one set of methods, over another: e.g., people may have a greater or lesser facility for handling mathematics, group dynamics, emotional conflict, etc. If people have different personalities which affect their paradigm allegiances, then it will no doubt take a great effort for them to learn new methods outside the set that they 'naturally' feel comfortable with (Stumpf and Dunbar, 1991; Mingers and Brocklesby, 1996). Clearly, psychological resistance is an issue that needs to be addressed.

11.4 Proposed Solutions

Over the last ten years, these problems have been addressed by a variety of authors, particularly those writing from a Critical Systems Thinking (CST) perspective. By far the greatest focus has been on the paradigm problem, primarily because this was identified well before the others in the management systems literature. Over the coming pages I will review some of the main contributions to the CST debate that have dealt with these issues. I will then present a new model of learning that I believe takes the debate one stage further.

11.4.1 Meta-Paradigmatic Thinking

To address the paradigm problem, Flood (1989a, 1990), Jackson (1990, 1991, 1993c) and Flood and Jackson (1991a,b) draw upon Habermas's (1972) theory of knowledge-constitutive interests (reviewed more fully in Chapter 10). In brief, this is the idea that, as a species, all human beings have a *technical* interest in work, a *practical* interest in achieving mutual understanding, and an *emancipatory* interest in freedom from oppressive power relationships. Flood and Jackson align the three waves of systems thinking (see Chapter 9) with the three interests: first wave (quantitative, modelling) methodologies are used to support the technical interest; second wave (qualitative, debate-orientated) methodologies are viewed as supporting the practical interest; and third wave (confrontative, boundary-challenging) methodologies are seen as supporting the emancipatory interest (see Chapter 10 for further details of these alignments).

Most importantly, Flood (1990) says that CST, in its use of the theory of knowledge-constitutive interests, is "meta-paradigmatic"—governing the use of other paradigms. Jackson has been engaged in a long-running debate over the paradigm problem with various authors, and has developed his position over the years in response to their comments. In 1993 he suggested that CST does not have to

> "decide whether the issues, or problems, or systems of concern are 'in the world' or whether they are in the minds of those conducting and participating in the analysis. As is demanded by its radical complementarism, since it embraces methodologies with varying

> ontological and epistemological presuppositions, it is agnostic on this matter" (Jackson, 1993c, p.292).[137]

And,

> "....systems methodologies can be related to different paradigms, each of which will constitute and frame social reality in its own way. Nevertheless, rather than these paradigms being incommensurable, it is possible to see them as complementary on the grounds of the three essential human interests identified by Habermas—the technical, the practical, and the emancipatory. The paradigms should guide knowledge production and the systems methodologies should be put to work, in an informed manner, in the service of appropriate human interests" (Jackson, 1993c, pp.290-291).

Essentially, Flood and Jackson 'solve' the paradigm problem by the use of a meta-theory which guides the practical use of the sub-paradigms.

However, in my view, there is a significant contradiction in this idea. Flood's (1990) claim that CST is "meta-paradigmatic" is undermined by the assumptions made in Habermas's (1972) theory of knowledge-constitutive interests. These assumptions are alien to, and incommensurate with, assumptions made by the proponents of the various systems paradigms that Flood and Jackson try to contextualise. Therefore, by accepting Habermas's theory, Flood actually sets up new paradigmatic assumptions: he does not rise above the paradigm debate at all. For further details of this argument, see Midgley (1989a,b, 1996c).

Clearly, I do not regard this 'solution' to the paradigm problem to be credible, and indeed both Flood and Jackson have now turned their backs on it too (Flood and Romm, 1996a; Jackson, 1999). It should also be noted that Flood and Jackson's use of Habermas's (1972) theory of knowledge-constitutive interests does not address the cultural and psychological problems either. Therefore, I suggest that it is appropriate to abandon this line of argument.

11.4.2 Towards a New Paradigm

When Flood and Jackson first proposed this "meta-paradigmatic" thinking, I saw the above problem and wrote a critique of their position (Midgley, 1989b). Since then, I have consistently argued that it is

[137] This comment was actually made in defence of Total Systems Intervention (Flood and Jackson, 1991b)—a meta-methodology inspired, in part, by CST (see Chapter 10 for details). However, it is reasonable to suppose that Jackson would take the same view if he were discussing CST more generally.

impossible for any approach to methodological pluralism to be meta-paradigmatic given that it must inevitably make assumptions that are alien to, and incommensurate with, assumptions made by the proponents of the various paradigms that methods are drawn from. Far from being meta-paradigmatic, I suggest that those engaging with methodological pluralism are trying to establish the foundations for a *new* paradigm (Midgley, 1989a,b, 1990a, 1992a, 1996c). Of course, pluralists can still learn from other paradigms (Gregory, 1992), but this learning is always geared to the enhancement of one's own paradigmatic position—there is no pretence that other people's methodological ideas are used in exactly the manner that their creators intended.

It is because I do not believe that paradigmatic thinking can be transcended that I stress the mixing of *methods*, not methodologies. In Chapters 9 and 10, I argued that we can learn from other methodologies to aid the on-going construction of our own, and we can detach methods from their original methodological principles in order to use them in new ways (seen through the eyes of our own methodology). This is now a widely accepted way of thinking about methodological pluralism in both the CST and operational research communities (e.g., Gregory, 1992, 1996a,b; Flood, 1995a; Flood and Romm, 1996a; Mingers and Brocklesby, 1996; Yolles, 1996, 1999b; Brocklesby, 1997; Mingers, 1997c; Munlo, 1997).

Clearly, this argument addresses the paradigm problem: there is no need to claim that we are operating across paradigms—we just have to acknowledge that we are setting up a new position which encourages learning about ideas from other paradigms, but reinterpreted in our own terms. However, it does not explicitly address the cultural or psychological problems (identified by Brocklesby and colleagues) because its production pre-dates their identification. Nevertheless, Brocklesby (1997) builds his own argument that psychological barriers can be overcome using my approach. His thinking runs as follows:

The psychological barrier to multi-paradigm thinking exists because of the demands of moving between fundamentally different sets of assumptions. It is difficult enough, when wedded to one paradigm, to accept the possibility that another one has anything valid to offer—but moving freely between two or more paradigms, changing one's assumptions as one goes, is infinitely harder. For example, at one moment it requires a person to believe that there is a real world that s/he can know, and at the next s/he may need to deny this 'basic fact' altogether! This is simply contradictory. However, it is a different kettle of fish to say that we can develop a *new* set of paradigmatic assumptions that embraces the best of several old sets:

> "Whereas multi-paradigm multimethodology would have an agent move from one paradigm to another depending on which

> methodology, or part thereof, is being used at any moment, an alternative possibility has methodologies originating in different paradigms being employed in the service of a *new* paradigm. The defining feature of such a paradigm is that it can dissolve the competing objective-subjective duality of the original paradigms by incorporating these perspectives within a broader ontological framework. It is not appropriate to delve into this matter here, save to point out that, in the MS [management science] context, this option was first raised by Midgley (1989a, 1990b, 1992a[138]). In Mingers (1995), and Mingers and Brocklesby (1996), it was developed further through reference to the work of philosophers such as Bhaskar and Giddens. It seems to me that putting various methodologies to work in support of this new paradigm is desirable because it provides a way of avoiding having to choose between.... the.... existing paradigms, or having to constantly adjust one's assumptions as one moves between them. This option.... provides the authority to throw away the old rule books and play by new rules. The new rules circumvent the need to be constantly adjusting one's philosophical position depending upon which methodology or technique is being used at any moment in time, which, as we have seen, can create difficulties" (Brocklesby, 1997, p.211, emphasis in the original).[139]

It should be clear that Brocklesby's proposed solution to the psychological problem (establishing a new paradigm with a broad ontological framework) is what I have been working on for the last ten years, and is actually one of the core missions of this book.

11.4.3 Paradigm (In)commensurability

Another pair of authors taking up the challenge of dealing with the paradigm problem are Flood and Romm (1995b, 1996a). They acknowledge the argument advanced by myself and others that any attempt to embrace methodological pluralism will involve the intervener making assumptions that other methodologists may not agree with. It is therefore very difficult to suggest that there is genuine commensurability between paradigms: there is no position outside the paradigm debate from which to achieve this commensurability.

Nevertheless, like Flood and Jackson, Flood and Romm insist that it *is* still possible to contextualise other ways of thinking from a pluralist perspective. People may thereby choose the 'most appropriate' approach to each intervention, depending on perceptions

---

[138] The annotation of these references has been changed from the original to reflect my own use of the letters 'a' and 'b'.

[139] I disagree with Brocklesby's (1997) use of the term 'methodologies' for the reasons stated earlier in this chapter, and in Chapter 5. My preference is to talk about *methods* in this context, but I suspect that Brocklesby is just following the usual convention in operational research and management science of using the two terms interchangeably.

of the circumstances and the wishes of the agent(s) involved. Because Flood and Romm (1995b) see both the paradigmatic nature of pluralistic practice *and* the possibility of contextualising ideas from other paradigms, they refuse to talk about either paradigm commensurability *or* incommensurability. Instead, they express the irony of the problem with the phrase "paradigm (in)commensurability" (note the 'in' is bracketed).

My own view is that resorting to a phrase like "paradigm (in)commensurability" expresses the irony of the paradigm problem very well, but it does not take us any further in dealing with the three problems (philosophical, cultural and psychological) that provide the focus for this chapter. It merely indicates the 'bluntness' of the language of paradigms (as used by authors up to 1996[140]) in helping us deal with the relationships between our own ideas and the ideas of others. Let me explain.

When Kuhn (1962) first popularised the term 'paradigm', his insights were revelatory for many philosophers of science: previously, science had been seen as an activity that allowed incremental progress by continually developing our store of knowledge. However, this older view did not take account of the experiences of scientists who often found themselves involved in lengthy theoretical debates with others. People trying to introduce new thinking encountered great resistance: old ideas were often defended by their advocates for many years. When Kuhn suggested that different groups of scientists make different paradigmatic assumptions, and that one view eventually *replaces* the other (rather than simply building upon it), this seemed to explain the difficulties people experienced in convincing others of their point of view: scientists wedded to established ideas were defensive because accepting the arguments of others could signal the annihilation of their own work. I suggest that the language of paradigms has been very important because of the light it has thrown on how scientific communities function.

Nevertheless, the first indications of its bluntness were identified quite early on. For instance, Masterman (1970) points out that Kuhn uses the term 'paradigm' in a large number of different ways—Kuhn cannot cover every angle without doing so. While the language of paradigms has certainly generated insights, and I have used it regularly myself, I wish to argue that its inadequacies become transparent when we think about *individual learning*. For paradigms to change, it must be possible

[140] I have used 1996 as a point of reference because this is when Yolles first published his work on paradigms, which (in my opinion) overcomes the problem of 'bluntness' that I have identified. This will be reviewed later.

for individual agents to propose new ideas that step outside old paradigmatic assumptions. The question is, what kind of 'paradigm' is operational when an individual breaks the paradigmatic mould? And, in the context of methodological pluralism, what is the status of the work of an individual who proposes a position which draws on ideas from other paradigms? Tsoukas (1993a) claims that an individual cannot give birth to a paradigm: a paradigm is only born when the individual's ideas have become widely accepted. What then is the relationship between paradigms and the thinking of individuals? Gregory (1992) and Yolles (1996, 1999b) have both addressed these questions, so their work is reviewed next.

11.4.4 Critically Appreciating Alien Paradigms

Like me, Gregory (1992) insists that it is impossible to transcend the paradigm debate: each attempt to do so must inevitably involve interveners in making new paradigmatic assumptions. However, she advances our thinking by examining the *nature of communication* between people based in different paradigms. Every time one person listens to another whose thinking is based in another paradigm, he or she can only interpret what they are saying through his or her own terms of reference. However, this does not mean communication is impossible—just that care is needed not to be either dismissive or to think that full understanding has been achieved. If care is taken to *appreciate* the other, in the knowledge that full understanding in the other's own terms is impossible, then one's own learning about methodology and methods can be enhanced. This way of thinking advances the debate because it allows us to see paradigms in relation to the perspectives of *individual agents*. Learning through the appreciation of others' viewpoints can feed back, via communication, to transform one's own paradigm.

Gregory's (1992) approach not only deals with the paradigm problem, it also addresses the problem of psychological resistance to methodological pluralism. This is because the primary emphasis is on *learning*: for the agent to start learning, there is no need for him or her to have full knowledge of a multitude of methods and methodologies. There is only a need for a *critical attitude*: a preparedness to listen to others when we encounter them, and a willingness to research new approaches when the need arises. Of course, there can be no absolutely *objective* need for new approaches, but processes of self-reflection, dialogue with others, observation of circumstances, and ideology critique can help to highlight the limitations of one's current armoury of methods and suggest alternative research avenues for exploration.

11.4.5 Virtual Paradigms

Like Gregory (1992), Yolles (1996, 1999b) also addresses the paradigm problem by shifting the focus to the level of the individual agent. He argues that, while paradigms are formalised sets of shared assumptions held in common by groups[141] (not just research communities, but also organisations), individual agents can establish *virtual* paradigms: that is, they can work out a set of assumptions through which 'reality' and ideas (including methods and methodologies) from other paradigms can be interpreted. A 'virtual paradigm' may be temporary (like a working hypothesis or model), or be developed over the longer term. A virtual paradigm may also become a true paradigm if others begin to share the assumptions: when it comes to be shared right across an organisation or community, the paradigm's transition from 'virtual' to 'true' status can be said to be complete. Yolles (1996) puts it like this:

> "....if paradigms are to be compared and coordinated.... [this can only be done] through the creation of a virtual paradigm because (1) without a paradigm, nothing can be said about reality, and (2) new language shows that a new paradigm has been created.... Its creation is dependent on the modeller[142], to whom it is totally relative. Different modellers may define different virtual paradigms, and classify situations in a modelling space according to the paradigm that they choose through which to see. In due course, however, if it becomes accepted by a group and if norms develop that modellers use in order to classify situations, then the paradigm loses its status as *virtual*" (Yolles, 1996, pp.568-569, emphasis in the original).

In my view, Yolles has dealt with the paradigm problem in a useful and interesting manner, building on the observation that any vision of pluralism must be paradigmatic. Indeed, Yolles's idea of a virtual paradigm helps me escape from Tsoukas's (1993a) pertinent criticism of my own writing on the paradigm issue: Tsoukas argues that an individual or small group cannot claim to set up a new paradigm—it is always a large group or community phenomenon. However, using Yolles's language, an individual or small group *can* claim to establish a *virtual* paradigm. Indeed, in the case of CST, I would argue that the basic premises are now sufficiently widely shared to claim that it has moved from 'virtual' to 'true' status.

---

[141] According to Yolles (1996), a paradigm can be distinguished from a *weltanschauung* (or worldview). A weltanschauung is a set of implicit assumptions shared by a social group, but a paradigm results from the *formalisation* of some or all of these assumptions.

[142] Yolles (1996) talks about "modellers", but this can be translated into 'agents' for the purposes of this book.

Unlike the paradigm problem, Yolles does not explicitly address the psychological and cultural problems. However, an attitude to them can be inferred from his writings. I surmise that Yolles would have the same answer as Gregory (1992) to the psychological problem: that on-going learning about methodology and methods at the individual level, via the establishment of virtual paradigms, is a means to overcome the psychological barriers to methodological pluralism that are partly the result of an unrealistic expectation that interveners should come into the world ready-equipped with a full armoury of methods.

However, perhaps Yolles's thinking is most relevant in relation to the cultural problem. While Gregory and I have not addressed this at all, and Mingers and Brocklesby (1996) talk in very general terms about changing the education of future generations of academics, Yolles at least clarifies the mechanism through which cultural change towards a more pluralistic practice might come about: the establishment of a virtual paradigm that begins to gain wide-spread support, until it becomes a fully-fledged paradigm in its own right which others can commit themselves to. Indeed, it is not beyond the realms of possibility to have a *variety* of pluralist paradigms which people can choose between [as Mingers and Brocklesby (1996) and Mingers (1997c) show, there are a number of virtual paradigms out there which have the potential to grow into something more].[143]

11.5 A Model of Learning

Having reviewed some of the writings that have addressed the three challenges facing methodological pluralism, I can now move on to present the model of learning that builds on these writings, and ultimately offers a new methodological understanding. I intend to construct the model in a series of stages over the coming pages, adding greater complexity at each stage. When the model is complete, I will reflect back on the three challenges and discuss how the model does (or does not) address them.

I should note that this model represents an ideal learning practice that can be used for critical reflection at any level of agency: individuals can work towards this pattern of learning for themselves, as can small groups and whole communities of practitioners (who therefore constitute a research community). Indeed, the learning of

[143] It is important for me to reiterate that these views are merely *inferred* from Yolles's writings. It remains for him to clarify whether or not I have correctly identified his thinking on these matters.

many individuals and groups may contribute to community learning. In writing the text below, I have tended to use language associated with individual learning, but it would only take a minor linguistic adjustment to argue the case for group and/or community learning too.

I should also be clear that, when I say that this is an *ideal* learning practice, I mean that it is a model of good practice to aim towards, but should not be seen as something that can be operationalised all in one go. It requires the development of many different skills over time. It would be unrealistic to specify a minimum set of skills to start systemic intervention [as Mingers and Brocklesby (1996) argue, most of us are products of education systems that limit the scope of skills development]—but agents do need to be willing to learn as they practice. Furthermore, even when they are committed to learning over time, individuals will tend to have predilections for particular types of learning: e.g., active or reflective, abstract or concrete (Kolb, 1984). While they may make some advances in the areas they are less comfortable with, it will inevitably be the case that their methodology and practice will have strengths and weaknesses. For this reason, there are advantages to building learning at the team, organisational and/or community level rather than just at the level of the individual: a team of individuals working together to improve their systemic intervention practice through mutual learning can complement and support one another (Gregory, 2000).

Let us start by reflecting back on the model I introduced in Chapter 9 (Figure 9.1, p.168), which shows the two levels of 'methodology' and 'methods'. We see that there are several 'isolationist' methodologies (according to Jackson, 1987a, those that proscribe all but a narrow range of methods) and a pluralist methodology. The latter has a wide range of methods associated with it, some of which may be drawn from isolationist paradigms, but their use comes to be seen through the 'lens' of the pluralist methodology. Learning about the existence of different methods, and their possible strengths and weaknesses, needs to be an on-going process: one can start with just a couple of methods and proceed from there. Proponents of a pluralist methodology may also learn from other methodologies, including isolationist ones (see Chapter 9 for details).

11.5.1 *Continuity and Discontinuity*

Let us now look at how learning about methodologies and methods comes about. We can see the pluralist methodology as a virtual paradigm: essentially it is associated with the activities of an *agent* (whether an individual or a group). If the agent is a relatively large group, constituting a research community, one could say that the paradigm is 'true', not 'virtual'—but here I will continue to refer to it as a virtual paradigm, if only not to be presumptuous (there are many different visions of methodological pluralism, some more widely shared than others, and it is always going to be open to debate when the transition from 'virtual' to 'true' status has taken place). Because learning is an on-going process, the armoury of methods will grow and develop as the agent becomes more and more experienced at systemic intervention. This is a relatively straight-forward kind of skills acquisition.

However, learning also takes place at the level of methodology. An important assumption I make is that a pluralist methodology (virtual paradigm) is dynamic, not static. If it is possible to learn from others, then it is necessary for a methodology to be evolved on an on-going basis. Therefore, we must oppose the usual practice in academia of building a methodology like a castle and then defending it against enemies who want to tear down the castle walls. People with this kind of attitude see the modification of a methodology as a sign of weakness.[144] I view it as a strength, as long as learning is part of a process of construction in which ideas change in relation to both practical experience, dialogue with others, and theoretical reflection. Building a methodology is more like constructing a house, where extensions can be added, internal walls demolished, rooms redecorated, etc., to enhance both its function and the experience of living in it. A methodology should be useful in terms of how it allows for the interpretation of methods and practice, both for its creator and for others wishing to learn from it. This is different from the uninformed vacillation of someone who is so unsure of what they believe that every new idea is swallowed wholesale (see later). Constructing a methodology is a much more considered process, *but is still essentially dynamic*.

Importantly, if the methodology (virtual paradigm) changes on an on-going basis, *there are always going to be tensions and discontinuities between different aspects of it that have been introduced at different times under different circumstances*. In this sense, the methodology can

[144] I once met an academic, who shall remain nameless, who rubbed his hands with glee when Checkland and Scholes (1990) revised Soft Systems Methodology (SSM) after a nine year period of reflection (see Chapters 9 and 15 for longer discussions of SSM). This person, who disliked SSM, said that Checkland was making a fundamental mistake admitting that there was any need for improvement. I thought then, and still think now, that Checkland showed more courage and integrity than his 'opponent' in making this admission, and the proof of the pudding is in the eating: the publication of Checkland and Scholes (1990) led to a whole new wave of systems research which enhanced, not diminished, Checkland's reputation.

be described as a 'fragmentary whole' (a deliberately paradoxical concept). It is the task of the agent, as part of his/her/their on-going learning, to balance two potentially contradictory activities: maintaining coherence and introducing new ideas. If there is too much emphasis on listening to new ideas, and these are not brought into a coherent perspective, then there will be no methodology to speak of—just a fragmentary set of theories, principles and rules for practice. The result will be interventions which jump from one impulse to another in a seemingly haphazard manner, with the risk of confusion for others involved in, and/or affected by, the agent's activities. Certainly, the agent who falls prey to fragmentation is likely to be influenced by all the new fads that come along, regardless of their worth, because s/he lacks a reasonably coherent set of ideas to critique new proposals against.[145] In addition, s/he will not have a consistent language to communicate insights to others, so learning is unlikely to be passed from one generation to another. Jackson (1987a) criticises atheoretical pragmatists[146] for this kind of fragmentary thinking: while pragmatists welcome the idea of a pluralistic use of methods, they turn their backs on theory, and thereby lose coherence. Finally, the agent who contradicts him or herself on a regular basis, without a coherent story to explain the contradictions, will lose credibility in the eyes of others (Aronson, 1976)—and is also likely to experience an unpleasant feeling of dissonance if s/he eventually realises s/he is thinking, talking and acting in a contradictory manner (Festinger, 1957).[147]

Conversely if, in the interests of internal coherence, an agent closes off to influences from ideas other than his/her/their own, learning at the methodological level will be minimal at best. The likely outcome

[145] See Jackson (1995) for a critique of management fads and an argument in favour of systems thinking.

[146] This use of the label 'pragmatist' does not refer to the Pragmatist movement which was active at the turn of the 20th Century (e.g., James, 1904; Pierce, 1934; Dewey, 1946). The Pragmatists were essentially *pro*-theory—or at least pro *useful* theory. Rather, 'pragmatist' (with a lower case 'p') is a degraded use of the term referring to interveners who are anti-theory. See Chapter 5 for a further discussion of the two uses of the same term.

[147] Festinger (1957) and Aronson (1976) both assume that it is 'natural' for human beings to wish to reduce 'unpleasant' dissonance between contradictory ideas. Of course, some authors (e.g., Taket and White, 1993), argue that this is a cultural phenomenon, not a natural one, and we should not be constrained by the Western prohibition of logical contradiction. I certainly believe that welcoming a degree of contradiction is necessary, as is acting to resolve contradictions (no new ideas could develop if there were not a tension between these two activities), but it is interesting to observe that, in Taket and White's writings, contradictions are given an acceptable status by a narrative explaining their historical prohibition. Ironically, Aronson (1976) identifies this technique of making a contradiction rational through the use of an explanatory narrative as one very effective means of reducing cognitive dissonance!

will be an impoverished methodology (virtual paradigm) which is self-justifying: if practice is always interpreted through the same methodological idea, then evidence that the methodology is impoverished will simply not be seen by the agent—practice needs to be interpreted through more than one methodological idea for potential problems to be surfaced effectively (Romm, 1996). To give an example, if a methodology focuses on the benefits of restructuring organisations, and no other rationale for intervention is entertained, then the possibility that restructuring may introduce coercion into an organisation, or may prevent constructive communication between employees, is unlikely to be made visible, let alone be addressed as part of intervention. Of course, an impoverished, self-justifying virtual paradigm is unlikely to be seen as useful by others, so will not become widely shared. It is in the interests of agents wishing to persuade others of the value of their ideas to listen to the concerns of others, interpret them, and reflect them back as part of the agents' own methodology.[148]

Maintaining the 'right' balance between coherence and openness to new ideas is not always easy, but in my own experience development goes in cycles. I will explain in more detail below. Although this feels right for me, it may not work for everyone, so I am providing it as an *example* of how the balance may be struck, not as a prescription. My own development of methodology is cyclical in the sense that I tend to go through (sometimes overlapping) periods of openness to new ideas, followed by periods of revision and consolidation. Some of these revisions follow from theoretical research, and others follow from reflections on my practice. Very often, when I am in a period of consolidation (building and communicating theoretical coherence), I may come across a valuable new idea, but I consciously refrain from integrating it into the whole unless it seriously undermines some aspect of what I am currently doing (in which case integration is urgent)—I wait until I enter another period of openness and revision.

Each time I take in new ideas, they are obviously interpreted through my own conceptual schema ('irrelevancies'[149] are filtered out and language reinterpreted), so some of the work of integration is

[148] It is certainly a motivation for most academics to have their work listened to and used by others. However, this may be less important for non-academic agents who are perhaps more likely to be content with constructing a virtual paradigm (methodology) solely for his/her/their own use. Communicating the ideas to others in order to make the virtual paradigm real may not be an issue at all—although communication about methodology is often still necessary just to enable others who are affected by an agent's interventions to understand the principles the agent is operating with.

[149] Of course, what appears to be an irrelevance to me might be vitally important to someone else, and vice versa.

already done at an unconscious level. This is simply a result of being a situated agent with a history of knowledge and language-use guiding my interpretations. Nevertheless, I still become aware of contradictions, discontinuities ('gaps' between ideas that need to be filled) and subtle changes in the use of language. The process of revision and consolidation smoothes out the contradictions by changing and thereby harmonising ideas, creates new theory to fill the gaps, and alters my use of language where necessary.

Some of this cyclical learning happens fairly automatically in the process of engaging in systemic intervention, talking with others, reading, and writing for publication. However, I also make conscious use of certain events to 'force' myself to open up to new ideas or consolidate a variety of semi-coherent theories into a whole. One example is conference invitations which I tend to use as challenges to explore new theory: knowing that one has to make a credible public presentation of the new ideas within a couple of weeks concentrates the mind wonderfully! The model of boundaries and marginalisation presented in Chapter 7 (Figure 7.3), for instance, was first created in just two weeks prior to a conference (Midgley, 1991b), and this resulted from a conscious decision to read Douglas (1966) to see if it contained insights for systemic intervention. In 1991, I had not yet fully integrated that thinking into the larger methodological whole. The integration happened at a later date during a period of consolidation. Again, I used an invitation—this time to write the introductory chapter to a book on Critical Systems Thinking (Midgley, 1996c)—to 'force' myself to produce a coherent whole that would have what I hoped would be a persuasive effect. Indeed, writing the book you are now holding in your hands represents the largest effort of consolidation I have ever engaged in: to produce this book I have had to draw together a wide variety of ideas developed through philosophical, methodological and practical reflections undertaken in different contexts over a number of years—and the process of harmonisation has been extraordinarily complex. I leave it to you to judge whether or not it has been successful.[150]

Before moving on, I wish to make one final point which relates to the comment I made earlier about the fact that agency and learning can be seen at various levels: at the level of the individual, the group, and/or the research community. It is obviously the case that the more diverse the research community, the more likely there is to be

[150] Earlier I said that, during periods of consolidation, I may encounter new ideas which have to wait to be integrated into the whole. This has been the case when writing the current book. I am now sitting on an idea that I am sure will send my research into a new and exciting direction (see the section entitled "The Politics of Systemic Intervention" in Chapter 18)—but hopefully more on that in future years.

burgeoning discontinuity which cannot easily be balanced by integrative research: many people in a widely diverse research community will see attempts to unify the (virtual) paradigm as a 'political' imposition by one group or individual on others. There are two possible solutions to this problem: either (i) discontinuity can be accepted at the community level and integration can be pursued at the individual and small group levels only; or (ii) an integrative theory (or narrative) can be sought which makes a virtue of discontinuity at the level of the research community.

The first of these strategies is likely to lead, over time, to the fragmentation and dissipation of the research community—but in a peaceful way, rather than through the violent disruption that would result from a minority group trying to impose a non-consensual coherence. The second strategy, finding a theory that makes a virtue of discontinuity at the community level (while continuing to balance discontinuity and coherence at the group and individual levels), is paradoxical: making a virtue of discontinuity requires *an integrative and coherent theory that can be consensually accepted*. If such a theory can be found, then a common agenda is preserved, along with the identity of the research community. My own view is that, as research communities grow and develop, they tend towards greater diversity, so the successful spread of ideas is continued through a tactical switch from a balance of coherence and discontinuity at the community level to a greater emphasis on the value of discontinuity—but without losing the idea that there is still a common agenda, expressed in part through the inclusion of a theory that values the discontinuity. In such a situation, diversity is unlikely to lead to 'political' in-fighting because most of the pressure for coherence is transferred to sub-community levels, yet enough commonality remains for the research community to continue to exist and enable learning.[151]

---

[151] It is interesting to note that this is exactly the transition that occurred in the Critical Systems Thinking (CST) research community. When only a small group were advocating CST, the balance between coherence and discontinuity was easy to maintain. However, by the mid-1990s, over 100 writers had contributed to the CST literature (and I assume that there were many more readers), introducing a great deal of theoretical and methodological variety. I therefore published several papers (Midgley, 1995b, 1996c) arguing that CST is a *debate around common themes* rather than a set of 'commitments' (the language of 'commitments' was used by Flood and Jackson, 1991a,b): this was both a theoretically necessary shift (given the increasing focus on dialogue in our research community) and a consciously devised, strategically motivated move to keep the research community intact. In my view this move was successful, as the CST community has continued to grow, and we currently have no significant problems with in-fighting (as far as I am aware).

11.5.2 Philosophical Reflections

Having clarified what it means for a pluralist methodology to be a 'fragmentary whole', I can now add the next layer of complexity into the model: reflections on philosophy. Given that I have argued the case for the relevance of philosophy to methodology (Chapter 2), it should come as no surprise that I propose that the construction of a methodology should be informed by philosophical reflections. The development of a philosophical position very much mirrors the development of a methodological one: a philosophical position can be a fragmentary whole which can take in and interpret ideas from other peoples' philosophies.

I have heard it said amongst systems methodologists and operational research practitioners (even some who share my views on the value of philosophy) that a cursory reproduction of a particular writer's ideas 'will do'. It is obviously the case that the systems intervener who bridges philosophy, methodology and practice will not be able to put as much time into philosophical reflections as a full-time philosopher. In my view, this is a good thing: as I argued in Chapter 5, there is a danger of getting caught in a narrow philosophical discourse and forgetting about the possible moral implications for action. However, using philosophical ideas in a purely *instrumental* fashion—bending them in any which way to suit one's methodological purposes—should be regarded with caution. For example, Valero-Silva (1996) criticises writers who 'instrumentalise' the work of Foucault (e.g., 1980, 1984a): they simply ignore the fact that Foucault argued vociferously against purely instrumental, uncritical thinking. It is not enough to find a philosopher whose work superficially seems to support a methodology and throw in his or her name to add spurious credibility to otherwise unsupported ideas: it is important that philosophical and methodological reflections *inform each other*.

Fundamentally, the reason for exploring philosophy (from a methodological point of view) is to ask penetrating questions about the assumptions that methodologies make—about their connections with other (in this case philosophical) discourses that flow through and influence social action.[152] Again, I return to the work of Spash (1997) for

[152] Note that I am not claiming that philosophical reflection reveals the 'underpinnings' of methodology: as I explained in Chapter 2, I see the relationship between philosophy, methodology and practice as non-hierarchical. Nevertheless, there are connections between the three discourses that, when exposed, allow the agent to see a whole system of mutually supportive argumentation. Sometimes it can be quite surprising what supports a supposedly innocuous idea (like cost-benefit analysis), and revealing existing connections gives agents the freedom to critique these and make new, more desirable connections.

a good example: by exposing the utilitarian philosophy implicit in supposedly 'neutral' cost-benefit analyses, Spash demonstrates their hidden bias in favour of profit-orientated thinking and against environmental conservation (see Chapter 2 for details).

11.5.3 Reflections on Practice

The next layer of complexity comes from reflections upon *practice*. It seems to me pointless to explore methodology for its own sake (see Chapter 12 for an extended argument): methodology only has meaning in relation to interventions. The actual feedback from practice to methodology happens when the agent makes connections between the methodology and the various experiences (and discourses about the agent's own and others' experiences) that surface during practice. Initially, the latter may not present themselves as methodological, but reflection may reveal their methodological importance. An example is Checkland and Scholes's (1990) addition of a "stream of cultural inquiry" to Soft Systems Methodology (in Checkland's, 1981, previous version this was absent). It was added partly because of Checkland's experience that intervention could be frustrated by the 'political' activities of participants and the contingencies of organisational cultures.[153] Supposedly non-methodological insights may create quite significant changes in methodology when their relevance for the latter is realised (see Flood, 1990, for a description of two methodological/ philosophical "breaks" he made as a result of becoming more aware of the implications of his ideas for practice).

Earlier, I mentioned Romm's (1996) argument that reflections on practice need to be undertaken using multiple methodological positions, otherwise evidence of problems might not be revealed. This is important if one's methodology is not to become self-justifying, and it is part of learning from other methodological positions: to ask how other methodologists might evaluate one's practice, and then to question whether they have a point that should be taken on board in terms of one's own methodological development, selection/design of methods, and/or future practice.[154] For example, when I started out in community

[153] I suspect that another motivation was the need to deal with the critique of SSM by Mingers (1980, 1984) and Jackson (1982) for failing to deal adequately with power issues. Without asking Checkland himself, it is not possible to know whether this critique or the experiences of dealing with 'political' participation was the prime motivator for adding the "stream of cultural inquiry", or whether it was a balance between the two.

[154] It should be acknowledged that another person's methodology can only be interpreted from one's own standpoint: there can be no objective interpretations of methodologies, but that is not to say that learning by looking through the eyes of others (however imperfectly) is impossible (Gregory, 1992).

operational research, I was relatively unconcerned about whether or not the other participants in my interventions learned how to use the methods I was using. It was only by reading and listening to Checkland and Scholes (1990), who argue that participants should learn their language of Soft Systems Methodology in order to free themselves from dependence on an external 'expert', that I noticed that people did indeed often become dependent on my presence in my own interventions (which, in my view, was not a healthy state of affairs). However, I have not taken on Checkland and Scholes's position wholesale: I now argue that there are *some* situations where the methods should be handed over to participants, particularly when there is a need for the empowerment of the various agents involved, and there are other more immediately task-orientated situations in which communicating the language of intervention might be less important. This discussion of the transmission of methods is just one instance of how reflections on practice in the light of different methodological ideas can be helpful.

11.5.4 Reflections on Theory

Another layer of complexity is introduced when we consider the value of reflecting on theory during interventions—not just theory surrounding the use of methods (i.e., methodology), which is obviously relevant, but theories about the world around us. The perceived relevance of particular theories during intervention may affect the choice of boundaries and thereby the design of methods. Reflection on, and discussion of, theories may need to take place 'on the spot' as part of an intervention, but theoretical learning may also take place outside the context of particular interventions. In my own case, I try to use 'spare' moments (e.g., on long train journeys) to do some reading in order to keep in touch with selected theoretical debates that I believe may be relevant to future interventions.

An important issue with regard to the use of theory is whether it is legitimate to draw upon several theories making contradictory assumptions within the same intervention (or indeed, over time in different interventions). If one takes the view that theories should fit together like pieces of a jigsaw puzzle to create a picture that is as near as it is possible to get to an accurate representation of reality, then theoretical contradictions are a problem. However, in Chapter 8, I argued against this 'cumulative' view of theory and knowledge. Instead, if theories come to be judged in terms of their utility for given purposes (and of course they can *interact* with purposes, shifting the agent's understanding of an intervention), contradictions need not be a problem. Let me provide a practical example.

Cordoba *et al* (2000) describe their use of the theory of autopoiesis partnered with boundary critique to inform their actions during an intervention in which they supported a Colombian University in participative, strategic information systems planning. At the point at which their paper was written, the intervention was incomplete. Subsequently (although the details have not yet been written up in the literature), they drew upon Foucault's (1984a-c) understanding of the connections between power, knowledge and the actions of 'subjects' (agents) to explain some problems that they encountered, and to suggest alternative paths for the intervention. While an argument has been advanced claiming that the theory of autopoiesis could be improved by the addition of Foucault's understanding of power (Vélez, 1999), it is undoubtedly the case that some of Maturana and Varela's (1992) ideas in the theory of autopoiesis would have to be modified to accommodate it. Similarly, Vega (1999) has proposed the use of boundary critique in the context of applying Foucauldian theory to the evaluation of social justice in health care—but he has had to challenge and re-write some of the theory of boundary critique (particularly Ulrich's, 1983, universalism) to make it compatible with Foucault's position. Therefore I would say that, in principle, it would probably be possible to harmonise the three theoretical ideas flowing into Cordoba *et al*'s intervention, but this harmonisation has not yet been systematically and fully undertaken (only some aspects have so far been tackled). Nevertheless, *in terms of the purposes of the intervention* that Cordoba *et al* undertook, the disjunctions between the three theories created no problems.

However, were a fresh intervention to be undertaken with the purpose of synthesising Foucault's view of power and knowledge with the theories of boundary critique and autopoiesis in order to make a new contribution to critical theory, these disjunctions would no doubt assume paramount importance. Indeed, the success of the intervention would depend on whether a 'seamless' theoretical product could be produced—it is unlikely that anything else would be persuasive to the academic community. It is the purposes of the intervention, seen in relation to agents' perceptions of the context (which affect how the purposes are defined), that are crucial in guiding how theory is used in an intervention.[155] Learning about theory therefore plays a part in the

[155] However, it should not be assumed that theoretical contradictions are only a problem in relation to academic debates, and that in other contexts they are perfectly acceptable. It may be quite important in some non-academic contexts to demonstrate that a strong, unitary theoretical position is being employed. Conversely, in exploratory academic discussions where the purpose is to creatively surface ideas (rather than refine existing ones), a good degree of contradiction can be tolerated.

model being developed in this chapter, but the extent to which harmony between theoretical ideas is needed will vary from intervention to intervention.

11.5.5 Espoused Methodology and Methodology in Use

The final layer of complexity then enters the picture when we consider what else effective reflection on practice might involve. Argyris and Schön (1974, 1985) and Schön (1983) make a very useful distinction between *espoused theory* and *theory in use*, which I will shortly adapt for inclusion in my model of learning. 'Espoused theory' is what an agent *says* that they use by way of theory, methodology, principles, etc., and 'theory in use' is what they *actually* use. Argyris and Schön suggest that the gap between 'espoused theory' and 'theory in use' can sometimes be quite wide, and the agent is almost always unaware of this fact. This is because theories in use tend to be learned in an unconscious, non-intellectual fashion, while espoused theories are the subject of conscious reflection. Indeed, if people try to 'force' agents to see that their words and deeds are contradictory, these agents are likely to resist the insight. Therefore, Argyris and Schön say that agents need support to explore the gap between espoused theory and theory in use—and this support should take the form of facilitated dialogue forums and space for self-reflection. In essence, theories in use need to be brought into consciousness to be subjected to the same scrutiny as espoused theory.[156]

Having said that Argyris and Schön's distinction between espoused theory and theory in use is helpful, I should nevertheless declare my concern that the concept of 'theory in use' can be interpreted in a naively objectivist fashion. I want to make clear my own position on this. By definition, a 'theory in use' is contrasted with an 'espoused theory', so there is bound to be an (initial) disagreement over what the theory in use actually is. Therefore, a theory in use can never be absolutely objective: it is always an interpretation offered by someone other than the intervening agent. The problem with treating it as objective is that there is an assumption that the agent must be wrong, and that the interpretation offered by an observer other than the agent has to be

[156] In the early 1980s, I had the good fortune to attend a training course run by a French Canadian woman who was a student of Freud's in the last few years of his life. Unfortunately, her name now escapes me. Interestingly, she claimed that, in reflecting on his life's work, Freud realised that he had concentrated on exploring the concept of the unconscious to such an extent that he had neglected the substantial ability of the conscious mind to change unconsciously motivated behaviour. This woman had made it her own life's work to take this idea forward, and in some ways Argyris and Schön's project of exposing theory in use to conscious reflection is along the same lines.

right. In contrast, I prefer to enable a dialogue between the agent and others affected by an intervention, and ultimately it is the responsibility of all the parties in that dialogue to make up their own minds about what is the 'theory in use'.

Now, in the paragraph before last, I said that I would *adapt* Argyris and Schön's concepts for use in my model. This is because I find the word 'theory' to be too specific: for the purposes of this discussion, we are interested in methodology and methods (with theory being seen as an aspect of methodology). Therefore, I prefer to talk about the evaluation of *espoused methodology*, meaning evaluation against stakeholders' interpretations of *methodology in use*.[157]

At this point I should admit that it took me a good while to really understand the importance of this kind of evaluation: for several years I undertook no post-operative evaluations of my interventions at all, other than personal reflections and occasional conversations with colleagues. As a result, I suspect my earlier intervention work was less sensitive to the effects of unconsciously learned assumptions than I might have realised. In the mid-1990s I began to take in the importance of dialogical reflection on interventions, and now I always try to hold at least one debriefing session after an intervention has been completed. During this session, amongst other things, I ask questions about whether any of my actions contradicted my expressed intentions.

However, it has only been in the last year that I have taken the need for this kind of evaluation seriously enough to do more than just post-operative evaluation: in my last couple of interventions, I have held regular sessions during the process to reflect on issues of methodology and practice, and I have also used questionnaires to elicit anonymous views that people might otherwise not have been willing to discuss openly.[158]

I have detailed my own failings in this regard partly to make clear that the model I have developed, although (in my view) useful to enable learning about methodology and methods, was not constructed in one go: it was developed over time, with some aspects (like the evaluation of espoused methodology) only being introduced after many years. The slow construction of the model essentially represents *learning about learning*. Acknowledging that my own learning (about

---

[157] In Chapter 7, I proposed a stakeholder theory which defines stakeholders more widely than is normally the case—including people whom agents think *ought* to be involved, as well as those already affected by, and/or involved in, a problematic situation.

[158] Here I need to acknowledge a debt to Alan Boyd, one of my research students at the University of Hull, who started to use questionnaires in this way before me, and convinced me of their utility.

methodology, and about learning itself) has taken place over time leads me to reiterate my earlier point that agents should not attempt to implement the whole model in one go: it is better to work at it incrementally, viewing the model as an ideal, otherwise the task looks too daunting. However, by benefiting from my own learning and the learning of others who have written about methodology, somebody who is newly interested in systemic intervention might learn much more quickly than I did! Also, by reflecting critically on the adequacy of my model (and others), you may produce new insights that, if communicated through books and papers, may result in more general improvements to systemic intervention.

11.6 Reflections on the Three Challenges

Earlier in this chapter I said that I have presented this model as my response to three challenges to methodological pluralism: the 'paradigm problem' (how can we mix methods drawn from a variety of incommensurate paradigms without getting into a philosophical muddle?); the 'psychological problem' (how can we minimise psychological resistance to methodological pluralism?); and the 'cultural problem' (will the intellectual climate enable ideas about pluralism to be taken on board?). Below, I show how the model does (or does not) take a position on each of these challenges.

11.6.1 Addressing the Paradigm Problem

My answer to the paradigm problem was clarified quite early on in this chapter: I do not believe it is possible to be 'meta-paradigmatic'—methodological pluralism involves us in setting up a new position which encourages learning about ideas from other paradigms, but reinterpreted in our own terms. This new position can be seen as a 'virtual' paradigm (Yolles, 1996), owned by an individual or small group—or, if it is shared sufficiently widely, it can be called a 'true' paradigm.

11.6.2 Minimising the Psychological Problem

My answer to the psychological problem of resistance to methodological pluralism, largely because of the wide span of knowledge it appears to require from the intervener, is to stress learning over time, starting from the knowledge base the intervener has at the point at which s/he realises the value of mixing methods. If this knowledge base consists of no more than one or two ideas from a single paradigm, then *that's a start*—s/he can reach out and begin learning from there. Even learning about an appropriate model of learning can be undertaken over time—but hopefully the model I have presented in this chapter (and others in the literature) will be helpful in this regard, as will more general writings on methodology (see, for example, Chapters 5-10).

11.6.3 Dealing with the Cultural Problem

The one challenge that the model in this chapter does *not* address is the cultural problem. Mingers and Brocklesby (1996) express doubts about whether the culture is right amongst academics and practitioners of systemic intervention for more than a minority to accept methodological pluralism. They talk about the need to establish new kinds of education programs to promote this kind of thinking, and this is certainly already happening: for example, the University of Hull offers an MA Management Systems degree specifically designed to introduce students to a wide range of methodologies and methods; to think about the theory and practice of their pluralistic use (e.g., by reflecting on ideas produced under the banner of Critical Systems Thinking); and to apply them in situations of direct relevance to themselves.

However, the bottom line in terms of cultural acceptance is whether or not methodological pluralism is perceived as adding value to people's current intervention practices (and practices, like scientific experimentation, which I regard as interventionary but others might not). I am personally convinced of this added value, especially if there is no expectation that interveners should enter the world with a widely informed, ready-made set of methods. These can be picked up through an on-going process of learning. Indeed, amongst a great many interveners the need for pluralism is no longer controversial: it is well established. The question is, what *type* of pluralism? My own plea is for a *critical* pluralism[159] that is theoretically informed and gives boundary critique a central role in intervention practice. In other words, my plea is fundamentally for methodological pluralism to be seen as an attribute of *systemic intervention*.

[159] The term 'critical pluralism' was first proposed by Mingers and Brocklesby (1996).

11.7 Conclusion

In this chapter I raised three problems which I suggested that proponents of methodological pluralism have to address if their practice is to be seen as credible: the paradigm problem, psychological resistance, and a lack of cultural readiness to accept pluralism. I then reviewed the works of a variety of authors who have tackled these problems, and found some useful ideas to take forward into a *model of learning*. This addresses the paradigm problem by making it clear that no pluralist methodology can exist without making its own paradigmatic assumptions. It deals with psychological resistance by talking in terms of *learning*, starting from wherever the agent is currently situated (there is no need to assume that a large knowledge base is needed to begin practising methodological pluralism). However, this model does not deal with the question of whether the time is right, culturally speaking, for methodological pluralism. In a sense, this is the task of the whole book, not just the model in isolation: I believe the case is strong that a systemic and pluralistic intervention adds significant value compared with other forms of intervention that do not concern themselves with boundary critique and methodological pluralism. I hope that this added value will become even more apparent in Section Three of the book, which presents a series of examples of systemic intervention in action.

III

Practice

12

Why Practice?

In this third section of the book, having discussed philosophy and methodology, I will now concentrate on practice—specifically, my own systemic intervention practice in the area of Community Operational Research (see Chapter 13 for an introduction to this), which I will use to provide illustrations of the methodological ideas that I have outlined in the preceding chapters. However, before doing this, I need to ask a key question (the same question that I asked about philosophy in Chapter 2, and methodology in Chapter 5)—why practice? Why be concerned about the implementation of any of the ideas I have described? Why not just put them forward as academic thoughts, and leave it to others to work out if they have any practical value?

In order to begin to answer these questions, I need to clarify the relationship between engaging in practice and writing about it, as it is this relationship that makes practice different from philosophy and methodology.

12.1 Practice and Discourses about *Practice*

There is engagement *in* practice (intervention for particular purposes) and there are discourses *about* practice, which take the form of written material and verbal conversations describing practice. Inevitably, the former cannot be understood and communicated except by interpretation through the latter. Equally inevitably, however, discourses *about* practice are meaningless if not related to engagement *in* practice. In contrast, philosophy and methodology are both primarily discursive: while they may (and in my view should) be informed by practice and its discourses, it is still possible (although I would argue inadvisable) to write about philosophy and methodology without practical engagement in anything other than philosophical and methodological ideas.

In this chapter, I am advocating engagement in both practice *and* discourses about practice. The latter are, of course, closely related to methodology, in that methodological ideas can influence the way that practice is both described and understood. I have already made the case for engaging with methodology (Chapter 5), so will not say much more about this here. What I will do, however, is concentrate on why engagement *in* practice is so important.

12.2 Why Practice?

Essentially, there are two different groups who may ask, why practice? The first are philosophers who like to keep their philosophy 'pure'. The second are methodologists (usually academics) who prefer not to get their hands dirty by moving from the *theory* of practice (methodology) to practice itself. I will not spend long dealing with the first of these groups (the 'pure' philosophers), as my answer to them is the same as my answer in defence of methodology (Chapter 5). However, I will refresh the reader's memory of my argument. Then I will use the bulk of the space in this chapter to highlight the limitations of 'pure' methodology.

12.3 The Argument against 'Pure' Philosophy

The argument for engaging in practice is the same as for talking about methodology. It is a moral argument. Given the scale of injustice, cruelty and greed in the world, is it really enough to indulge in thinking purely 'for its own sake'? Philosophers can choose between activity that is interesting and self-gratifying, but is ultimately little more than this, and activity that creates changes in the world which people other than philosophers may value (and which is *also* interesting and self-gratifying, but perhaps in a different way than 'pure' philosophy). Bearing in mind that philosophy and practice are both forms of *intervention* (see Chapter 6), we can rephrase this as a choice between intervention to change philosophical discourse, or intervention to change the conditions of life that people and other sentient beings experience.

In a way, posing the question as a simplistic choice between 'pure' philosophy and practice is just a rhetorical device to confront the reader with the fact that, if they choose the path of 'pure' philosophy, they are implicitly choosing *not* to follow other paths that may have more effect in terms of challenging injustice, cruelty, unsustainable lifestyles, etc. However, the choice need not (and indeed, I argue *should* not) be thought of as a binary one, between 'pure' philosophy and 'pure' (unreflective) practice. There are two senses in which the choice should be regarded as more complex:

First, it should be clear that, throughout this book, I have argued that philosophy, methodology and practice are *all* necessary for systemic intervention to flourish, and that each one of them should inform the other. Therefore, to engage in practice does not mean *abandoning* philosophy and methodology—it simply means allowing insights to flow between the three. Cutting off one area (practice) in which important insights may be generated in order to concentrate purely on philosophy and methodology is problematic, as is engaging in practice without philosophical and methodological reflection (see Chapter 5).

The second sense in which the choice between philosophy and practice is more complex than implied by this binary opposition follows from the observation made earlier that taking the path of 'pure' philosophy also involves a form of intervention—intervention in philosophical discourse. In this sense, it is itself a kind of practice, albeit a limited one compared with the many other forms of practice it is possible to engage in. Therefore, it would be contradictory to deny the value of philosophical intervention—but its value becomes most apparent when philosophy is allowed to inform methodology and (other) practice, and when the latter is allowed to feed back to inform philosophy.

Now, some people who embrace the arguments for seeing philosophy as an applied discipline may reply by saying that they agree that they may make a contribution by allowing their philosophy to inform methodology (and vice versa), but this should not mean that they should have to get their hands dirty by engaging in practice, beyond the (limited) activity of discussing philosophical and methodological ideas. Certainly, in Chapter 5 I argued that a principle means by which philosophers can make a meaningful social contribution is through methodology—but it is now time to widen the boundaries still further. The difficulty is that methodology formed in the absence of practice can be problematic, not to say dangerous! Here, we may enter the argument in favour of practice that is aimed at (mostly academic) methodologists who prefer not to get their hands dirty. This argument is also relevant to philosophers who may accept an extension of their remit to methodology, but resist going any further.

12.4 The Arguments against 'Pure' Methodology

One argument against 'pure' methodology (that is, theoretical understanding about methods, untainted by reflections on experiences of application) is quite simple: without engaging in practice, methodologists are not able to test for themselves whether methodological ideas work as suggested by their authors. Linked to this is a second argument: given that all methodological ideas are theoretical, and all theories are partial (they are 'ways of seeing' that assume particular boundary judgements, as described in Chapter 8), an unwillingness to engage in practice prevents methodologists from becoming aware of possible limitations of theories that can only be highlighted by understanding practice through discourses other than purely methodological ones. Let me go into some more detail about these issues, as they are not as straight-forward as they might at first appear.

Testing methodological ideas in practice is vital. I suspect—but this is only my interpretation, and there may be other views—that one reason for the controversy surrounding the System of Systems Methodologies (Jackson and Keys, 1984; Jackson, 1987b), described in Chapter 10, is that Jackson and Keys first developed it as a teaching aid, not as a support for intervention practice. Essentially, they were working at a purely theoretical level, trying to show students that different systems methodologies make different assumptions, and can be divided into four paradigms. For these teaching purposes the System of Systems Methodologies worked very well, but its creators then went further and suggested that it might also be a useful guide to methodology choice for intervention. Many people (including myself in those early days of the emergence of the third wave of systems thinking) hailed this as a breakthrough: on the face of it the System of Systems Methodologies seemed to be acceptable to both first and second wave systems thinkers (who were engaged in a paradigmatic 'war', and therefore tended to regard one another as profane). It was an attempt to show that all the main systems paradigms have a unique and complementary role to play in support of systems and operational research practice.[160]

[160] I say *on the face of it* the System of Systems Methodologies seemed to be acceptable to both first and second wave systems thinkers because, when I and many others with an intuitive (but not yet theoretically elaborated) commitment to methodological pluralism first read Jackson and Keys (1984), we believed that their argument would end the paradigmatic war. However, in the longer term, it turned out that many first and second wave systems thinkers resented having their ideas portrayed as part of a larger framework of methodologies. Therefore, it is fair to say that a new 'army' entered the war—this army being the paradigmatic perspective that would later be called Critical Systems Thinking (CST). Refer back to Chapter 11 for a further discussion of the paradigmatic nature of CST.

However, it was my own reflections on using the framework *in practice* (e.g., Midgley, 1988, 1989b, 1990a) that highlighted the problem of trying to select whole methodologies instead of trying to be more flexible and responsive by mixing methods. Similarly, Dutt (1994) reached the same conclusion. Without these practical reflections, the debate might never have moved on. Indeed, the idea of methodological pluralism in the management systems and sciences might have been still-born if there had been no willingness to allow methodology and practice to inform one another.

Of course, as we saw in Chapter 11, one problem with the idea of learning about methodology from practice is highlighted by Romm (1996): if practice is just looked at through one methodological 'lens', then only evidence supporting that methodology is likely to be seen. One answer to this is to look at practice through multiple methodological 'lenses'—acknowledging, of course, that one's reading of a methodology is not impartial: it is filtered through the intervener's interpretive framework of ideas so, while learning is still possible, the exercise can never be perfect (Gregory, 1992).

Also, I argued in Chapter 11 that a key route to learning through practice is for agents to consider the methodological implications of non-methodological descriptions about the experiences of the agent and others. In my view, reflection on these is crucial to the successful development of methodology. It is possible to follow Romm's (1996) advice and look at practice through a variety of methodological 'lenses' (and I believe this is valuable), but ultimately it may be the case that none of the methodological lenses make satisfactory sense of the perspectives of local participants in an intervention (or on-lookers who might have a different view). It is perfectly possible that their experiences, which are unlikely to be communicated in the form of methodological discourse, may transform one's own methodology if they are respected as salient.

An example of this kind of learning is my own participation, on several occasions at the beginning of my career, in participative group

---

My own feeling is that the 'war' is now effectively over. While it was very active and 'political' in the 1980s and early 1990s, I now experience fewer and fewer aggressive incidences. This is partly because the third wave systems thinkers are no longer 'upstarts' at the beginning of their careers (many now have positions of academic leadership), and the third wave is firmly established. It is also because methodological pluralism, which was once so controversial because of the dominance of neo-positivist thinking right into the 1980s (which resulted in the proscription of all but a narrow range of methods), is now well accepted by many interveners (if not traditional scientists who think in terms of 'observation'). This acceptance has been partnered by a breakdown in the hegemony of neo-positivist ideology (at least in the research communities which are explicitly engaged in intervention) and a proliferation of alternatives.

work with people with mental health problems: when staff and service users participated together, most found the experience valuable, but a minority of users found it quite distressing. Indeed, I witnessed one person have a panic attack when he was asked to contribute to an open discussion. Actually, this is not just a problem that affects people with mental health problems: anyone who is very shy or who lacks confidence in their own ideas can find the experience distressing. In 1997, this led me to argue that methodologists should be aware that, in some situations, and with some people, methods that have been designed to promote participation may actually be oppressive (Midgley, 1997c): there is a need to be sensitive to this when designing modes of participation. On occasion, I still work with people with mental health problems, but I take these issues into account when designing methods. In one case, for example, I worked with individuals and amalgamated their insights into a report (Midgley and Milne, 1995), while in another I worked with service users and professionals in separate groups (Cohen and Midgley, 1994; Chapter 16 in this volume). I would not entirely rule out the possibility of bringing staff and service users together, but I would be careful to make sure that participation is entirely voluntary; that the service users have space to develop their thinking separately from staff as well as with their participation; and that people know and trust each other enough to be comfortable in speaking publicly.

Without engagement in practice, it is simply not possible to have this kind of learning. Indeed, this learning is *essential* if unwanted and unanticipated side-effects of the implementation of methodological ideas are to be identified and addressed. It is for this reason that I claimed earlier that refusing to get your hands dirty, but still working on methodology (especially in an academic context where lecturers are responsible for communicating methodological ideas to students), can be positively dangerous. Unwittingly, oppressive practices may be passed from one generation to the next unless methodologists are willing to engage in practice and listen to others' experiences of the effects of their interventions.

12.5 Conclusion

In this chapter I have argued against both philosophical and methodological purism in favour of engagement in practice *alongside* philosophical and methodological reflection. Practice is about intervention to change the conditions of life that people and other sentient beings experience—philosophy, isolated from methodology and practice, cannot *directly* challenge injustice, cruelty and greed. Also, methodology that is developed without any relationship with practice can only be subject to a limited form of theoretical testing, and cannot be informed by the many non-methodological discourses about agents' and others' experiences of intervention.

In Chapters 14-17 I will detail a number of my own interventions, each of which illustrates one or more of the methodological ideas presented in Chapters 6-11. Before this, however, I need to provide some background about the discipline (Community Operational Research) that my practice is based in. This will be done next, in Chapter 13.

13

Community Operational Research

All of the examples of systemic intervention I will present in this third section of the book come from my own Community Operational Research practice ('Community OR' for short). To give a very broad definition, Community OR is intervention in the service of community development: working for improvement by dealing with issues that have a perceived negative effect on either the whole of, or sections of, local communities. As Community OR is a child of the wider Operational Research (OR) movement, most Community OR practitioners draw on OR and management systems methods—although (unsurprisingly, given the subject matter of this book) I argue that well developed methods are not enough on their own to enable someone to undertake Community OR in a satisfactory manner: Community OR needs to be a *systemic* intervention practice if agents wish to minimise the occurrence of unforeseen consequences of intervention. Boundary critique is essential to enable agents to reflect on the complexities of the issues they are trying to deal with, including different views surfaced by multiple stakeholders (see Chapter 7 for more details). Flexibility in the use of methods is also important, as complex twists and turns are often experienced during Community OR interventions, where new and unexpected dimensions to issues can be surfaced and have to be addressed in an on-going manner (see Chapters 9-11 for the arguments in favour of a pluralistic use of methods).

Having declared that all of my examples of systemic intervention (to be presented in Chapters 14 to 17) are drawn from my Community OR practice, I should be clear that this does *not* imply that systemic intervention, as I have described it, is *only* of use in Community OR. Far from it. I hope it is clear that boundary critique; theoretical and methodological pluralism; mixing methods; and on-going learning about methodology are relevant across the board for all kinds of intervention—including some kinds, like those practised in the traditional sciences, which are not usually thought of as interventionary at all.[161] However, it is inevitably the case that interveners specialise in addressing certain kinds of issue, even if their systems practice strays across disciplinary boundaries.[162] This happens partly because people have specific interests that they wish to pursue; partly because the roles they assume in organisations may require them to specialise; and partly because, as people gain a reputation for dealing with certain kinds of issue, they can get 'type-caste' (whether they want to be or not).[163]

In this short chapter I provide more details of Community OR so that my own specialism is made clear. I then make some more general comments about professional identities and disciplinary knowledge: I recognise that, within Western culture, professionals are subject to disciplinary constraints, but I highlight the need to be critical of these (which involves clarifying the constraints and engaging with them in a strategic manner).[164]

---

[161] While I argue that the ideas expressed earlier in this book have a wider application than Community OR alone, it is inevitably the case that the limitations of my practical specialisation will have been reflected in my presentations of philosophy and methodology. I therefore invite readers with different practical specialisations to reflect on what would have to be added to my methodology to cater for different contexts. I could say 'what would have to be *different* from my methodology' (rather than *added* to it), but I believe that even some radically different forms of practice, like the use of scientific methods (which I have had several years experience of applying within the discipline of psychology), would be enhanced by regarding them as systemic intervention.

[162] See Midgley (1998) for an argument for transdisciplinary systems research that nevertheless still allows specialisation.

[163] For several years I was 'type-caste' as a disability and mental health researcher, partly because I used to work in the Rehabilitation Resource Centre (RRC) at City University (London), so disability projects came my way as a matter of course; partly because I studied Psychology and worked as a Residential Social Worker in a therapeutic community before discovering systems thinking, so people wanting interventions in mental health services sought me out in preference to more general disability researchers; and partly because, even when I had left the RRC, my reputation in the field resulted in it being much easier to secure work in this area than others in which I had relatively little experience. It took a conscious effort over a period of years, working on more general social issues, to establish a new reputation in Community OR. At the present time I am also trying to extend into the field of environmental planning and management (but without leaving Community OR behind), so I anticipate a further period in which I have to establish my credibility.

[164] In my view, these constraints are not absolutely inevitable, but they are so well institutionalised in Western societies that it will undoubtedly take many years of strategic action to free ourselves from them. Incidentally, this discussion of disciplinary constraints is not meant to imply that only 'professionals' can engage in systemic intervention. Later, I will clarify this point further.

13.1 *The Origins of Community OR*

As we saw in Chapter 9, Operational Research (OR) was born in the Second World War: it was an inter-disciplinary, applied science originally harnessed to make the war effort more efficient and effective. After the war, the methods developed by operational researchers were transferred into industry, with some success. Although it had its origins very much in the quantitative, applied science tradition, in the 1960s and 1970s many OR practitioners embraced a paradigm shift which took them on a similar path to the one followed by second wave systems thinkers: they developed a variety of problem structuring methods which were essentially about facilitating debate around possible models for action (as opposed to the expert-led, quantitative modelling of scenarios).[165] It was around this time that a number of exponents of OR realised that their methods (both quantitative and qualitative) might be useful for community development as well as in the more usual business and public sector contexts. In the United States, OR practitioners have worked with community groups since the late 1960s (e.g., Ackoff, 1970) and in the UK since the mid-1970s (e.g., Noad and King, 1977; Trist and Burgess, 1978).

Although this is the case, it was not until 1986 that the term 'Community OR' was first coined (Rosenhead, 1986)—and labelling Community OR in this way facilitated a significant expansion of the number of community-based interventions. Funding from the Operational Research Society was secured to establish a research centre (the Community OR Unit) at Northern College in the UK, and this opened in 1988.[166] In 1987, the Community OR Network (a study group of the Operational Research Society) was formed (with over 300 members), and the Centre for Community OR was opened at the University of Hull (later to be merged into the Centre for Systems Studies, where I am based, resulting in a massive expansion of Community OR activity at Hull). Since 1987, there have been Community OR streams at many UK and international conferences, and two conferences have been dedicated solely to Community OR. More detailed histories of the institutional development of the Community OR movement can be found elsewhere (Parry and Mingers, 1991; Mar Molinero, 1992; Ritchie, 1994; Ritchie *et al*, 1994).

---

[165] Refer back to Chapter 9 for further details.

[166] For anybody wishing to contact the Community OR Unit, I should note that it is no longer at Northern College: it is now at the Lincoln Campus of the University of Lincolnshire and Humberside.

13.2 Motivations for Involvement in Community OR

Several papers have been written about people's motivations for getting involved in Community OR. Wong and Mingers (1994) surveyed a number of practitioners, and concluded that the following motivations are most common: 'doing good' in the community; broadening the client base of OR (see also Rosenhead, 1986); broadening and developing the set of methods available to OR practitioners (see also Rosenhead, 1986; Keys, 1987); promoting methodological pluralism (see also Jackson, 1988); the challenge of a new experience; practising new techniques; trying out old techniques in a new environment; promoting community OR as a discipline; gaining a qualification (or some other professional reason); and practising OR in the locality in which one lives.

However, it is noticeable that there is relatively little documentation about the *political* motivations people have had in building the institutions of Community OR and engaging in its practice, except for some quite general comments about 'doing good' (see above).[167] This is despite the fact that many of us who are active in the Community OR research community know that some people's motivations have included the desire to promote socialist revolution; to serve God; to develop participative democracy; to halt the advance of capitalist ideology; to reconstitute civil society; or several of these at once. For this reason, Midgley and Ochoa-Arias (1999) argue that Community OR practitioners should be explicit about the normative vision(s) of community they wish to promote, rather than hiding their political interests behind the rather more 'mundane' motivations surfaced in Wong and Mingers's (1994) survey.

Midgley and Ochoa-Arias (1999) discuss eight different forms of community politics: two forms of liberalism; two variants on Marxism; and four kinds of communitarianism (see the original literature for definitions). They argue that different forms of Community OR support different visions of community, and if practitioners fail to reflect on these then it is likely that they will slip uncritically into a form of Community OR that automatically supports the cultural norm (in most countries where Community OR is practised, this is welfare liberalism). Nevertheless, Midgley and Ochoa-Arias do not suggest that practitioners are *bound* to choose between these eight political positions: they may also use the power of reflective thought to develop something different. It is my view that the understanding of systemic

[167] Indeed, there are only a handful of writers in OR more generally who have publicly expressed their own political motivations, perhaps the best known being Rosenhead and Thunhurst (1982) and Rosenhead (1986, 1987) who argue for the creation of a "workers' science"—but these papers do not relate this "worker's science" to Community OR.

intervention presented in this book can provide the basis for a new political position that is not wholly liberal, Marxist or communitarian, but enables the transcendence of these categories. However, this is an argument for another day.[168]

13.3 The Breadth of Community OR Practice

Community OR practice is actually very varied. Some intervention work is carried out solely with community groups (the term 'community group' is used in the UK to denote a group of people who are organising or campaigning in their local community on a voluntary basis without significant funds). See Mar Molinero (1993), White and Taket (1994) and Wilsdon (1994) for some examples. However, Community OR interventions have also been conducted with the British National Health Service (e.g., Taket, 1994a); voluntary organisations with paid staff (e.g., Gregory and Jackson, 1992a,b); local government bodies (e.g., Vahl, 1994; Midgley *et al*, 1996); and multi-agency groups (e.g., Barr and Vangen, 1994; Cohen and Midgley, 1994; Pindar, 1994; Midgley and Milne, 1995; Gregory and Midgley, 2000). See Ritchie *et al* (1994) for 26 case studies that reflect some of the breadth of Community OR activity.

Early Community OR tended to follow the model of mainstream OR, having a 'client' whom the intervener serves—except in the case of Community OR the client was likely to be a community group or voluntary organisation rather than a business or public sector organisation. However, Midgley *et al* (1994) argue that it is more appropriate for Community OR practitioners to see themselves as working on an *issue* as opposed to serving a client: this allows a range of organisational and non-organisational participants to be swept in, depending on the nature of the issue, and it ensures that no one agent (even a community group and/or the intervener him/herself) can dominate the intervention in an unaccountable manner. All participants are encouraged to engage in critical reflection on what should be done.

While there are different views about the extent to which Community OR can legitimately engage with the agendas of public sector and business organisations, as opposed to the agendas that emerge solely from communities of people outside these organisations (whether they are organised into community groups or not), it is arguably the case that the majority of practitioners now recognise the need to engage with a *network* of organisational and non-organisational interests in order to

[168] It is my intention to write another book on this subject, probably in two or three years time.

address issues of concern to members of local communities. However, as I see it, a common commitment of all Community OR practitioners is to a practice of *meaningful* community participation: if public sector and/or business organisations are involved (especially if they are funding the intervention), then the agenda must take seriously the views of people outside these organisations (preferably through their direct participation, but at the very least through a genuinely open process of consultation). For example, community operational researchers may get involved in the planning and evaluation of statutory services—but, for this to be called Community OR, the users of these services (and possibly other people affected by them) have to play an influential role.[169]

In terms of methods, generally speaking there seems to be a greater emphasis on the use of problem structuring and second wave systems approaches than on first wave ideas and the more quantitative OR techniques. This is partly because of the emphasis on participation, but also because community issues are often perceived as more complex and multi-faceted (with a variety of stakeholder perspectives impacting upon them) than business problems (Jackson, 1988).[170] However, there is certainly no exclusion of quantitative methods, although the focus on participation means that the intervener has to think carefully about the process of application to be used: some quantitative methods lend themselves to an expert-led style of intervention which has to be countered, or balanced out by the complementary use of different methods to ensure that the intervener does not make other participants dependent upon him or her (Midgley *et al*, 1994). See Thunhurst *et al* (1992a,b) for examples of the use of quantitative methods as part of participative Community OR practice.

---

[169] Arguably, the only exception to this is when an organisation wishes to rethink its activities in ways that will enable greater participation in the future. On occasion, I have taken on projects with narrow boundaries of participation in the knowledge that one of the explicit purposes being pursued is to consider this kind of reorientation, which will widen participation for the future.

[170] This is not to say that business problems are never perceived as complex and multi-faceted (many obviously are)—just that community issues are rarely defined in a narrowly focused manner. Arguably, the profit motive for business makes it acceptable, in some people's eyes, to use one heavily dominant output measure—financial return. This makes simplistic definitions of problems more frequent in business practice, although (from a systemic point of view) these can often take attention away from other significant matters: e.g., social and environmental issues that are not seen as having an immediate or easily quantifiable financial impact.

13.4 *The Community Operational Researcher as an Agent*

Before closing this discussion of Community OR, it is important for me to highlight a key aspect of Community OR practice: in the vast majority of cases, the practitioner is invited in by an organisation or group to offer his or her services to help address a particular issue.[171] Therefore, the *agency of the practitioner* is pivotal to the intervention, even if this agency is exercised to generate wide-spread participation, ultimately resulting in *group* agency for change.

In Chapter 6, I argued for a critical attitude to the boundaries of agency: the agent may be seen as an individual or group (defined in accordance with the theories that are drawn upon) acting under the influence of a wider knowledge generating system, which may be bounded in many possible ways. This is obviously relevant to Community OR practice, in that the thoughts and actions of the agent who is invited in by an organisation or group to engage in intervention can indeed be seen as influenced by many possible knowledge generating systems. However, while the boundaries of the knowledge generating system(s) the agent may be part of are massively variable, the identity of the agent him/herself is less so (given the current constraints of Western culture). As Community OR is a *discipline* (which I define as a heterogeneous body of knowledge used by an identifiable professional community), the key agent is inevitably seen as a 'Community OR practitioner'. While this may have a variety of meanings, depending on the Community OR theory (or theories) being used to understand the identity of the practitioner, the variety is constrained: it is strongly influenced by the theoretical variety available within the discipline that prescribes possible identities; the desires (or lack of them) of practitioners to draw on other forms of knowledge to build new identities; the normative forces within the disciplinary community which act to legitimise, marginalise or exclude certain theories and identities; and the normative forces in wider society which put pressure on the disciplinary community and the individual practitioner. See Foucault (1980) for some fascinating comments on disciplinary knowledge.

I am making an issue of this for two reasons. First, while the idea of being critical about the boundaries of agency is vital, when we recognise that an important boundary of the wider knowledge generating system is disciplinary, and this results in a primary focus on

---

[171] There are exceptions to this, such as Taket's (1994b) use of OR techniques with a feminist collective of which she was already a member: in this case, she felt that there was no need to declare the fact that she was practising Community OR. However, this is a relatively rare way of operating.

the actions of professional agents (in this case Community OR practitioners), it is essential to be aware of the consequences of this: there will inevitably be disciplinary constraints on the extent to which some critiques of the identity of the agent will be regarded as legitimate. Nevertheless, there *are* still opportunities for critique: as I said earlier, disciplines are not homogeneous, and (in most Western societies) neither are the wider social forces impacting upon them. However, to take advantage of the opportunities in a critical and systemic manner requires the interplay between disciplinary constraints and social forces to be made visible so that strategic arguments for new practices and ways of looking at identity and intervention can be constructed.

While I would hope that many of the theoretical and practical ideas presented in this book have a critical edge to them, I cannot claim, in my write-ups of practice, to have moved beyond a primary focus on the actions of the Community OR practitioner. There are two sides to this admission. On the one hand, there is still considerable scope for exploring the roles of other kinds of agent (e.g., groups, organisations, communities and ecosystems) in Community OR and other intervention practices. On the other hand, there is a positive side to keeping the role of the professional agent visible: professionals undoubtedly have the capacity to influence the course of interventions in ways that may or may not be acceptable to other stakeholders, so de-emphasising the focus on professional agency to the point where professional identities become invisible could be dangerous. Therefore, I make no apology for my primary focus in Chapters 14-17 on my own agency in interventions, but I do acknowledge that this focus could be complemented in future writings by thinking in terms of other boundaries of agency.[172]

The second reason for making an issue of these disciplinary constraints is to explain the limits of my own intervention practice (in my role as Community OR practitioner) detailed through the examples in Chapters 14 to 17. I have argued that a wide range of practices, including supposedly observational science, can and should be seen as interventionary (Chapter 6). However, my own intervention practice does not span the full breadth of these practices. As I made clear in my discussion of methodological variety (Chapter 9), it is impossible for

[172] Reynolds (1998) provides an interesting analysis of how professional expertise can be 'democratised' during interventions: he argues that professionals should open their expertise to lay analysis, and should work with others to set the agenda of interventions in a co-operative manner. It is important to note that doing this requires the agency of the professional to remain visible (Reynolds, 1999). In the view of both Reynolds (1999) and myself, it is far too early to proclaim the "death of the expert" (White and Taket, 1993).

any one person to be aware of the full variety of methodologies and methods available. My own practical focus has primarily been on the use of management systems and OR methods, although I sweep in a more limited knowledge of other methodological ideas too (e.g., applied social science and action research). As these methodological ideas are channelled into my Community OR practice, the extent of my interventions is inevitably subject to limitations. Ultimately, these limitations can be seen as a function of the dynamic interplay between the disciplinary Community OR community (acting in relation to the knowledge base of Community OR) and my own critical engagements with the boundaries of my practice and my personal and professional identity. I do not pretend to be all-encompassing in my own practice, even though I advocate the use of a plurality of intervention methods and have a desire to continually push out the boundaries of my engagements (and thereby the possibilities for defining the disciplinary and professional identity of the Community OR practitioner). However, I do claim that I embrace more variety of methods in my practice than some other interveners who choose to work solely with isolationist methodologies.

I suggest that other people, acting with other professional identities and drawing upon different disciplinary knowledge bases, will be similarly constrained, albeit in different ways. However, this does not invalidate the idea of pushing out the boundaries of practice and engaging in systemic intervention. It simply means that engaging in systemic intervention entails a learning process (see Chapter 11); it means recognising the constraints imposed by professional identities so that conscious decisions can be taken about whether and when to challenge them; and it also means becoming aware of the disciplinary legitimation processes that act to marginalise or outlaw certain innovations, making it necessary to think strategically about how new ideas are best communicated to others.

Finally, I should make one last (but very important) comment. I have focused in this section on professional identity and disciplinary constraints, primarily because these issues are crucial to my own situation as a Community OR practitioner. However, *it should not be assumed that the arguments of this book are only of relevance to professionals engaging in disciplinary practice.* Any other human agent (individual, group or organisation) may engage in systemic intervention, however they define themselves. For such agents, the arguments of this book should still be relevant, even if they are not subject to specifically disciplinary constraints. Nevertheless, just because they are not part of a disciplinary community does not mean they are *un*constrained. The constraints will simply be different, and the need to identify them in order to liberate the potential for critique will be just as important.

13.5 Conclusion

In this short chapter, I have introduced the reader to Community OR, which I practice as systemic intervention in the service of community development. All the examples of systemic interventions I provide in Chapters 14 to 17 are from Community OR projects. The discipline of Community OR assumes a focus on the agency of the professional intervener, which is why it is possible for me to talk about the practice represented in Chapters 14 to 17 as *my* practice. There is, of course, a positive side to this: professional agency should not become invisible while the professional has such an influential role in society, although I acknowledge that the roles of other forms of agency could and should be further explored in future writings.

I also need to reiterate that, just because I have focused my own practice primarily on Community OR, this does not imply that the arguments in Sections One and Two of this book are only relevant to Community OR contexts. I believe that they have much broader applicability. Nevertheless, any disciplinary specialism will impose its own restrictions on practice, and it is important that people work to raise their awareness of these so that they can push the boundaries of their professional identities and practices and thereby enable increasingly critical and flexible forms of systemic intervention to emerge.

14

Developing Housing Services for Older People

In this and the next three chapters I will present examples of my systemic intervention practice. The first example (this chapter) will illustrate boundary critique, while the other three will each illustrate different aspects of the choice/design of methods (methodological pluralism). Of course, in all four interventions (and the many others I have engaged in over the years), *both* boundary critique and methodological pluralism were involved: the fact that the interventions have all been used to illustrate different aspects of methodology is merely a matter of emphasis, not substantive difference. Indeed, it will become evident later in this chapter that a key way in which boundary critique can be pursued is through the *design of methods* to allow both Community OR practitioner(s) and other stakeholders to explore, and choose between, boundaries and associated values.

14.1 Boundary Critique

A full methodological discussion of boundary critique, the practice of which is illustrated shortly, was provided in Chapter 7. To remind the reader, Churchman (e.g., 1970) introduced the fundamental idea that the boundaries of analysis are crucial in determining how improvement will be defined during systemic intervention, and hence what actions will be taken. He also argued that pushing out the boundaries to make intervention more inclusive may well involve sweeping in new stakeholders. Then Ulrich (1983) built on this by pointing out the need to rationally justify the setting of boundaries: he suggested, following Habermas (1976), that rationality is dialogical. Therefore, if boundaries are to be established rationally, they should be defined in dialogue by all those involved in and affected by the intervention. I then built on the work of Ulrich by examining the systemic forces that work to stabilise conflictual situations, and produced a model (Figure 7.3) of marginalisation processes that can be used to inform critical reflection during interventions. This focuses on how marginalised elements (people and/or issues) may assume a 'sacred' or 'profane' status, allowing action to be taken by those ascribing this status without resolving the wider ethical conflict that these ascriptions reflect. Yolles (1999a, 2000) has since elaborated this model. Of course, there are some differences of opinion between Churchman, Ulrich, Yolles and myself over philosophy and methodology, but the development of boundary critique (albeit seen from slightly different angles) has been a common theoretical interest that we all argue is of significant value to systemic intervention.

14.2 The Initial Remit of the Intervention

The intervention I shall use to illustrate boundary critique was commissioned by the Joseph Rowntree Foundation[173], which is a UK-based charity that funds research for social benefit.[174] Rowntree invited the submission of research proposals to discover how information from assessments of the housing needs of individual older people (which are always conducted before service delivery to determine what actual services should be provided) could most effectively be aggregated and used in the development of housing policy. As I was already aware (from my involvement in previous projects) that information from assessments is rarely used in this way, I started with the idea that there may well be significant problems obstructing the aggregation process. My initial proposal, which was accepted by Rowntree, was to conduct two phases of research: in phase one, I was to interview stakeholders in two geographical areas to identify problems of assessment, information provision and planning. I then proposed working in partnership with stakeholders in the second phase to design improvements to information provision.

The boundaries of *what* I would look at were therefore quite clearly defined: the problem identification phase would focus quite

[173] While the Joseph Rowntree Foundation supported the project reported upon in this chapter, the material presented here represents the views of my intervention team, not necessarily those of the Foundation.

[174] I have chosen this particular intervention to illustrate boundary critique, rather than one of the many others I have conducted, because it involved dealing with some particularly complex boundary issues.

broadly on assessment, information provision and planning, while the later design phase would narrow down on information provision, seeing assessment and planning problems in relation to this. In terms of *whose views* were to be included, I deliberately left it vague, saying that 'a broad range of stakeholders' would be interviewed and involved in the design process. However, I made it clear that one group who would definitely be involved would be older people—the clients of the housing services. This had to be stipulated in advance for two reasons. First, the exclusion of service users could well have resulted in a design proposal that failed to meet their needs. In this respect, the decision to include the views of older people reflected both Churchman's general 'sweeping in' of stakeholder concerns, and Ulrich's inclusion of the 'affected' along with the 'involved'. Second, clients of service systems very often have their views marginalised and made profane. This allows professional discourses to maintain a sacred status and thereby dominate the business of service delivery (see also Thompson, 1995). It was important for local government officials to know right from the start that the intervention would promote the views of clients alongside other views. The decision to involve older people therefore reflected my theoretical understanding of marginalisation (and, of course, the disciplinary focus of Community OR on meaningful community involvement).

I was funded for an 18 month period to conduct the intervention, and brought together a team of three Community OR practitioners to undertake the work: Isaac Munlo (a Ph.D. student under my supervision who had recently graduated with an MA in Management Systems, and who had substantial experience of rural development in Africa); Mandy Brown (a colleague in the Centre for Systems Studies at the University of Hull who had experience in applying a wide range of systems methods in both business and community contexts); and myself. Rowntree also convened an Advisory Group (made up of housing managers, specialists in the development of services for older people, and other Community OR practitioners) to oversee the project.

14.3 Phase One: Identifying Problems

Commitment was obtained from the Housing and Social Services Departments[175] in two geographical areas to 'sponsor' the intervention:

[175] In the UK, Housing Departments are part of the local government apparatus. They provide low-rent, publicly owned accommodation, and specialist housing tailored to the needs of specific groups (e.g., older people who are no longer able to live independently). Social Services Departments, which are also part of local government, provide a wide variety of non-financial services, such as aids and adaptations for people with disabilities,

that is, to act on our behalf in making first contacts with other key agencies. This gave us some initial points of entry, but there was a need to define in more detail who should be interviewed in the problem identification phase. We already knew that a diverse variety of agencies were involved in housing for older people (local and regional government, the National Health Service, housing associations, voluntary organisations, private building companies, etc.)—far too many to identify right from the start. We therefore used a method devised by Midgley and Milne (1995) for 'rolling out' the boundaries of who was to be interviewed. Starting with those categories of people whom we could easily identify, we asked interviewees to name others whom we should talk to, either because they were also stakeholders in the system, or because they had a different view to the interviewee. We also used some of Ulrich's (1983) questions from the methodology of Critical Systems Heuristics to identify stakeholders. In particular, asking who is involved in, or affected by, the interviewee's activities helped to reveal stakeholders whom we might not otherwise have identified. In addition, we asked for examples of specific decisions regarding assessment, information provision or planning that people had made, and we 'mapped' the subjectively perceived variables that had been assessed in the decision-making process using Cognitive Mapping (Eden, 1988). Stakeholders involved in, and/or affected by, interviewees' activities were identified through analyses of the maps. The boundaries defining who was to be interviewed were finally established when interviewees stopped bringing up new names of relevant agencies or stakeholder groups. Altogether, 131 people were interviewed.

After about twenty interviews had been conducted, a significant issue arose with regard to the initial remit of the project. A relatively wide boundary had been established for the first phase (looking at problems of assessment, information provision and planning), but in the second phase the boundaries were supposed to narrow so that the focus would primarily be on finding solutions to the problems of information provision. However, it became very clear that the problems people were identifying in the areas of assessment and multi-agency planning were so important that to narrow the focus in the design phase to issues of information provision alone would mean ignoring the concerns of many stakeholders. In terms of assessment, many older people were worried about a perceived mismatch between what they requested in

child protection, day care for people with mental health problems, etc. There is some overlap with health and welfare services, but Social Services are financed through local taxes rather than general taxation or national insurance.

assessments and what was actually provided in the way of housing services, and this was an issue that would be marginalised by the focus in the second phase on information only. Also, managers claimed that there were significant difficulties with multi-agency working. If the agencies found co-operation in the areas of policy making and planning problematic, then the issue of providing better information was really of secondary importance. We therefore faced an ethical dilemma: continue along the same path, which would be 'safe' in the sense that it had been agreed with the funding body, or widen the boundaries in the design phase to ensure that the resulting proposals for change would deal with the larger problems that stakeholders had identified. We tackled this dilemma by convening a meeting of our Advisory Group at the Rowntree offices, at which the ethical consequences of adopting the various boundaries were explored. It was agreed that the boundaries of the intervention should indeed be widened.[176]

The perception of this dilemma as an essentially ethical one reflects Churchman's insight that 'improvements' look very different through the eyes of different stakeholders: had we not rolled out the boundaries of who we interviewed, some of the key problems of assessment and planning might never have become visible, and the resulting 'improvements' would have seemed irrelevant to most stakeholders. Indeed, to have focused on information alone might have intensified some of the problems in the assessment system. One of our key findings was that needs that could not be met given current resources and spending priorities were not even recorded during assessments, making them invisible (and legislation prevented any changes to this practice). To have aggregated the information from assessments for use in planning would have presented a false picture to policy makers of a set of services meeting everybody's needs (Midgley *et al*, 1997). The narrow boundary around information issues would therefore only have generated an improvement in the eyes of those wanting to close down debate on the appropriate level of public spending on housing services for older people. To everybody else, this would have been the very opposite of an improvement.

Having widened the boundaries of the project, we continued interviewing stakeholders. We then extracted all the information about problems in the areas of assessment, information provision and

---

[176] Ormerod (1999) says we were fortunate that we were allowed to widen the brief of the project when we came across problems that were not originally taken into account, and that in another situation we might have been asked to 'deliver as promised'. I must say that I always try to make it clear from the beginning of a project that, if unforeseen issues emerge, the course of the project might need to be changed. My experience is that managers generally appreciate why this should be the case.

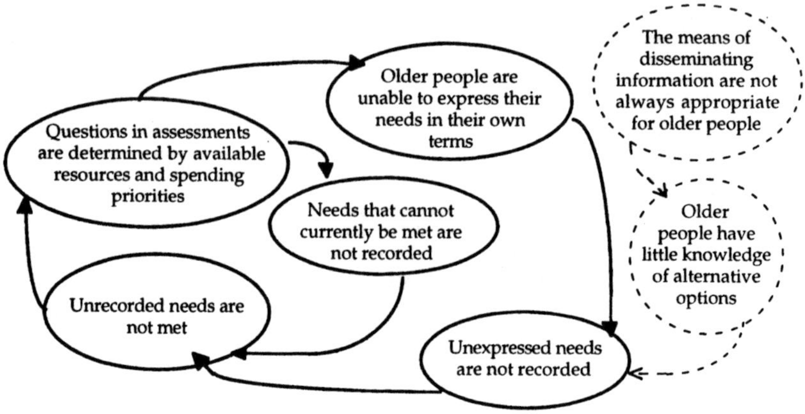

FIGURE 14.1: Problem map of assessment issues in one geographical area (after Midgley et al, *1997)*

planning from the interview transcripts. Once we had a list of key problems, we were in a position to look at how these related together. Taking each problem in turn, we asked if and how each of the other problems impacted upon it. Over the course of a day we developed two 'maps' (one for each geographical area) showing relationships between issues (see Midgley *et al*, 1997, for specific details of our method). We then proposed to the Housing and Social Services Departments who had sponsored our intervention that the resulting 'problem maps' should be used as the basis for presentations of our findings in workshops to decide what should be done in the second phase of the project to produce recommendations for improvement. One example of just part of a problem map is shown in Figure 14.1 (the whole maps, which are reproduced in full in Midgley *et al*, 1997, are about three times as large).

14.4 Designing the Methods for the Second Phase

At this point, due to personnel changes, we lost the commitment of the agencies in one of the geographical areas. Despite repeated attempts to make contact with officers to set up a meeting, we received no response. However, in the other area, the Housing and Social Services Departments (who had acted as our sponsors) were keen to go ahead. A workshop was set up to present our findings and discuss ways forward.

In planning this workshop, another boundary issue surfaced.

Because it was already clear from the maps that some sensitive issues concerning service provision and inter-agency relationships were to be discussed, the Housing and Social Services Departments made it clear that they wanted their own managers to be the only people to participate. This was understandable given the fact that the first phase of the intervention had been solely problem-focused: the statutory agencies were afraid of publicity being given to weaknesses in the system without any counter-balancing information about strengths. However, this meant that, unless we were very careful, other agencies, and indeed the older people themselves, could have had their concerns marginalised. Once again we were faced with an ethical dilemma. In terms of the theory of boundary critique, there was a risk that our intervention could reinforce the marginalisation, and hence the profane status, of service users—a risk that we perceived as significant, as one of the problems we had identified through our interviews was resistance on the part of a minority of professionals to hearing users' views.

We handled this ethical dilemma by accepting that the workshop would only be attended by managers from Housing and Social Services, but we also looked at ways in which we could prevent the concerns of other stakeholders being marginalised. We came up with a strategy to achieve this. First, we looked at our own strengths and weaknesses as a team and assigned ourselves different roles: one of us gave the presentation (which contained the information that we anticipated would be seen as controversial), and also chaired the discussion on ways forward; a second person paid particular attention to emotional interactions in order to address conflicts between the person making the presentation and agency representatives; and the third one of us acted as an advocate for other stakeholders who were not directly represented. In addition, we decided to make a conscious effort to ask the managers from Housing and Social Services to place themselves in the shoes of other stakeholders and speak on their behalf. Although the boundaries surrounding who was involved in the workshop were quite narrow, we felt that it would still be possible to ensure that wider concerns were addressed.

When we held the workshop, our belief that it would still be possible to introduce wider concerns by facilitating debate in the manner described above turned out to be justified. A key principle to guide the design of methods for the second phase of the intervention was established at the meeting: any recommendations for improvement that were to be generated through our intervention should be based on a vision of the ideal service produced by stakeholders, including older people and their carers. 'Ideal', in this context, means the best *possible* service, not one that is unattainably perfect (Ackoff, 1981).

Following some facilitated discussion of the problems in the system, and possible ways to address them in the second phase of the project, we asked for ten minutes in which to develop a more concrete proposal for the intervention methods to be used. This was readily granted, and we held our discussion in front of the other participants so as to sustain the atmosphere of mutual trust that had by now developed. Our sudden lapse into systems jargon produced laughter all round! We then provided an impromptu thirty minute presentation of our proposal in plain English, which was modified through further discussions.[177] The final result was a proposal that has been written up in Midgley and Munlo (1996).

This technique of debating methods in the presence of participants, making a joke of the jargon, and ensuring translation into plain English, was designed to prevent, as far as possible, the marginalisation of the Housing and Social Services managers in the process of designing the methods. As professional interveners, we needed to introduce our 'expertise' into the situation: the other participants did not have the same knowledge of systems and operational research methods. However, it was important that this should be done in such a manner as to enhance the resources of the whole group rather than to enforce our own 'sacred' status as Community OR practitioners, thereby making participants feel profane, disempowered and resentful. To have done the latter could have jeopardised the entire second phase of the project: ultimately, the commitment of this group of participants would be needed for the implementation of recommendations for improvement. Here, our approach to decision-making about methods embodied the concern of both Churchman and Ulrich (as well as more recent writers, such as White and Taket, 1993, and Reynolds, 1998) that 'expertise' should be opened up to challenge, whether it is the expertise of a stakeholder group or the expertise of professional interveners. Our

[177] Ormerod (1999) argues that we need not have conducted a debate on systems methodology in front of participants which then had to be translated into plain English—we could have talked in plain English from the start. He has a point, and in other circumstances I would have done this. However, the idea of taking time out to discuss methodology amongst ourselves was my own suggestion stemming from my knowledge that this was the first time that the three of us had worked together as a team. I was developing an idea for a set of methods in my mind, but felt that to have taken the time to spell these out in plain English would have committed the team before I was sure that we were in agreement. Talking in jargon, and using abbreviations (CSH, IP, VSM, etc.), is the quickest means I know of checking that the whole team is on the same wavelength. We explained why this was necessary to the other participants; asked permission before doing it; and the result was that it was treated as a big joke. Of course, we then made a plain English presentation straight away. Given the necessity to balance the need for the team to harmonise its understandings against the other participants' need for accessibility to decision-making, I think we achieved a fair compromise.

approach also reflected my own recognition of the need to counter processes of marginalisation: this time it was the potential marginalisation of the managers by our Community OR team rather than the marginalisation of service users by managers.

14.5 Phase Two: Designing Improvements

The second phase of our intervention is described over the coming pages. We used the methods that had been agreed with the Housing and Social Services managers in the workshop described above, but one adjustment was made: the intervention was to be based on a series of further workshops with stakeholders, but the number of these had to be reduced to allow us to finish the project within the deadline set by the Joseph Rowntree Foundation. This did not involve reducing the number of participants because we simply combined groups together whom we originally intended to work with separately, thus creating double-size groups (which were still small enough to provide a reasonably intimate atmosphere). Agreement was sought from the Housing and Social Services Departments before the change was made. It turned out that holding fewer workshops suited them because it made it easier for them to guarantee the participation of key players.

The first phase of our intervention (described above) concluded that the problems faced in the housing system for older people were highly interactive (see Midgley *et al*, 1997, for details): it would be impossible to design an effective solution to one problem without considering the effects of all the others. In consequence, the second phase started by taking an *overview* of how the situation could be improved. We began by looking at what stakeholders thought the desired properties of an *ideal* housing system should be, and went on to support managers from the statutory agencies in designing a form of multi-agency organisation that could deliver services which would work towards this ideal. As Ackoff (1981) makes clear, the word 'ideal' means the best *possible* practice, given current technology and the need to maintain viable and adaptable organisations.

To provide some confidential space, especially for users and carers, it was initially decided to run three separate workshops. The first was conducted with older people in receipt of housing services; the second with carers and representatives from relevant community groups and voluntary organisations; and the third drew together managers and front-line professionals from the statutory agencies involved in housing. By beginning with these three stakeholder groups, the boundaries of participation were immediately widened beyond the small gathering of managers who had been involved with us in designing the methods for the second phase. The inclusion of a variety of stakeholders in this manner reflected the insight of Churchman that a more comprehensive understanding can be generated by considering a variety of views than by working with a small group in isolation. It also reflected Ulrich's idea that both the views of the involved *and* the affected need to be accounted for if the results of planning are to be considered normatively acceptable.

The provision of confidential space for the various stakeholder groups allowed users and carers to develop their views outside the hearing of professionals before communicating them to others. This practice, which is common to a number of systems and OR approaches (e.g., Mason and Mitroff, 1981; Beer, 1994; Gregory *et al*, 1994; Midgley, 1997c), was designed to mitigate against the usual tendency for professional discourses to dominate, making the 'ordinary language' of service users profane. It therefore reflected my understanding of the need to be aware of, and act to counter, processes of marginalisation.

To generate visions of the ideal housing service for older people, and identify a possible form of organisation that could begin to work towards these, we synergised methods drawn from three different methodologies: Ulrich's (1983) Critical Systems Heuristics; Ackoff's (1981) Idealised Design (which is part of his larger methodology of Interactive Planning); and Beer's (1979, 1981, 1985) Viable System Modelling. The theory of synergising methods (part of the 'creative design of methods') was described in detail in Chapter 10, and will not be discussed further here.

To stimulate discussion about the ideal service system, we prepared a list of questions in advance. These were based on Ulrich's (1983) Critical Systems Heuristics, and were specifically designed to explore the boundaries of proposals for improvement. However, we had two reservations concerning Ulrich's original set of questions: they are not all phrased in plain English, and their generic nature makes them relatively abstract. We therefore used a modified set that had been altered to improve their accessibility (Cohen and Midgley, 1994), and further developed them so that they were specifically related to housing for older people. These questions were used with each stakeholder group to generate a list of "desired properties" (Ackoff, 1981) of the ideal housing system.

To facilitate the creative exploration of ideas, participants were asked to imagine that all housing occupied by older people, and all the agencies responsible for it, had disappeared from the face of the Earth, and that it was the group's job to design a replacement system. To prevent participants from becoming utopian, the following three rules (developed by Ackoff, 1981) were provided for people to use:

- Participants' ideas had to be *technologically feasible*. They could not propose something that it would be impossible to implement given current technology. For example, magic pills that made everyone 21 years old forever were not allowed!

- Their ideas had to be *viable*. Participants were allowed to disregard start-up costs, but, if implemented, their thinking had to be realistically sustainable by the agencies identified as having responsibility for making the system work.

- Their ideas had to be *adaptable*. That is, participants were required to think about how control could be exercised over the system to ensure that it does what it is supposed to do, and has the capacity to change to meet new circumstances.

This process of generating ideas, which Ackoff (1981) calls "idealised design", has the effect of promoting creativity by suspending belief in the usual taken-for-granted boundaries of what is considered to be possible. In my view, this challenge to the boundaries of possibility was enhanced by the integration of questions from Critical Systems Heuristics (Ulrich, 1983), as these raise boundary issues that might not otherwise have been considered. However, it would be a mistake to say that idealised design, conducted in this manner, is *un*bounded. Boundaries are explicitly set in terms of feasibility, viability and adaptability—exactly what is feasible, viable and adaptable being a matter for debate. Boundaries about what is possible will inevitably remain, but they will be set more widely, and will be justified more rigorously, than they might have been during a conventional conversation about design. The idea of being explicit about boundaries, and seeking to widen and justify them (without losing the practical focus), reflects the basic thinking of Churchman, Ulrich and myself that it is impossible for plans *not* to be based on bounded (i.e., limited) understandings. The inevitability of limited understanding, but the possibility of reflecting on and overcoming particular limitations when they are identified, is the prime reason why the critique of boundary judgements is necessary.

The outputs from the workshops were three long lists of desired properties. Importantly, there was substantial agreement between the stakeholder groups on the characteristics of the ideal housing system for older people. Indeed, there were only a handful of disagreements, and (in our judgement) none were so large that they represented obstacles to designing improved services. This was not expected: we had anticipated the need to convene a multi-stakeholder workshop, using a method like Strategic Assumption Surfacing and Testing (Mason and Mitroff, 1981) which subjects alternative strategies to oppositional debate before a synthesis of ideas is sought, in order to sort out substantial differences of opinion between stakeholders. As the differences were minimal, we felt justified in moving straight on to synthesise one single vision from the three without further multi-stakeholder involvement.

We therefore produced a first draft of the synthesis ourselves. Desired properties were listed on a flip chart using coloured pens, with different colours indicating which stakeholder group(s) had made particular points. The areas of disagreement were also highlighted. This was then presented in a workshop to a group of managers from the statutory agencies (not the same group of managers from the Housing and Social Services Departments who had worked with us on the design of the methods, although two members from that first group were present—it was a larger group drawn from these agencies and the Health Purchaser[178]). The managers were first required to discuss, and express an informed opinion on, the few disagreements between stakeholders that were outstanding. In dealing with each in turn, we advocated for the position(s) of stakeholders who were not represented. The managers took all points of view seriously, often debating them at length, and the resolutions they arrived at certainly reflected the concerns of users and carers as well as their own. The final list of desired properties (together with details of the disagreements and how they were resolved) can be found in Midgley *et al* (1997). In terms of the theory of boundary critique, our advocacy technique was designed to mitigate against the potential consequence of restricting participation to managers only (this potential consequence being the dismissal of the viewpoints of other stakeholders), and therefore reflected our concern to avoid marginalisation.

---

[178] In the UK public sector health system, there is a split between 'purchasers' and 'providers'. In local areas, one purchasing organisation buys health care from many providers which compete to provide the most efficient and effective services.

14.5.1 *Designing Organisational Delivery*

Having finalised the list of desired properties, the managers were then able to move onto their main task—designing a desirable and feasible form of organisation that would be capable of delivering the kind of housing system described in the previous workshops, and which would address the problems highlighted in the first phase of the intervention.[179]

First, we presented a cybernetic model of good organisation (the Viable System Model, originally developed by Beer, 1979, 1981, 1985) which we suggested could be used as a template for the design. This model proposes that, for an organisation to become and remain viable in a complex and rapidly changing environment, it must carry out each of the following five functions:

- *Operations*: the provision of products or services that address particular needs in the organisation's environment.
- *Co-ordination* of operational units, facilitating communications between them and ensuring that they work effectively together.
- *Support and control*, especially with regard to distributing resources, providing training, gathering and distributing information about quality, etc.

[179] Ormerod (1999) points out that the reason for excluding service users from the workshop at which the findings from the first phase were presented did not apply in the later design workshops, so we should have included them. However, we were still not in a position to invite users in. At that first (contentious) workshop, we agreed the set of methods which would involve service users and carers (as well as other stakeholders) in determining the 'desired properties' of the ideal system. If there were significant disagreements, then we said we would hold a multi-stakeholder workshop to debate and resolve them (in the event of the disagreements being minor, which in fact they were, we agreed that this stage could be omitted). Finally, we agreed that the resulting vision of the ideal service system would be used by managers from the statutory agencies to design an organisational structure that could deliver it in the future. Given various sensitivities expressed at that meeting, we felt that this was a reasonable compromise. Most importantly, we had established the principle that the older people would be involved at the outset in determining the 'desired properties' of the system. This is in contrast to the common practice of user consultation where design ideas, produced by 'experts', are presented to a user forum with the expectation that they will be welcomed (if they are not, then it is usually too late to do much about it). It should also be kept in mind that, by this time, we had gained some experience of working with the managers from the statutory agencies, and it was our judgement that they would respect the views expressed by the older people when they produced their organisational design. Indeed, we were right about this: they were painstaking in their efforts to accommodate all the 'desired properties' produced in the earlier stakeholder workshops—actually, much more so than I expected.

- *Intelligence*: the forecasting of future needs, opportunities and threats. This involves a comparison between the external requirements placed upon the organisation and its internal capacity.

- *Policy making*: setting long-term goals and objectives.

According to the model, the key to effective organisation is not only to make sure that all five functions exist, but also to ensure that communications between functions are appropriate and effective. Together, these functions manage the information and decision flows necessary for effective organisation, and consequently each function is of equal importance.

This model was chosen because it contrasts with the usual hierarchical and multi-agency structures used in organisations. It was already clear from the earlier workshops that the organisation to be designed was going to have to be multi-agency and co-operative in character, and thus a hierarchical structure would have been inappropriate: developing a hierarchy would have meant placing a minority of agencies in a privileged position with regard to policy making, and information flow would then have been largely vertical (to and from the 'top' agencies). Such a structure does not aid co-operation, as the privileged position occupied by the 'top' agencies breeds resentment in other agencies who are forced to conform to the goals set for them. An alternative was also required to the usual practices of multi-agency working where senior managers come together to set policy, but nothing else is done to support joint working (here information flow between the agencies is horizontal, but very limited): in the first phase of the intervention this practice had already been identified as highly problematic (see Midgley *et al*, 1997, for details). In contrast to both the hierarchical and traditional multi-agency structures, the Viable System Model emphasises the need for a diversity of operational units (in this case housing services for older people), together with co-ordination of service delivery and planning. Information flows *within* the agencies would therefore remain largely vertical (although in the ideal world the Viable System Model might be applied within all the individual agencies too), but these would be co-ordinated through multi-layered horizontal information flows *between* agencies through the functions of co-ordination, support/control, intelligence and policy making.

In presenting the Viable System Model, we were aware that we were introducing something into the intervention that, if we were not very careful, could be seen as 'sacred' expertise. If the model were to be perceived as sacred, then the knowledge brought in by the managers could come to be regarded as profane in comparison. The result could have either been dependence on our own expertise, or resentment on the part of the managers who would have felt the need to rubbish our ideas in order to reassert their own sacred status. It was therefore very important for us to explain the model in plain English, encourage critical comments about it, and give the managers a genuine opportunity to decide against using it. When we did this, the managers saw the benefits of the model in comparison with alternatives. They appropriated the model as their own through a process of critique where they identified and challenged a remaining element of hierarchy in the way (following Beer) we had diagrammed it. In Beer's diagramming technique, policy making is at the top, while intelligence, support/control, co-ordination and operations are below it, and the environment (including the clients of the organisation) are to one side. The managers turned the diagram around 90 degrees so that the clients were at the top, and the other functions were at the same level below. After this critique had been conducted, and we had accepted its validity, the knowledge embodied in the model was easily taken on board by the managers. Here again it is clear that the theory of boundary critique informed the approach taken, helping us identify and avoid the possible negative consequences of imposing our own expertise on the intervention.

After they had revised and accepted the Viable System Model, the managers were asked to use it as a template to generate a vision of an organisational means for delivering housing services to older people according to the specifications set by stakeholders. They did this by generating ideas for how each of the five functions should be organised, continually cross-referencing with the list of desired properties of the ideal housing system developed previously by stakeholders. When their design seemed complete, they then rigorously reviewed the list of desired properties of the ideal housing system once again. The proposal for multi-agency organisational development was required to deliver all the desired properties, either directly (by instituting them in its design) or indirectly (by providing an organisational means by which they could be discussed and realised in the future). This review process represented a final safeguard against the possibility that managers would design the multi-agency system to suit their own interests, thereby marginalising all the concerns expressed in the work that the other stakeholders had already undertaken. Once again, this reflects the insight of Churchman, Ulrich and myself that boundaries can become narrow when participation is restricted: the possible negative effects of limiting participation to managers from the statutory agencies were minimised by close adherence to the principle, established in previous discussions, of basing the organisational design firmly on the earlier, wider stakeholder involvement.

14.6 The Design of Methods

The above case study provides many examples of the theory of boundary critique informing practice. At virtually every stage of the intervention, boundary questions were crucial. In reflecting upon how these questions were answered, however, it is possible to make a key observation. The translation of theory into practice is principally achieved through the design of methods. In this sense, as I suggested earlier, separating the discussion of boundary critique (Chapter 7 and this chapter) from the creative design of methods (Chapters 10 and 15-17) is rather artificial, but necessary given the linear nature of written text. Let us briefly look at the case study again to illustrate the importance of the design of methods.

First, in phase one of the intervention (concerned with identifying problems in the housing system for older people), we needed to talk with a diverse variety of stakeholders, many of whom were initially unknown to us. We therefore needed a method for exploring who should be 'swept in'. We based the design of this method on Midgley and Milne's (1995) idea of a 'rolling program' of interviews, with each interviewee recommending others until no new stakeholders could be identified. However, we built on this by integrating questions from Critical Systems Heuristics (Ulrich, 1983) into the interviews. These were specifically designed to help identify those affected by interviewees' activities, thus rolling back the boundaries of inclusion in the intervention. We also used Cognitive Mapping (Eden, 1988) to look at specific decisions taken by interviewees, once more allowing the identification of people involved in, or affected by, interviewees' activities. Clearly, by sweeping in new stakeholders, we were also sweeping in new perspectives on the situation.[180]

[180] Ormerod (1999) points out that our technique of 'rolling out' the boundaries is attractive, but time-consuming. This is indeed the case. Munlo (1997) examines the strengths and weaknesses of our Problem Mapping method (which the 'rolling out' technique is part of) and concludes that this is one of its most significant weaknesses. When I first used it (Midgley and Milne, 1995), I was fairly sure that the boundaries would not expand so far that the project would become unmanageable (altogether, thirty four organisations were swept in). In the Rowntree project I was aware that more organisations would need to be involved, and it is obviously the case that many interventions have to be completed within time constraints that will not allow this kind of process to be used. Nevertheless, when the time is available, I think Problem Mapping has something useful to offer as a method for

Second, when it became apparent that the original boundaries of the study would exclude the concerns of key stakeholders, we established a debate with our Advisory Group on the ethics of the various possible boundaries that could be employed. This can be classed as the use of a method in the sense that a series of techniques were used to organise the debate so that an appropriate way forward could be identified: we convened the meeting, clarified the options (revealing ethical consequences), discussed the pros and cons of these, and reached agreement.

Third, once the first stage of the intervention was complete and we had produced the 'problem maps', we needed to develop a method that would allow us to present our work to a very limited set of stakeholders (who were seeking to control access to the results) without marginalising the concerns of others. We achieved this by using a facilitation and advocacy method that specifically allowed for the introduction of other stakeholder concerns into the debate; helped us handle the negative emotions this generated; and prevented the intervention team from marginalising other participants when designing methods to be used in the second (solution-focused) phase of the work.

Fourth, when moving on to the second phase, we needed to ensure that the concerns of service users were not marginalised. We therefore started with separate stakeholder groups to allow the older people (and others) confidential space to develop their own views. We also used a synergy of Critical Systems Heuristics (Ulrich, 1983) and Interactive Planning (Ackoff, 1981), which facilitated both the widening and explicit justification of boundary judgements.

Fifth, when embarking upon the multi-agency organisational design with managers from the statutory agencies, we discussed the disagreements between stakeholder groups. Here we advocated for other stakeholders who were not present to ensure that their concerns were taken seriously.

Sixth, when selecting a model upon which to base the design, we presented our preferred model, compared it with alternatives, encouraged criticisms, and finally handed over the choice of model to participants. Thereby, the participants came to own the model and were not marginalised by our expertise.

Seventh, as we moved from the initial involvement of a wide range of stakeholders to the final involvement of just a few committed managers, we needed to find a way to preserve the concerns of those who were no longer participating. We did this by designing the problem

---

sweeping in stakeholder views and highlighting systemic relationships between problems.

solving methods so that the normative basis for the work (in the form of the 'desired properties' of an ideal housing system for older people) could be set by a wide range of stakeholders, leaving only the organisational response (using the Viable System Model as a template) to be designed by the managers alone. As an added safeguard, we ensured that, when the managers produced their design for a multi-agency organisation, they reflected on the desired properties in a rigorous manner to ensure that each one was dealt with adequately.

We see from these examples that the means by which insights from the theory of boundary critique can be introduced into interventions is primarily through the selection or design of methods. Some methods specifically facilitate the exploration of boundary issues, while others enforce particular boundary judgements by including or excluding people and/or their concerns from the process of intervention. Importantly, in undertaking this work, we found that no previously existing method that we were aware of was able to handle all the boundary issues faced. Therefore, the focus was very much on the *creative design* of methods (refer back to Chapter 10 for more details).

14.7 Conclusion

In this chapter I have detailed an intervention to support the development of housing services for older people, thereby illustrating the practice of boundary critique. At just about every point in the intervention, boundary issues were crucial—and these issues were addressed through the creative design of methods, informed by the theoretical writings of Churchman, Ulrich and myself.[181] Of course, such an elaborate exploration of boundaries might not be seen as appropriate in all situations (see the intervention presented in Chapter 15, for example, where the extent of boundary critique was more limited), but my experience of Community OR practice is that a careful consideration and participative exploration of boundary issues is indispensable in the vast majority of situations: without it, there would be a tendency to deal with issues at a relatively superficial level, taking account of only a limited range of stakeholder views.

In the next chapter, I discuss an aspect of the creative design of methods: reflection upon a variety of methods, narrowing down to just one (instead of synergising several methods, which is more usual). The intervention that illustrates this is my involvement with a multi-

[181] At the time the intervention was undertaken, Yolles had not yet started work on developing boundary critique.

agency group seeking to design a counselling service to be activated in the event of a disaster.

15

Planning for Disaster

This and the next two chapters illustrate the creative design of methods. Let me briefly remind the reader of what I said in Chapter 10 about this subject. The creative design of methods involves the development of a dynamic set of interrelated questions, expressing the purposes of the agent(s) concerned. Each of these questions/purposes might need to be addressed using a different method, or part of a method. A synergy is generated that allows each purpose to be addressed as part of a whole *system* of purposes. It is important to note that, in generating the purposes, the need for critical thinking and debate about boundary judgements is crucial (see also Chapters 7 and 14). To know which method(s) from the wide variety available it might be appropriate to use in any particular situation, the agent(s) involved in an intervention may consider the purposes expressed in the system of questions, and then inquire into the purposes, principles, associated theories, ideological assumptions and examples of past practice of various methods in order to design an appropriate path for action. Of course, some purposes might not be expressed explicitly: an action can be intuitive, but in such a situation conscious reflection can reveal that a question *could* have been asked, and a purpose expressed, leading to that action.[182]

In terms of writing up interventions, I (and others) who use the creative design of methods tend not to labour the presentation of questions. Rather, I prefer to focus on the *purposes* expressed in the questions: how they are arrived at in local situations; why they are important to the agent(s) concerned; and how they are pursued in terms of the choice and/or design of methods.

---

[182] This kind of reflection is useful because it is not always the case that intuitive reactions are the best. Sometimes, when things go wrong, it is worth clarifying the intuitive assumptions that were made so as to enable learning for the future. Also, when intuitive decision making is successful, it needs to be articulated so that it can be translated into theory and passed on to others.

15.1 Choosing Methods from Just One Source

It should be clear from the above summary that, in most of my interventions, the focus is on mixing methods—and in Chapters 16 and 17 I present two examples of intervention where mixing methods was of central importance. However, in this chapter, I discuss one of the few interventions that I have been involved in where I felt that it was appropriate to draw upon methods from just one source (other than some initial boundary critique to clarify stakeholder perceptions of the situation). This source was Soft Systems Methodology (SSM) (Checkland, 1981; Checkland and Scholes, 1990). I have chosen to dedicate a chapter to this intervention because I don't want to give the impression that interveners *must* mix methods: there are some situations that can be handled quite satisfactorily using just one.

15.2 Planning for Disaster

This chapter describes a set of workshops, facilitated by Wendy Gregory and I, where representatives from nineteen agencies in a County in the North of England came together in six one-day workshops (spread over one month) to plan the basis for a counselling service that could be activated in the event of a disaster.

I will begin by describing our initial contacts with the multi-agency group wishing to undertake this planning (some boundary critique was involved here), and will then move on to discuss how and why we decided to recommend the methods from SSM. Having provided the necessary background, I will proceed to detail our intervention, focusing in particular on the difficulties we experienced in co-ordinating debate (resulting in some modifications of the SSM methods), and the learning outcomes generated from the process. I will then reflect on the intervention in the light of feedback from participants. I will argue that our work can be considered successful according to four criteria (generated from the priorities of participants and the stated purposes of SSM).

15.3 Initial Contacts with the Multi-Agency Group

Our first introduction to the possibility of undertaking this project came when we were approached by the Chair of the multi-agency working party that had been set up to develop plans for a counselling service that could be activated in the event of a disaster. We were told that the working party had been meeting for about eighteen months, but had achieved very little in terms of comprehensive planning. This was because the situation was perceived as so complex that participants found themselves unable to come up with an effective plan that satisfied all their requirements. In particular, they knew that their disaster response had to be multi-agency, because no one organisation had the resources to go it alone, but they anticipated significant problems in harmonising their different views and working practices. The idea of the intervention was to support the working party in generating a sufficiently cohesive plan to warrant putting an application for funding in to Social Services.[183]

15.4 Boundary Critique

We said that we would like to explore the situation before deciding if and how we might help—essentially, we wanted to engage in some boundary critique. It is important for me to note that the boundary critique conducted in this intervention was less elaborate than in the intervention reported in Chapter 14. There are three reasons for this: (i) time constraints (this intervention needed to be short and sharp, while we had eighteen months to conduct the other one); (ii) the stakeholders were already quite clearly defined; and (iii) there was no prospect of the kind of 'user involvement'[184] normally undertaken when Community OR practitioners get involved in service planning and/or evaluation—there had not been a major disaster in the locality in the recent past, and the logistics of accessing service users from other localities to participate in the planning rendered user involvement impossible given the time constraints. We did not view this as a problem, as we could not foresee any major disagreements between people over the need for a post-disaster counselling service—it was just a question of *what kind* of service. Also, amongst the nineteen participating agencies were a number of voluntary organisations as well as statutory services, so we were reasonably confident that a wide range of viewpoints would be represented.

[183] Social Services Departments, which are a part of UK local government, provide a wide variety of non-financial services to disadvantaged people (see Footnote 175 in Chapter 14 for details).

[184] The term 'user involvement' is commonly employed in health and welfare organisations to refer to the involvement in planning, evaluation and/or management of the people who use their services. Calling people 'users' is generally considered preferable to 'customers' (which implies a commercial relationship) or 'clients' (which implies a dependent therapeutic relationship).

In systemic intervention practice there is nearly always a need to balance the desire to undertake a penetrating boundary critique against the desire to get a practical outcome relatively quickly. However, there is no easy answer to the question of how this balance should be achieved: in each intervention, the relevant agent(s) need to weigh up the pros and cons. In some situations it is obvious that boundary critique is necessary if marginalised issues and/or stakeholders are to be considered, and if there is a push for a 'quick and dirty' job in these situations then I resist it. Conversely, there are other situations where there has already been wide-spread involvement in setting the agenda, and the goals of the intervention are relatively uncontroversial (as was the case in the disaster planning project reported in this chapter), and in such scenarios it would be unnecessarily time-consuming to do much more than check out people's perceptions to ensure that there are no major, as-yet-unspoken issues bubbling below the surface of the presenting situation.

In the case of this disaster planning intervention, we were sent written details of the work undertaken so far, together with information about the constitution of the working party. The agencies represented were as follows: four Health Authorities[185]; the Ambulance Service; the Fire Service; the Police; the Police Welfare Service; Victim Support (a voluntary organisation linked with the Police, offering counselling to people affected by crime); the Samaritans (a voluntary organisation offering crisis counselling to individuals); CRUSE (a voluntary organisation offering bereavement counselling); a local Association of Counsellors; the Emergency Psychological Service; the Council of Churches; two University departments; Emergency Planning (County Council); and Social Services (County Council).

We were then invited to a meeting of the working party at which we were expected to make a proposal for providing support. In between receiving the written material and attending the meeting, we made contact with a selection of working party representatives to arrange individual interviews. The purpose of this was to see if their reasons for wanting an intervention (if indeed they did want one) mirrored those given by their Chair. We were particularly keen to talk with people from the voluntary sector, as the Chair was a Social Services

[185] In Chapter 14, Footnote 178, I said that, in the UK public sector health system, there is a split between 'purchasers' and 'providers'. In local areas, one purchasing organisation buys health care from many providers which compete to provide the most efficient and effective services. However, this intervention was conducted just before the purchaser/provider split was introduced. The 'Health Authorities' described in this chapter were responsible for both planning and providing health services in a non-competitive environment.

representative, and the idea was that the group should eventually apply for funding from Social Services. This put the statutory agencies (especially Social Services) in a pivotal position, and we wanted to know if this caused any problems in terms of the participation of other groups. Here we were giving people some confidential space so that, if there were issues of marginalisation, they might be discussed in a relatively 'safe' environment.

These initial interviews revealed that there *was* a difference of view between the voluntary and statutory agencies. The latter wanted to develop a 'professional' service, with selected employees in various caring professions paid a retainer to be on call in the event of a disaster. In contrast, the voluntary organisations envisaged a 'volunteer' service, with the names of a larger number of potential, unpaid counsellors kept on file to be contacted when necessary. However, the voluntary agency representatives agreed with the Chair that, whichever vision was followed, the planning task was so complex that they really did need support. Also, they felt that it was important to get *something* done; the 'professional' service would be better than nothing. Importantly, everybody we talked with said that open communication was possible in the group: nobody said that they feared representing their views, or felt that the results of the planning had been determined in advance by any single agency. We were therefore reasonably confident that there were no significant issues of marginalisation within the multi-agency group.

15.5 Choosing Soft Systems Methodology

Based on these preliminary discussions, we developed a proposal to put to the working party. In terms of the creative design of methods, our central question was, what are the main characteristics of this situation? It seemed abundantly clear that there was a primary need to structure the problems and to facilitate debate (to enable people to move beyond the 'mess' of issues and viewpoints they saw at that time). However, we made the judgement that any use of methods involving ourselves in an expert role with regard to the production of a design would have been highly problematic, and there were two reasons for this. First, disaster planning is a specialised activity of which we had no prior knowledge. The expertise resided in the agencies. Second, we felt that, given the large number of stakeholder organisations involved, any proposals that were developed without their participation would not have engendered their commitment. Indeed, this is a common problem in systems and OR practice: the need for commitment through participation is one of the strongest arguments for the use of 'second wave' systems methods or problem structuring approaches (Rosenhead, 1989a). There was no evidence of problems that might have prevented a debate-orientated method from working, or which would have made debate a sham (such as coercion or insincerity on the part of those controlling resources), so we proceeded with this line of thinking.

The question then was, which problem structuring method(s) should we use? Because of the difference of opinion between the voluntary and statutory agency representatives, we were tempted to employ a method like Mason and Mitroff's (1981) Strategic Assumption Surfacing and Testing (SAST). This asks people with different ideas about strategy to justify their position to others before a synthesis of viewpoints is sought. However, to be used effectively, SAST really needs the strategic options to be clearly defined in advance (Jackson and Keys, 1984), and the complexity of the situation made this condition difficult to satisfy. In addition, the likely outcome would have been a general direction which the working party could pursue, but little concrete detail. An alternative approach might have been Ackoff's (1981) Interactive Planning (IP). This asks participants to generate a list of 'desired properties' of the ideal system (in this case, the ideal disaster response counselling system), which participants can then use to generate more concrete plans. However, most of the examples of IP in practice seem to indicate that planning takes some considerable time (Ackoff, 1981), and even in situations where planning is truncated, the result is often a plan that needs to be worked towards over many years (Cohen and Midgley, 1994). With this project, it was clear that plans needed to be generated quickly so that an application for funding could be submitted within a period of months, and the application had to detail activities that could be implemented in the subsequent one or two years (the longer the period of implementation needed, the more likely it would be that a disaster would occur before the response system was in place).

After exploring the various options above, and some combinations of options[186], through dialogue between ourselves, Wendy Gregory and I focused on the possibility of using the methods from Soft Systems Methodology (SSM). SSM has the advantage of allowing participants in debate space to develop their ideas as they go along, thereby supporting them in getting to grips with complexity. The outputs, in terms of the identification of specific activities that need to be undertaken, can also be quite detailed. We felt that this would be useful given the short term goal of submitting a funding application. Finally,

[186] I will not go into excessive detail by listing these.

it was obvious that we were only ever going to be able to help the group 'kick start' their planning activities, and that a lot more work would need to be done once we had left and the funding application had been submitted: in common with some other second wave systems methods and problem structuring approaches, SSM claims to provide participants with a planning 'language' that people can use to guide their discussions over the longer term—SSM is not only useful for one-off interventions (Checkland, 1981; Checkland and Scholes, 1990). We felt that this 'language' could help prevent the working party from getting into a similar 'stuck' situation in future.

Before finalising our choice, however, we wanted to consider one final issue. SSM has been criticised by a number of authors for being 'regulative': reinforcing the political status quo rather than allowing the possibility of 'radical' change (Mingers, 1980, 1984; Jackson, 1982; Munro, 1999). This is partly because it is based on the idea of equal participation by stakeholders with a variety of views, when (in many situations) there are actually power relationships ('power', in this context, meaning relationships between authority figures and subordinates, employers and employees, etc.) obstructing free and open communication. Also, Checkland (1981) insists that proposals for change must be both systemically desirable (not just desirable on their own, but as workable parts of wider systems) and *feasible*: that is, they must be implementable by those identified as the key actors in local situations. Mingers and Jackson both claim that these requirements limit the scope of changes that can be proposed, effectively removing the possibility of political 'consciousness raising' where people discuss the desirability of change on a wider scale than local actors might be able to bring about on their own.

When we discussed this critique of the ideology implicit in SSM, we agreed that it has some validity (although we took issue with aspects of the detail in the arguments of Checkland's critics[187]). However, given the nature of the task (planning the development of a post-disaster counselling service), which was not the subject of political controversy (disaster planning would be necessary whatever form of government was in place), and given that everybody felt that they could speak openly and that nobody was pre-determining the outcome, we did not feel that the critique of the ideology of SSM was particularly relevant. Indeed, if anything, a regulative approach would be to the multi-agency group's advantage: their need was to

[187] In my view, the requirement for outcomes to be both desirable and feasible need not *always* result in a regulative intervention: in the hands of some groups, it might be most desirable and perfectly feasible to plan consciousness raising activities to prepare the way for 'radical' social change.

produce detailed service plans in a relatively short time, not to question the political status quo. I can envisage situations in which disaster planning *could* become a significant political issue: for example, I have heard that in Venezuela the government is cutting back State support for disaster planning, putting the onus on ordinary citizens to resource it. However, this was not our situation, and I have yet to meet anyone in the UK who seriously questions whether disaster relief should be a statutory responsibility. We therefore went ahead and developed a proposal based on using the methods from SSM, which we presented to the working party.

At the meeting with the working party, the participants readily accepted our idea. However, we made it clear that we wanted to set up a series of full-day workshops so that people could clear their minds of everyday concerns and thereby be more creative. We also wanted to use the first couple of workshops to train participants in the use of SSM so that they would not be wholly reliant on us for support. This caused a serious problem for the group. They had wanted to work in relatively short evening meetings, and had not anticipated our desire for training. After a lengthy discussion, they decided that it *would* be possible to have full-day workshops, as long as there were no more than three blocks of two days each. This inevitably meant abandoning the idea of training participants beforehand, as we anticipated that the group would need the whole six days for planning. This is a compromise we were prepared to make on the grounds that participants would gain some familiarity with the language of SSM while engaged in the process of problem structuring.

We also explored one other issue at the meeting, and that was whether there were any further agencies, not yet represented on the working party, who could usefully be involved. This question was a last check on the boundaries of participation: missing a key stakeholder with a different perspective on the situation could have posed problems in the longer term, as that person or group could have access to critical information affecting the potential success of the whole operation. The group reflected on their membership, but couldn't think of anyone else who should be involved.

15.6 Soft Systems Methodology

In Chapter 9, I provided a very brief introduction to SSM. Below, I flesh out the basic idea of SSM and its methods so that the reader can see what they require of agents involved in intervention. However, I would still recommend consulting the original literature for more details if an application is going to be attempted. Key texts are Checkland (1981) and Checkland and Scholes (1990). In my view, it is preferable to use these two books rather than the burgeoning secondary literature because, as Checkland (1993) points out, much of the work by other authors is of variable quality. For criticisms of SSM, see Mingers (1980, 1984), Jackson (1982), Romm (1995a) and Munro (1999).

SSM encourages participants in the intervention to generate issues through on-going explorations of their perceptions, allowing people to model desirable future human activity. Given the necessary commitment from individuals involved in and affected by possible changes, these models of future human activity can be used as a basis for guiding actual human activity in the world. However, to ensure that the models will indeed be useful, it is necessary for participants to relate them back to their perceptions of their current situation. In this way, possibilities for change are tested for feasibility. At one time, Checkland (1981) talked about relating models of future human activity back to the "real world", but in a recent paper with Tsouvalis (Tsouvalis and Checkland, 1996) he makes it clear that this is merely management-speak for *perceived* real world.

Checkland and Scholes (1990) describe two 'modes' of using SSM: mode 1 is the use of their specific methods, while mode 2 is the expression of their methodological idea (comparing models of future human activity with perceptions of the current situation) through the use of any methods the intervener cares to draw upon. The suggestion is that, as interveners get more practiced, they can move away from mode 1 towards mode 2 applications. In our intervention, we were concerned with mode 1 alone: use of the methods as described by Checkland and Scholes (1990).

To maximise the accessibility of the argument, I will follow Checkland (1981) in describing the methods of SSM as if they fit into seven stages. However, in doing so, I should acknowledge that presenting SSM as a series of stages encourages the reader to make the erroneous assumption that it is a simple set of techniques to be operationalised in a linear sequence. In fact, although Checkland himself presents SSM in stages, he always stresses its *iterative* nature—that is, one moves backwards and forwards between the various stages as and when necessary.

SSM is usually (but not always) conducted in a workshop format involving a general group discussion. The methods ask workshop participants to (iteratively) follow the stages set out in Figure 15.1 [taken from Checkland (1975, 1981) and slightly simplified for purposes of brevity]. This figure is briefly explained in the text below (the numbers in brackets refer to the numbers in Figure 15.1), but the reader

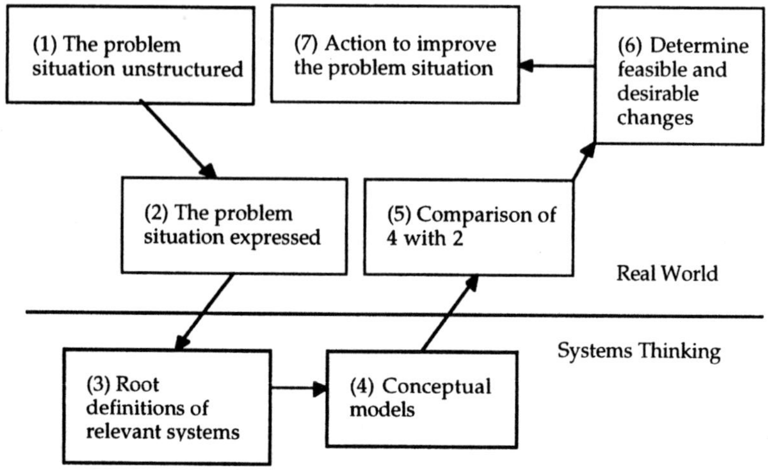

***FIGURE 15.1**: Soft systems methodology (after Checkland, 1975, 1981)*

should consult the original literature on SSM for more details.

First, the problem situation is considered in an unstructured form (1). Then the participants are required to express their understanding in a 'rich picture' (2). A rich picture is a visual representation of the situation people currently find themselves in. It is usually a mess of drawings and arrows showing the interconnections between the various facets of the situation. This is, if you like, a map of the (perceived) 'real world'.

Next, it is necessary to identify possible 'relevant systems' that might be designed to improve the situation (3). These have to be precisely defined to ensure common understanding amongst participants, and to do this a 'root definition' can be compiled for each relevant system. This is a statement of what the relevant system is all about. The adequacy of the root definition can be tested by use of a mnemonic, CATWOE, each letter of which refers to a different facet of the relevant system that should be explored in creating the root definition. CATWOE stands for: **C**ustomers (those who might be harmed as well as beneficiaries), **A**ctors (those who will be involved in making the system work), **T**ransformation process (an identification of a 'raw material' that the system will transform into an 'end product'; e.g., volunteers may be transformed into trained counsellors by a training system), **W**eltanschauung (the world view underlying the wish to make a transformation), **O**wners (those who have the power to stop the system from working), and **E**nvironmental constraints (things that have to be taken as given by the system). By exploring the possible customers, actors, transformation process, weltanschauung, owners and environmental constraints of a relevant system, participants in debate begin to develop common understandings and concretise the root definition.

A 'conceptual model' is then produced for each relevant system (4). A conceptual model is a 'map' of the human activities that would need to be undertaken if the system were to become operational. Activities are first listed, then arrows are used to link them to show which ones need to be done first, and how the activities support one another. A conceptual model does not express the full complexity of the necessary activities, it simply highlights key points that may act as a focus for debate (Checkland, 1997). Once a conceptual model is complete, a comparison can be made with the rich picture to make a judgement about whether those activities would indeed make a difference to the problem situation (5). Then an action plan for making desirable and feasible changes can be developed (6), leading to action for improvement (7).

15.7 The Process of Application

Below, I detail our process of application. In writing up this intervention I have followed the 'stages' as detailed above, but (as Checkland and Scholes, 1990, indicate should be the case) the group moved iteratively between the stages, comparing, contrasting and changing things, so that in the end everything, from the identification of relevant systems through to the design of specific activities, came to be harmonised.

We found it necessary to modify the method of SSM in a number of ways. The first of these was a decision to begin the workshops with an exercise designed specifically to explore the nature of a 'disaster' prior to embarking on the production of rich pictures.

15.7.1 Exploring Disasters

We felt that a preliminary exercise was necessary because we had seen references to definitions of disasters in the written material we had been sent, and we had also spoken to several people who had claimed that it was important to be able to define what a disaster actually is before an appropriate counselling response could be identified. The written definitions usually said something like "more than four people killed in any one incident". Our immediate thought was, does this mean that 200 people injured in a football stadium is *not* a disaster? Or that a serious radioactive leak from a nuclear power station, where nobody is killed outright but deaths are expected in the long term, is not a disaster? It seemed to us that disasters are so varied that they are impossible to characterise with a single phrase of use to everyone. If we were right, then this would be important. It would have been easy to have started the SSM method with a restricted definition of what the group was dealing with, only to find that a counselling service had been designed that was unresponsive to disasters that fell outside the official definition. If we were wrong on this, and the group *could* define a disaster in a water-tight manner, then we were happy to go with it.

Participants were asked to work in pairs to identify a real disaster and list its defining properties on a poster. Following this, people circulated around the room reading all the posters before coming together in a large group to discuss the results. The participants recognised that there was no single definition of what constituted a disaster, although common themes emerged. Some participants still wanted to try to pin a definition down, but the majority agreed with our own first thoughts that a restrictive definition might prevent the design of a sufficiently flexible system. It was therefore decided to keep an open mind on what constitutes a disaster, at least until more understanding had been generated.

15.7.2 Producing Rich Pictures

We then moved on to the first part of the method of SSM, production of rich pictures. At this point we divided the participants into two small groups, working in separate rooms. We felt that this was necessary because of the size of the larger group (nineteen people), which resulted in a rather 'strained' atmosphere in the disaster exploration exercise (described above), with a number of people not actively contributing to discussions. We were the ones who decided who was to go in which group, as we wanted to ensure that both groups contained representatives from the statutory agencies, the emergency services and the voluntary sector.

We held a feedback session once the rich pictures had been completed, where a representative from each group presented their work to everybody else (unfortunately, the drawings have not been preserved, so cannot be reproduced here). The problem situation described in each of the pictures was quite similar, although both groups had identified complexities that the other had not thought of. These were left on the walls throughout the workshops as reminders of the 'mess' which the groups were trying to handle. They were also altered periodically throughout the workshops as peoples' ideas were clarified.

At an emotional level, the group were at a low point when they had finished their rich pictures, and we were not surprised at this. They felt that they had identified a lot of interrelated problems, simply confirming their worry that the whole thing might be far too complex to deal with. Nevertheless, we were able to make clear our own belief that the next stage of the process, identifying relevant systems, would begin to address their concerns.

15.7.3 Identifying Relevant Systems

The following day, participants went back into their separate groups and were asked to think about what discrete (but interrelated) systems would be necessary to create a counselling network and ensure that it functions effectively in the event of a disaster. At the beginning of the process of identifying relevant systems, the groups appeared to be stuck with feelings of anger. They were frustrated that they had so far only identified problems (in the rich pictures), and in their eyes there seemed to be an unbridgeable gap between this and designing solutions. Several participants questioned the facilitators, suggesting that the methods we were using could not help. It took some considerable self-control not to give in and start the process of identifying relevant systems ourselves, but to have done this would have risked dependence on us for ideas, or alternatively would have set us up as scapegoats for the participants' frustration if our ideas had appeared inadequate. Instead, we asked them to reflect on the rich pictures again and try to identify some general themes (not necessarily focusing on local areas of the rich pictures) that might be addressed. Themes slowly began to emerge, and were listed on a flip-chart, but the groups were still unsure about how they could move from these to the identification of relevant systems. To get the process moving, we asked if anyone could identify just one thing (initially) that could be done to improve the situation. Once somebody had taken a risk and identified a possible relevant system, other ideas immediately followed. The mood turned from depression to excitement in a very short period of time.

Each small group produced a poster which described the relevant systems they felt were important, and they then presented this to the large group. The output from the two groups was noticeably different, and the relevant systems were therefore debated in the large group and a list was finalised. At this point participants asked for clarification about how long it would take to explore the relevant systems further (testing them using the CATWOE mnemonic and developing conceptual models). It became apparent that not all the relevant systems could be explored in the time available. Participants therefore chose what they felt were the immediately important ones, and committed to explore the others in their own time after the workshops had finished. Each group took three relevant systems for further exploration.

15.7.4 Exploring the Relevant Systems

The next task was to examine the relevant systems in more detail to ensure the development of a common understanding of what each was about. The groups used the CATWOE mnemonic for this purpose, but root definitions were not produced. This is because both the participants and ourselves felt that enough clarity would be engendered by the CATWOE exercise, and not much added value would be gained by having protracted discussions about the precise wording in root definitions. This phase involved much debate, and the mention of 'CATWOE' led to numerous quips about cats causing grief! These remained with us throughout the rest of the workshops. Needless to say, the interjection of this sort of humour helped the process, and we believe that it significantly improved the participants' grasp of what was required at this stage.

It was when we joined the two groups together again for feedback that the first major problem surfaced. While it had *appeared* that there was mutual understanding when everyone first decided that each small group would work on three relevant systems, the CATWOE exercise showed this to be an illusion. There were two types of problem. First, when representatives from the small groups presented their CATWOEs, there were significant disagreements over fundamental issues (such as what the transformation process was and who the actors should be) between the presenters and those who had not been involved in developing the CATWOEs. Second, while each of the relevant systems had appeared to be discrete when they were originally identified, they had been elaborated in such a manner that there were now major areas of overlap. Unsurprisingly given this situation, people from each of the small groups began to advocate for their own relevant systems against those of their 'competitors'.

Tempers became frayed at this point. Some participants made the comment that, having felt like they had made enormous progress identifying the relevant systems, they were now back at square one with too much complexity. Interestingly, a suggestion was made that the groups had ended up replicating the two separate visions that they had started with (a 'professional' versus a 'volunteer' service), despite our efforts to ensure that the groups were well mixed in terms of agency representation. It actually emerged at this point that the two visions were not the property of two separate camps (the statutory agencies and the voluntary sector) as we had originally been led to believe, but cut across agency boundaries. By sheer chance we had replicated the division in the constitution of the two groups. People went home at the end of this discussion in an angry and dispirited mood, and we were left with a big question to deal with through the creative design of methods: how could we address the situation at the next meeting?

15.7.5 Whole System Modelling

We felt that, uncomfortable though it may be, the only viable solution was for the group to face the problem head-on. It was under the considerable pressure of needing to find a way to facilitate this that we once again modified the SSM process. We decided to work with the whole group, taking each of the relevant systems they had defined (plus the ones that they were yet to think about in detail) and get them to do two things. First, to identify problems of disagreement and overlap, and deal with these in open debate. The goal was to reach accommodations and redefine the relevant systems in such a way that they became discrete once again. Second, to examine how the relevant systems, if created, would operate together holistically. This meant creating a model of the whole disaster response counselling system, showing links between key elements. I will discuss each of these aspects of the task in turn, starting with the business of redefining the relevant systems.

Emotionally speaking, the first couple of hours working on the redefinitions were exceptionally difficult. There was still a lot of anger in the room, and people were struggling just to understand the mess they had created, let alone deal with it. However, the breakthrough came when, at more or less the same moment, those championing the 'professional' vision realised that there was far too much work for a professional team to manage, while those pushing the 'voluntary' vision realised that it would be impossible to ensure a quality service if there was no centralised selection and training system. There was a sudden synergy. The resolution was to propose a core, professional team to manage the selection and training of a wider pool of volunteers.

Once this breakthrough had been achieved, the task was much easier. The relevant systems were redefined in line with the new thinking. As the process of redefinition began to gather steam, we introduced the idea of showing the links between the relevant systems to create a unified vision of the whole system. Doing this proved to be a

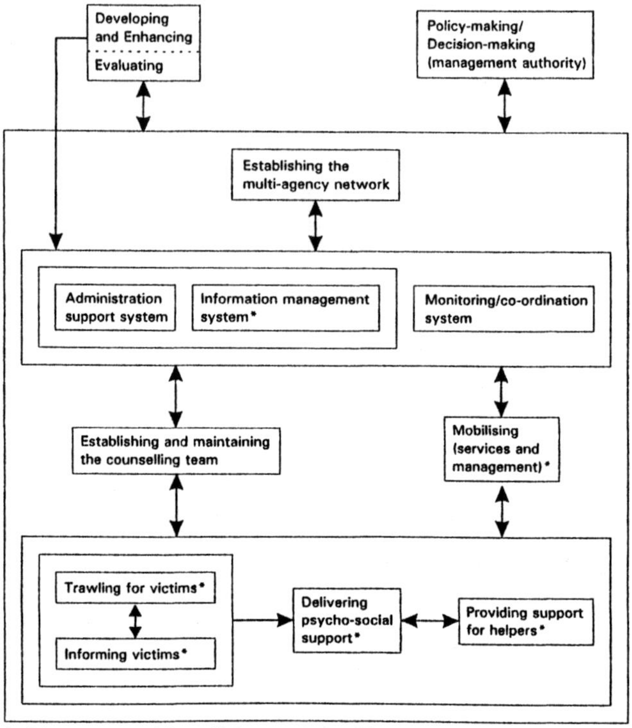

FIGURE 15.2: Whole system model of the disaster response system

turning point in the intervention. By the time they had developed a model of the whole proposed disaster-response counselling system, the atmosphere was electric. This was no doubt partly in contrast with the earlier feeling of negativity, but by the end of the day people were saying that they finally had a concrete vision of where they were heading, and a pride in its innovative nature. There was a palpable sense of achievement. The model of the whole system is presented in Figure 15.2.

One thing should be noted concerning this technique of modelling the whole system. In our workshop, as people began to draw links between the relevant systems, gaps became evident. Several new relevant systems had to be defined to fill these gaps and make the system complete. This is an important observation because it indicates that, up until that point, the relevant systems had been seen as parts without regard to the functioning of the whole. I therefore suggest that the addition of this new stage into the SSM process, which we called *whole system modelling* (Gregory and Midgley, 2000), represents an improvement to the systemic nature of SSM.

15.7.6 Conceptual Modelling

Having developed the whole system model, participants then reflected once more on the time constraints of the workshops. They decided that it would still only be feasible to explore six of the relevant systems in further detail. They split up into two groups, taking three relevant systems each. Once again, they made a commitment to explore the others after the workshops. For each of the six relevant systems, the participants set about developing conceptual models: models of related human activities that would need to be put into place if the relevant systems were to become a reality.

Now, in order to express the human activity systems in a parsimonious manner, so that a whole conceptual model can be comprehended at once, Checkland and Scholes (1990) suggest that each model should be restricted to 7 plus-or-minus 2 elements. This number was chosen because, according to Miller (1956, 1968), human short-term memory can only contend with 7 plus-or-minus 2 'chunks' of information. However, the workshop participants said that they found this guideline impossibly restrictive. We put Checkland and Scholes's (1990) point that each element of the conceptual model could be opened up, and a new conceptual model built to explain it, so that a hierarchy of models is created. However, they rebelled against this, seeing it as excessively complex and time consuming. We got them to try it out, but they continually insisted on getting down to the lowest level of detail straight away. After a short time we gave up trying to persuade them.

The six conceptual models that they eventually built had an interesting characteristic. In most cases the activities were organised into clusters, with relationships between the clusters made clear. One example in which the clustering is particularly noticeable is given in Figure 15.3. I will not go into detail about the content of this conceptual model. What is important is its form. There are four sectors in the diagram representing different arenas of action in time and space, and within each sector is a cluster of activities. Essentially, what the participants managed to do with this and the other models is generate *groups* of human activity systems, diagrammed in systemic relationship with one another.

The full set of conceptual models is presented in Gregory and Midgley (1999).

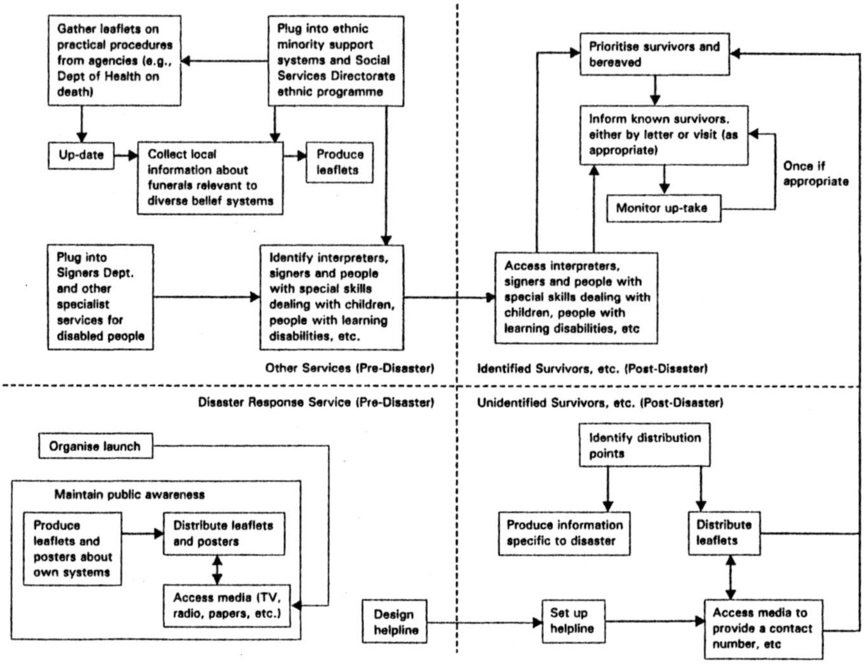

FIGURE 15.3: Conceptual model of 'informing victims'

15.7.7 Creating an Action Plan

The final stage in this project was the creation of an action plan. Checkland and Scholes (1990) suggest that this is a particularly important aspect of an intervention because it involves 'reality checking': relating the conceptual models back to perceptions of people's current situation. As participants reached the end of the conceptual modelling phase, some anxiety surfaced with regard to the practicality of the outcome. While they said that they had definitely found it useful modelling the human activity systems, as it made what needed to be done to create the disaster response counselling service more concrete, they were nevertheless afraid of the gap that still existed between these models and the practicalities of making the activities happen. They wanted to use their last workshop to identify how this was to be done, which was exactly what we had planned.

The method we initially used was the one that Checkland and Scholes (1990) claim is most widely practised in the application of SSM. The sets of human activities within each conceptual model were examined in terms of whether they already existed; who did (or should do) them; what the resource implications were; whether they linked with activities in other relevant systems; whether those links already existed; and who was, or should be, responsible for them. Other relevant comments were also recorded, and several new activities that would need to be developed were identified.

A great deal of detail was generated through this exercise, and this actually intensified the anxiety of participants. By the end of the morning, several people were saying that there was just too much to be done. An intense fear was expressed by one or two people that what they had planned was not really practical at all. This raised the question for us, how should we depart from the SSM action planning to address these anxieties? Our immediate response was intuitive: we asked participants to step back from the situation and revisit the original purpose of the exercise. This was to plan a large, multi-agency service from scratch so that funding could then be sought for its implementation. Further discussion clarified the fact that producing the application for funding was an immediate priority, and that the resources asked for would have to be sufficient to allow personnel to undertake the key activities identified through the workshops. Once this insight had been achieved, it became obvious that a more systematic prioritisation exercise was needed (the SSM action planning did not distinguish between urgent and longer term actions to be taken). The decision to undertake this exercise allowed participants to feel more confident and motivated once again.

The action plan was reviewed in order to add a time dimension to it. Actions were designated as short-, medium- or long-term. The whole list was then taken away, with the idea that the actions needed in the short-term would be placed on the agenda of the next unfacilitated meeting of the working party. This allowed participants to finish the workshop in an up-beat mood (focused rather than ecstatic), determined to take the work forward.

15.8 Learning Outcomes

Having detailed our process of application of SSM, I can now highlight the learning outcomes of this project for the participants. One form of learning can be found in the *content* of the conceptual models (and the whole system model), and another form of learning was about the *process* of working together.[188] I will not detail the content learning,

---

[188] This is not the same use of the terms 'content' and 'process' as in Chapter 4 when I presented my own understanding of process philosophy: in this chapter, I have used the

as this would side-track the reader from the main issue addressed in this chapter—that it is possible for the creative design of methods to result in the choice of methods from just one source (perhaps modified slightly in the light of local contingencies), and that a successful intervention can result. To read about the content learning, see Gregory and Midgley (1999, 2000).

The participants' learning about process can be assessed partly through the post-operative reflections undertaken by Wendy Gregory and I, and partly by discussing the results of a debriefing exercise we undertook immediately after the workshops had finished. This took the form of an open debate in which people talked about their impressions of the workshops. We also made telephone contact with the Chair of the group one month, one year and two years following completion.

At our debriefing session, people said that they had a much better understanding of the positions of other people, as well as what was needed to construct their counselling service, and had built far closer working relationships than they had had previously. There were numerous instances of learning about each others' practices which proved vital in enhancing mutual understanding. For example, some of the agencies were unaware of the details of plans that had already been worked out (by the County Council Emergency Planning Department, Social Services and the Health Authorities) for the provision of hospital and other facilities in the event of a disaster. The details were important because counselling would have to be provided alongside these other services.

There was also a great deal of learning about the different priorities of agencies. For instance, some people put great emphasis on the provision of psychological support to those on the front-line, such as the emergency services (Police, Fire and Ambulance). However, while two of the emergency services welcomed this, the representative of the third said that his agency had its own support system fulfilling a clearly specified function for the organisation: dealing with post-traumatic stress in order to minimise absenteeism from work. He felt that his senior management team would be against the idea of staff seeking external counselling from a system that may not take account of the organisation's wish to minimise absenteeism. Initially, this agency representative strongly resisted the idea of providing any psychological support to emergency services personnel, but an accommodation was reached when details were worked out about how

terms in the more usual 'lay' sense.

internal agency counselling services could interface with the wider system in a manner that would be acceptable to them.

At the debriefing session, everybody expressed pleasure at the fact that they had achieved some key accommodations (such as the one described above), and had thereby generated a unified vision of the service to be developed. Some of these accommodations were highly significant, dissolving major differences between agency representatives. Perhaps the most striking was the one described earlier in which those championing a 'professional' vision and those wanting a 'volunteer' service both realised that there were important problems with their positions. A moment of synergy resulted in an accommodation in which it was agreed that professionals would be used to train volunteers. This represents learning at two levels: learning about a new possibility for service provision, and learning that the group has the internal capacity to deal effectively with major differences of opinion in order to secure a result that can satisfy everybody. I suggest that having 'breakthrough' learning experiences such as this gives confidence to a group that it can handle future internal problems, and thereby encourages individuals to feel it is worthwhile committing energy to the group's collective work.

Apart from the task-orientated learning that takes place within workshops, Checkland and Scholes (1990) also talk about the value of people learning a problem-structuring language that they can use to facilitate further debates once the facilitator has left. We did not have high hopes of the group continuing to use SSM once the workshops were over because our original idea of training the participants in the method had been rejected. We had the distinct impression that the group wanted this to be a one-off intervention. However, in his telephone conversations with us, the Chair reported something interesting. One month after the workshops he stated that the group had found it very difficult indeed to complete the exercise (elaborating on the relevant systems not fully dealt with in the workshops) without us present as facilitators. He said that they had only been partially successful in this regard, and had become bogged down in debate. However, a year later, although the group was not *routinely* using SSM, the Chair reported that they occasionally drew upon aspects of it to clarify particularly sticky problems. Thus, for example, they might return to the output of the workshops for guidance when they felt that they had lost their way, or they might do an impromptu CATWOE to harmonise their understandings. Some limited learning about the use of SSM had therefore taken place.

15.9 Feedback on the Intervention

This intervention was completed several years ago, and I believe that sufficient time has passed to judge its success or failure. Judgements about the success of the intervention can be made in terms of the claims for the value of SSM made by Checkland and Scholes (1990), and also in terms of the priorities of the working party.

Checkland and Scholes (1990) stress that the *process* of an SSM intervention is as important as the content of any plans that are drawn up. It should foster a spirit of team learning, and facilitate accommodations between participants. As described above, the workshop participants unanimously expressed the view that important learning outcomes, including some key accommodations, had been achieved. One year on, the Chair of the working party said that the vision and team spirit had been maintained, partly through continued joking references to CATWOEs, etc., which reminded them all of the team-building 'ordeal' they had gone through.

Checkland and Scholes also talk about the value of providing a problem structuring language that people can use to facilitate future learning. The intervention was less successful in this regard, but (as indicated earlier) some limited assimilation of the language of SSM did take place.

The success of the intervention can also be judged in terms of the priorities of the working party. One major priority was that the workshops should support them in gaining enough clarity to make a strong bid for funding from Social Services. One year later, the Chair confirmed that a bid had been put together, and full funding for all their proposed activities had been granted. Implementation was therefore proceeding as planned.

Finally, there would be little point in spending time on systemic planning if the resulting system failed when it came to be used. After two years the Chair said that there had not been a disaster in the locality, but there had been one in a neighbouring County. One aspect of the group's planning related to the need to activate their system in the event that help was required elsewhere. This is indeed what happened in the case of the incident in the neighbouring County, and the Chair said that all ran smoothly. Indeed, the incident was reported on national TV, and special mention was made by journalists of the speedy and efficient psychological support made available to victims.

15.10 Conclusion

In this chapter, I described an intervention in which representatives drawn from nineteen different agencies came together to plan the development of a disaster-response counselling service. To provide support, the methods of SSM were used (and were modified to deal with various contingencies encountered during the intervention). In reflecting on the intervention in the light of feedback from participants, I argued that it was largely successful, both in terms of what one would expect given the stated purposes of SSM, and in terms of what the participant group hoped to achieve. This case study therefore illustrates that the creative design of methods *can* lead to the successful selection of methods from just one source: mixing methods is not an absolute imperative, although in my experience it is actually necessary in the vast majority of interventions. In the next two chapters, I give examples of interventions featuring the mixing of methods from a variety of methodological sources.

16

Planning and Evaluating Diversion from Custody for Mentally Disordered Offenders

Both this and the next chapter deal with mixing methods. I have devoted two chapters to this subject rather than one, because the two interventions I will describe are substantially different, and therefore illustrate the flexibility that methodological pluralism can give to interveners. Although they were both about support for multi-agency working (amongst other things), the intervention reported upon in this chapter drew upon several traditional social science methods (from the applied science and human relations traditions[189]), as well as second and third wave systems methods, and it involved the presentation of a summative report making recommendations for change. This is common practice in the applied social sciences, despite the strong criticism that summative reports of research results are most likely to make people defensive and unwilling to implement the recommendations (Cronbach, 1982). However, for reasons to be explained, it seemed like an appropriate approach in this instance. In contrast, in the intervention I will present in Chapter 17, this would undoubtedly have done a great deal more harm than good: what was needed in that instance was a strong focus on stakeholder participation and careful work with a variety of agencies to build bridges to enable change. The final report was a record of the activities undertaken, not a judgmental document making recommendations.

Another reason for presenting both interventions side by side is that the one in this chapter was undertaken in 1992-3, while the one in Chapter 17 was undertaken in 1998-9. In the intervening six year period, I believe my thinking about the use of methods matured considerably,

[189] These methods are usually described as observational, but I interpret them as interventionary (see Chapter 6).

and I hope that the extra subtleties I took into account in the 1998-9 intervention are apparent in comparison with my 1992-3 work. I do not want to make my personal learning a major feature of these chapters (I want to concentrate on the mixing of methods), so I will not dwell on this issue any further. Nevertheless, by comparing the two interventions, I hope the reader will get a feel for how learning about methodology and methods can indeed be an on-going process, as I argued in Chapter 11.

16.1 Diversion from Custody

The Diversion from Custody Project was a multi-agency initiative launched to keep people with mental health problems out of prison. The rationale for diversion from custody is as follows:

- People who have committed offences, and who also have mental health problems or a learning disability, require assessment, treatment and rehabilitation within a therapeutic environment;

- prisons cannot usually provide such an environment. Indeed, custody in prison can exacerbate the distress of people with mental health problems or learning disabilities; and therefore

- non-custodial assessment, treatment and rehabilitation is required.

According to Reed (1992) and Bynoe (1992), there are large numbers of people with mental health problems and learning disabilities on remand or serving prison sentences in the UK, and their diversion to more appropriate facilities has to be an urgent priority if their needs are to be adequately addressed.

It is common to call a person with a mental health problem and/or a learning disability who is caught up in the criminal justice system a 'mentally disordered offender'. As someone who has been engaged in a number of interventions in mental health settings, I am acutely aware that this is considered by many people to be a problematic term. There are two reasons. First, the term 'mentally disordered offender' contains an assumption that a person is actually guilty of an offence, despite the fact that some 'mentally disordered offenders' (like others in the criminal justice system) never reach trial, or are found not guilty when a trial takes place. Second, the implication is that a person with a mental health problem is indeed 'disordered'. There are movements amongst some mental health user groups, or 'survivor groups' as they are often called [indicating that their members have survived abuses perpetrated on them by psychiatric institutions (Campbell, 1987; Church, 1992)], to change the language of mental health/illness. They wish to remove terms which stigmatise people, or which imply that symptoms (like hearing voices and/or hallucinating) are necessarily bad and should be eliminated through the use of medication (some people prefer to learn to live with these experiences, and even learn from them). Saying that someone is 'disordered' is very definitely stigmatising. However, the term 'mentally disordered offender' is now very widely used, and we spent a great deal of time working with both staff and users of the Diversion from Custody Project to identify a more acceptable alternative, but to no avail. Despite general misgivings about the term, nobody could suggest an alternative that did not involve substituting the three words with a whole sentence. Therefore, in our evaluation report to the project (Cohen and Midgley, 1994), and in this chapter, I have reluctantly retained the term 'mentally disordered offender', but acknowledge its problematic connotations.

16.2 The Project

In June 1991, sixteen months before the launch of the Diversion from Custody Project, a multi-agency working party was set up, consisting of senior representatives of most of the key agencies involved in the criminal justice and mental health systems within the region: the Crown Prosecution Service; defence solicitors; five Health Authorities; two Prisons; the Police; the Probation Service; Social Services; the Clerks to the Magistrates Court; a local branch of MIND (a mental health voluntary organisation); and the Criminology Department of a University.

The working party undertook the following tasks:

- they organised a one-day conference on the issue of mentally disordered offenders being inappropriately remanded or sentenced to prison. This was intended to promote the creative exchange of ideas, and the conference also helped to improve networks of communication; and

- they prepared position statements for each of the agencies involved in the working party, showing where each stood in terms of their practices.

When the above conference was held (November, 1991), the working party formed itself into a 'steering group' in anticipation of the launch of the Project in 1992. As before, it was composed of senior representatives from the agencies involved in the Project, hence it signalled a commitment to multi-agency co-operation. The steering group acted as follows:

- as a conduit for information flow both between the various agencies, and between the full-time Diversion from Custody Project staff (to be recruited) and the agencies;

- as a way for the various agencies to keep up to date with developments in the Project;

- as a potential lever for changing or amending practices within the agencies in order that they could work better towards the aims of the Project; and

- to provide executive guidance to the Project. The steering group, which met every three months, was kept up to date with the progress of the Project by receiving reports from the management group (see below).

Certain members of the steering group also sat on the 'management group', which was directly involved in the day to day management of the staff team members. The members of the management group were the line managers of the team, hence they could give supervision and support, and make decisions regarding how the team members should act on behalf of their agencies. They could also be influential in amending the job descriptions, conditions of service, etc., of the team in order to further the aims of the Project. The management group was scheduled to meet once a month.

The 'diversion team' itself consisted of three workers seconded from the Health Authority, Social Services and the Probation Service. They were an approved social worker[190], a probation officer and a community psychiatric nurse. Their task was to work with the various agencies in the criminal justice and mental health systems to try to prevent mentally disordered offenders being remanded in custody or given prison sentences.

---

[190] An approved social worker is a social worker who is qualified to detain a person with a mental health problem against his or her will under the powers of the Mental Health Act, either for the protection of the public or for the person's own protection.

The Diversion from Custody Project was funded from a diverse variety of relatively short-term sources (most funding being for around two years). The staff of the Project had offices in a building owned by the mental health voluntary organisation represented on the steering group. In addition, the Director of this organisation chaired the steering group. It is common in the UK for voluntary organisations to take this kind of pivotal role in multi-agency initiatives, as they are regarded as having a more 'independent' agenda than any of the statutory agencies, and are therefore most widely trusted.[191]

16.3 Negotiating the Remit

I was asked to offer three months of team-building and 'operational planning' support after the launch of the Project to help it develop its working practices (here I have used the term 'operational planning' in distinction to 'strategic planning'—we were told that the strategic objectives of the Project were set, but planning was still needed at the operational level). I was also asked to evaluate the Project over a period of one year (starting once the outputs from the operational planning had been implemented), and to provide information systems (IS) support for record keeping and monitoring purposes. I wanted to take on this intervention because I believed (and still do believe) that a great deal of damage is done to people with mental health problems in prison, and anything I could contribute to improving that situation would be worthwhile.[192]

---

[191] Sometimes there can be significant conflicts between statutory agencies because they fulfil different functions in relation to the same users, which are not always viewed as complementary, and respond to the demands of different pieces of legislation designed at different periods of social policy development. In contrast, voluntary organisations are rarely tied in this manner.

[192] At the beginning of my career, I was employed for a short while as a residential social worker in a hostel for people with mental health problems. One night, a very confused young man broke into the hostel and started smashing things up. I called the Police, who took him away. It turned out that he was an ex-resident who remembered that the hostel had helped him previously, and he came back when he found that he was having another breakdown. However, he had lost the capacity to speak, and in his frustration he had started smashing our property. Unfortunately, I found out that my employer's policy was always to prosecute in situations like this, regardless of the circumstances. The case took months to come to trial, and in that time the young man was held on remand in prison. When it did eventually come to court, I had to give evidence. He was still unable to speak, and the Judge declared him unfit to stand trial. This meant that he would have to return to prison until his mental health improved! In effect, he was condemned to serve an indefinite prison sentence without being convicted of a crime. I was horrified by this state of affairs, but was prevented by my employer from withdrawing the charges, even though we were asked by the court to do so. This stayed on my conscience for a number of years, so when I was offered the opportunity to work with the Diversion from Custody Project, I

I brought together a small team of Community OR practitioners: Claire Cohen (a colleague in the Centre for Systems Studies who worked with me on the team-building and evaluation); Alison Savage (another colleague who undertook the IS planning alongside, and interfacing with, the operational planning); John Croston (an independent community operational researcher who designed a database for the Project's use); and myself (I was directly involved in the team-building and evaluation; facilitated one session of IS/operational planning; and monitored the development of the database). In this chapter I will focus mainly on the team-building, operational planning and evaluation, as these were the areas I was most involved in—only those aspects of the IS planning that closely interfaced with my own and Claire Cohen's work will be discussed.

My first contact was with the Chair of the steering group, who sent a letter inviting me to tender for the work. When I responded to the letter, a short brief was sent to me specifying the nature of the intervention required. I immediately asked to meet the Chair to discuss it further. I explained the basic principles of systemic intervention in plain English, and she was pleased with the approach I outlined—especially the commitment to methodological pluralism, which I stressed was intended to make the intervention responsive to local needs.

I was quite happy with most of the brief, as it was phrased in general language allowing quite a lot of room for manoeuvre to meet contingencies encountered along the way. However, I needed to raise two general issues. The first was the assumption in the brief that the staff of the Project were to work directly with users (people with mental health problems and/or learning disabilities) caught up in the criminal justice system. The staff team's role would be to negotiate with (and co-ordinate the various activities of) the relevant agencies to ensure that people did not end up in prison. My instinct was to ask whether their time would be better spent facilitating strategic and operational planning with the agencies to enable them to become more responsive to the needs of mentally disordered offenders. I thought that, if they could support the agencies in developing their policies and practices, then there would be no need for the team to be directly involved with users: the agencies themselves could take responsibility for finding appropriate alternatives to prison for mentally disordered offenders. I suggested that, as part of the operational planning support I was going to offer, I could work with the staff team on how they could facilitate

---

jumped at it. It was a chance to do something positive to ensure that others did not have to suffer like this man that I had been involved with.

wider strategic planning. While the Chair understood my logic, she made it clear that they had been granted funding for the Project on the understanding that it would indeed be working directly with users. Therefore, this aspect was non-negotiable.

However, she said that she would be happy to see the issue addressed as part of the evaluation once a year's worth of activities with users had been undertaken (I would have liked to have done this at the beginning of the evaluation, but she wanted it at the end of the year on the grounds that, if strategic planning indicated that something other than the established Project was needed, it could undermine work with users). Although not ideal from my point of view, this seemed like a reasonable compromise: as I saw it, if the funders were so closely focused on work with users that this aspect of the Project could not be changed, they might need to see the problems this threw up (if indeed it did throw up problems) before accepting my rationale. Also, I might be wrong to think that a strategic planning focus would be best: it could be that working with users would actually provide an effective vehicle for changing the policies and practices of the agencies. I therefore outlined (in plain English) how I might go about doing this aspect of the work at the end of the evaluation, making user involvement in strategic planning a central focus of it (see later in this chapter for details of my proposed methods). The Chair said that this would fit well with the agenda of the voluntary organisation hosting the Project, Social Services, and all of the Health Authorities which were trying to improve user involvement in their affairs.

Another issue was whether the evaluation would be summative (taking the form of a final report) or formative (focused on the provision of on-going feedback about performance).[193] Linked with this was a question about whether the primary focus would be on quantitative data, or the views of users about how involvement with the Project affected their lives. The quantitative/qualitative question was linked with the summative/formative issue because, if the primary focus was quantitative, then at least some of the evaluation would have to be summative: it would take most of the year to collect sufficient data to derive any meaningful conclusions.

It was made clear to me by the Chair that there was definitely a need for quantitative data presented in a summative report. There were two reasons: first, the evaluation would be used by the steering group to make a recommendation about whether funding for the project should be

[193] Scriven (1991) says that summative evaluations are conducted *on* services (suggesting an external, judgmental audience), while formative evaluations are conducted *for* services (suggesting an internal, managerial audience).

picked up by the various agencies once the current sources of money had run out; and second, if the project was judged a success, a report was needed that would present the findings in a form that would be seen as valuable by the Home Office[194] (which, at the time, tended to take a fairly traditional 'observational' line on research matters). This was to communicate the value of the Project so that it could be replicated in other areas of the country: the Diversion from Custody Project was being looked upon as a pilot initiative by some Home Office personnel, as it was quite an ambitious scheme compared with others of its kind.

I accepted the need for quantitative, summative research in this situation, but also argued that a successful service (judged in terms specified by agency representatives) that achieves results in a manner that is unacceptable to its users is rather suspect. She concurred, and added that it is not only important that the service is seen as valuable in the eyes of its users, but that the views of the many professionals in direct contact with the users (e.g., lawyers, psychiatrists, police officers, social workers, probation officers) are important too. Without the commitment of these people, wider change would be unlikely, and the Project would always be working reactively rather than proactively. We therefore had no difficulty agreeing that the evaluation should use both quantitative and qualitative data, and that qualitative data would also be gathered from a wide range of stakeholders through semi-structured interviews. Also, because some of the qualitative data could be collected on an on-going basis, we should set up formative processes to feed back information on performance at three different levels: to the staff team; to the management group; and to the steering group.

I asked to speak with the Project staff, service users and other members of the steering group before writing a final tender document which would meet the agreed specifications. I was explicit that I wanted to undertake this consultation because these stakeholders might have insights that would affect the choice of methods. Talking with the Project staff and other members of the steering group proved to be unproblematic, but meeting users at this early stage turned out to be impossible. The reason that team-building and operational planning support was being requested before the evaluation commenced was that the staff team had only just been recruited: they had not yet begun working with users, and did not have a clear idea how to start. Therefore, I had to develop my proposal for undertaking the intervention without input from mentally disordered offenders.

[194] The Home Office is the UK government department responsible for handling issues of criminal justice and national security.

However, I did not see this as a major problem because I had already clarified with the Chair how users could be involved in the evaluation through interviews and the strategic planning, allowing issues of importance to them to be surfaced and dealt with on an on-going basis. Discussions with the staff team revealed a substantial perceived need on their part for the team-building and operational planning support, but did not throw any new light on how the evaluation should be undertaken. Therefore, I went ahead and wrote a proposal based on the ideas already mentioned. I kept it general enough to allow flexibility in the precise choice of methods (within the agreed parameters), and made it explicit that, if unforeseen issues emerged, we should be able to re-negotiate our approach (as it turned out, however, this was unnecessary).

We submitted our tender, and were commissioned to undertake the intervention shortly thereafter.

16.4 Some Preliminary Boundary Critique

Prior to starting on the team-building and operational planning, we wanted to engage in some boundary critique to surface issues that might have an impact on how we worked with the staff team. A key question was whether there were any controversial issues that people could not talk openly about, or any significant antagonisms within the staff team that might prevent effective team-building. Another question was whether there were any disagreements over strategic objectives between members of the steering group that might compromise operational planning with the staff team. We were given the impression by the Chair of the steering group that all the agencies (and the individuals on the staff team) were in accord, but we thought we should double-check this.

Obviously, if we were looking for controversial issues and/or unspoken conflicts, it would not be appropriate to use methods that involved people speaking in the hearing of their colleagues. We therefore opted for confidential individual interviews, structured very loosely around some general topic headings (e.g., 'strategic objectives', 'relationships between agencies' and 'relationships between colleagues'). When I use the term 'confidential' to refer to these interviews, I mean it was clear that no information would be transmitted to others if the source of it might be identifiable: these interviews were for our own purposes to enable us to make appropriate suggestions for methods to support the team-building and operational planning.

The individual interviews also had the added advantage over group work of allowing the interviewees to get to know the evaluators on a one-to-one basis (at least superficially), providing a basis for establishing mutual trust. We interviewed all the members of the staff and management teams, as well as selected members of the steering group (selection being determined largely by availability).

When interviewing people it seemed that there was a consensus on broad objectives. However, a closer examination of the issues raised at the individual interviews revealed that many interviewees also harboured certain deeper, less willingly voiced concerns. Most of these were expressed in 'asides', and they centred around the management of the Project. This included the formal structures (the management group, the steering group, etc.); the ability of those structures to enable harmonious working between agencies; and the capacity and willingness of various agencies to co-operate and to amend their working practices as required. Essentially, while there appeared to be a consensus on aims and objectives, there was some scepticism about multi-agency working: most people had had experience of agencies defending their own boundaries of interest whilst disregarding the views of others, and were unsure about whether commitment to the Project ran deep enough to overcome this.

These points were worth noting since they revealed potential problem areas that might be significant for the future of both the Project and the intervention. Nevertheless, in relation to the proposed team-building and operational planning exercises, there were no immediate obstacles.

16.5 The Team-Building

When we interviewed the staff, they all said that a significant problem for them in working as a team was that, coming from different agencies, and having different disciplinary backgrounds (social work, probation and nursing), meant that they often used language in ways which obstructed effective communication. They all employed different terminologies: very often the same words meant different things to different people. We thought that it would be appropriate to begin the team-building activities by focusing in group work directly on this issue. There were four reasons for this: first, we were aware that it is indeed a substantive issue for many multi-agency teams[195]; second, it provided a

[195] See Watson (1997) for a review of multi-agency housing research which identifies a range of problems that multi-agency teams experience.

useful 'task-orientation' which would allow the team to develop trusting relationships at a more unconscious level (acclimatising and adapting to each other's verbal and non-verbal patterns of communication); third, it would be a signal that we, as professional interveners, took their expressed concerns seriously (enabling us to build a foundation of trust between the team and ourselves which would undoubtedly be of value as the intervention progressed); and fourth, using the correct terminology would be important for the IS planning, and in constructing the database. Here it is possible to see multiple layers of analysis of the situation (analysis of the expressed needs of the team; of relationships within the team; of relationships between the team and ourselves; and of the needs of the IS planners), resulting in a decision to engage in language-focused group work.

Reflection on our own intellectual resources[196] revealed that Claire Cohen and I knew of no 'formal' method(s) designed to pursue these multiple purposes.[197] Therefore, we designed our own method, which was agreed with the staff team and the Chair of the steering group before being enacted. First, we considered who should be involved. We felt that it would be important for any new terminology to follow from the objectives of the Project, so it would be inadvisable for the staff team to work on its own in case they defined objectives that were problematic for the management and steering groups (although broad objectives were already in existence, there was a consensus that much more detail was needed). Therefore, we invited the management group to work with the staff team in a half-day workshop to clarify objectives. For this purpose, it was crucial to explore the boundaries of the Project's activities (even though certain aspects of the boundaries were already given, such as the focus on working directly with users). To clarify objectives, we drew on some of the 'ought' questions from Critical Systems Heuristics (Ulrich, 1983; Chapter 7 of this book), but we did not work through them systematically: rather, we kept them in mind to inform our facilitation.

Having defined the objectives, a further series of half-day workshops were held with the staff team and management group

[196] This is the kind of questioning advocated by Mingers (1997a), as discussed in Chapter 10.

[197] This is not to say that no such method(s) exist, just that we were unaware of any (and a brief search of the literature did not help either). This is worthy of note because it highlights the point made in Chapter 11 that, in the absence of adequate intellectual resources, systemic intervention requires a willingness to research and experiment to enhance practitioner learning. If the intervener has no ready-made method to hand, s/he should not change the focus to fit with known methods: if the purposes being pursued are important, then they should be preserved. This is when the resourcefulness of the intervener comes into its own.

working in partnership. The participants took each objective in turn and questioned the terminology used in it (and in subsequent statements made to explain the objective). The idea was to agree specific meanings. Where differences were surfaced, the participants explored the significance of these (in terms of the purposes and values implied by the different meanings) in order to reach an accommodation. The definitions of key terms agreed by the participants can be found, along with the objectives of the Project, in Cohen and Midgley (1994).[198]

Some quite complex issues to do with the treatment of mentally disordered offenders were revealed in these discussions, and this helped to clarify the stance of the Project on a number of potentially controversial matters—arguably the most sensitive being the need to divert a person with a mental health problem from custody even if his or her psychiatrist says s/he is 'untreatable' (i.e., medication has no effect on his or her condition), and therefore refuses to admit him or her to hospital.[199] Interestingly, there were no major disagreements between participants that required a 'forced' accommodation—and, because the staff team were aware that they were taking positions on some controversial issues with which other stakeholders might disagree (e.g., psychiatrists), this helped to forge a common identity and a common set of values.

Of course, if the Project took too antagonistic a stance on some issues, this might undermine multi-agency working in the locality—which everyone was conscious was vital to the long-term success of the Project. On the subject of the challenge to psychiatry, it is interesting to note that the psychiatrists associated with the Project (on the steering group, and later working with the staff team in the Courts) were in agreement with the staff on the issue of the treatment of patients for whom medication had no effect. Therefore, the staff team and management group were able to take a controversial stance in the

---

[198] One aspect of the agreement on terminology was a decision not to use the term 'mentally disordered offender' (for the reasons stated earlier). However, this was rescinded when no succinct alternative could be identified.

[199] Psychiatrists in the UK tend to take the view that, just because someone has been diagnosed as having a mental health problem, this does not automatically give him or her the right to treatment (e.g., a hospital admission, or admission to some other residential or non-residential facility). On the whole (although there are exceptions to this rule), psychiatrists regard 'treatment' as the use of medication to suppress or eradicate symptoms. Some people's illnesses are resistant to the use of medication, and in these cases treatment is often refused. The staff and management teams were of the view that people who could not make use of medication should still be diverted from custody, and some other form of treatment should be devised—even if this was not yet of 'proven' scientific value (the worth of alternative treatments could be researched). In the case of potentially dangerous offenders, such treatment would need to be provided within a secure environment other than prison.

knowledge that they had allies in the psychiatrists' camp who were prepared to argue their case with others (and did so quite successfully on a number of occasions). Here we see the use of the model of boundaries and marginalisation presented in Chapter 7 (Figure 7.3): we were aware of the risks of supporting the staff in establishing their bounded identity as a team on the basis of making a value judgement about psychiatric practice—they could have begun to view psychiatrists as profane. As the psychiatric profession is pivotal in the mental health system, this would most likely have resulted in a backlash from the psychiatric establishment (the team itself would be regarded as profane by psychiatrists), leading to the marginalisation and nullification of the Project. However, in assessing the situation, we felt that the involvement of sympathetic psychiatrists in the Project would have two counter-balancing effects: (i) it would prevent the team from stereotyping all psychiatrists as profane; and (ii) it would ensure continued communication between the Project and the wider community of psychiatrists. As it turned out, we were right in making this assessment: the staff team was indeed able to gain a common identity through its value commitments without alienating the psychiatric establishment.

16.6 The Operational Planning

We envisaged that the operational planning should provide another focus for continued team-building within the staff team, and it should also support them in getting to grips with the complexity of their task: they had made it clear that they were unsure how to start their diversion activities.

At this point, Alison Savage (our IS planner) proposed to use the methods from Soft Systems Methodology (SSM) (Checkland, 1981; Checkland and Scholes, 1990; Chapter 15 of this book) as a lead-in to her IS planning. Several authors[200] have identified the utility of SSM for this purpose, and very specific links between SSM and other more technical IS methods, such as Jackson System Development,[201] have been devised (Savage and Mingers, 1996), which Alison Savage proposed to draw upon. I realised immediately that SSM would also be useful for addressing the other two purposes: team-building and working out how to start diverting mentally disordered offenders from custody. This is

[200] See, for example, Stowell (1985); Mingers (1988); Avison and Wood-Harper (1990); Prior (1990); Lewis (1993, 1994); Lander *et al* (1997); and Checkland and Holwell (1998).

[201] Jackson (1983); Sutcliffe (1988); Cameron (1989); Davies and Layzell (1993).

because SSM is based on the idea of evolutionary learning, where the process of exploration (which builds co-operative capacity) is as important as the content of the models that are produced (and this content is elaborated over time, allowing participants to get to grips with the complexity of their task). Given the opportunity presented by Alison Savage, and in the knowledge that there were no obvious barriers within the staff team to the use of SSM (such as coercion or an unwillingness to openly discuss differences of opinion), we amalgamated the operational and IS planning.

Alison Savage facilitated most of this, working with the staff team and management group together (in this sense, the SSM workshops were a continuation of the earlier team-building activities) and then went on to use Jackson System Development (JSD) to build a specification for the database. I came in towards the end of the SSM and before the JSD in order to facilitate some action planning following on from the rest of the SSM learning.

16.6.1 Outputs from the Soft Systems Methodology

In addition to the continuation of team-building, a vision of the main activities to be undertaken by the Project emerged from the SSM workshops. The principle focus was on the staff team's position as the 'hub' of a complex network of inter-agency relationships. They recognised that they would have to: (i) be a conduit for the flow of information on diversion between the various agencies; (ii) maintain momentum and enthusiasm for the Project within the various agencies; and (iii) facilitate the harmonisation of different practices.

Because a successful Diversion from Custody Project would require a significant evolution of attitudes and practices amongst the various agencies involved in or affected by it, the staff team and management group felt that it would fail unless its work with users could be complemented by a concerted public relations (PR) effort. In other words, the SSM workshops highlighted the Project's perception of itself as a proactive facilitator of change within other agencies. As an example of this heightened awareness, the following list of activities was identified as necessary during one of the SSM workshop sessions:

- gain and maintain the commitment of key individuals in the criminal justice system;
- generate a vision of what a local criminal justice system could be like that takes account of the needs of mentally disordered people;
- involve key people in generating this vision;

- demonstrate the effectiveness of the Project; and
- mobilise resources to this end.

Of course, it would be difficult for the staff team to be involved in both work with users *and* PR. Therefore, some separation of responsibilities was agreed. Formal PR became the main responsibility of the steering and management groups (particularly the Chair, who was strongly committed to championing the aims of the Project), and the staff team focused on work with service users. Nevertheless, it was recognised that the team would necessarily be involved in 'informal' PR in their day-to-day contacts with other professionals (i.e., they could build constructive relationships from the bottom up without ever needing to make formal presentations on diversion).

Further workshop sessions using SSM explored other aims of the Project. In so doing, team members developed a set of activities that needed to be performed, and they identified key people to be contacted and 'nurtured into' the Project.

In addition, these workshops helped the staff team define a template for their working day. They would start at 7.00am, visiting Police Stations to discover whether people with mental health problems were being detained in the cells. They would conduct assessments if necessary. They would then go to the Magistrates' Court at 10.00am in order to meet with the court probation officers and to find out whether people with mental health problems were appearing before the magistrates on that day. The rest of the day would be spent following up the cases identified; receiving information via fax or mobile phone regarding further cases; attending Police Stations again if required; and attempting to divert. Except for minor changes, this structure of the working day remained intact throughout the period we worked with the Project.

By the end of the SSM workshops, the staff team not only had a clear sense of identity and purpose, but had also worked out in detail how this purpose was to be pursued.

16.7 Designing the Evaluation Methods

We worked with the staff and management teams on team-building, operational planning and IS planning for just under three months, scheduling workshops on a twice-weekly basis. This phase of our involvement ended once the staff team felt ready to go out and work with users. At this point, the evaluation needed to begin. Alongside the planning activities, we built up the detail of the evaluation design so as to be ready to roll once work with users commenced.[202]

To remind the reader, we had already agreed to a summative evaluation (one based on the production of a final report, making recommendations to key stakeholders), but with elements of formative evaluation integrated into it (periodic feedback to the three hierarchical levels of the Project's management[203]). We had also agreed to both a quantitative analysis and qualitative assessment (particularly taking into account users' views). Finally, we had negotiated the idea of conducting a strategic planning exercise at the end of the evaluation period to determine whether the Project was orientated correctly (working directly with mentally disordered offenders as well as staff and management).

As will become apparent over the coming pages, we mostly used quantitative and qualitative methods from the applied science and human relations traditions for the evaluation, and we synergised two management systems methods to structure a participative, strategic planning exercise to round off our analysis and provide a platform upon which local agents could base future proposals for change.

16.8 *Quantitative Methods*

The first question to be answered was, how should we quantify 'success'? And a subsidiary question was, what important measures of success would *not* be quantifiable, and would therefore have to be assessed by other means?[204] We worked with the staff and management

[202] We were aware that the database would not be ready when work with users began, and everybody accepted the fact that data would have to be collected manually at first and then entered into the database when it went on-line. As it turned out, the database took longer to deliver than either the Project or I had anticipated: there were many bugs to be ironed out, especially as we were using what, at that time, was an entirely new technology: remote entry by staff in the field onto laptops, feeding into a desktop which then regularly up-dated the laptops (ensuring that everybody had data at their fingertips in remote locations that had been up-dated no more than 24 hours previously).

[203] Scriven (1991) says that formative evaluations can be designed as "early warning summative" evaluations (p.169). This means that they use an analytical design, and feed back information periodically before the final report is due. This is what we did for the Project, because it seemed the most appropriate option (see later). However, I do not agree with Scriven that this is typically the best way to conduct formative evaluations: in different circumstances, stakeholders may benefit from using participative, debate-orientated methods, and there may not be any need for analytical work or formal reporting.

[204] I have conducted a number of evaluations and have not yet come across a situation where (in my view, and in the view of other stakeholders) all relevant measures of success were quantifiable. For this reason alone, I believe it is necessary to use qualitative methods, or qualitative and quantitative methods together. We should not accept the neo-positivistic

teams to evolve answers to these questions in a participative manner (at this point there were still no users involved in the Project, so we could not get a view of success from the 'receiving end' of the service—this would be rectified later). We did this as part of the IS planning, as the database would need to be designed to accommodate the evaluation needs of the Project as well as the staff team's operational needs.

We held a discussion with the staff team and management group about the meaning of 'success'. The first issue that emerged was that there were many points in a mentally disordered offender's progress through the criminal justice system at which a diversion could be achieved: e.g., at the Police Station (either in consultations before charge or when discussing Police bail); at a court bail application; at trial (if there is a chance that a trial could be discontinued); at sentence (if the person is found guilty of an offence, then a pre-sentence report about their mental health problem could be influential); from Prison (if the person is already in there serving a sentence); and at appeal (if grounds can be found). The problem is, if diversion fails at the Police Station, but is later successful at bail application or sentence, should this be seen as a success or a failure? The team anticipated (and this was later confirmed by their experience with users) that multiple attempts at diversion would be needed for many users: should we therefore be taking as our unit of measurement a *diversion attempt*, accepting that there could be as many as ten attempts per individual, or should we only be looking at the *final* diversion attempt? After much debate, we decided on the latter—primarily because it was expected that diversion would be difficult, and the important thing was whether the *sum total* of the efforts of the staff team were successful, not whether an individual action taken along the way did or did not have the desired effect.

The next question was, if we have a set of figures expressed as proportions of people diverted at Police Stations, bail applications, sentence, and from Prison, how can we judge what constitutes a 'good' or a 'bad' result, both for each stage and overall? The Project and the evaluation team all agreed that it would be unethical to conduct a controlled experiment: to refuse a service to some users just in order to make a statistical comparison. Therefore, it was accepted that the quantitative results had to be judged qualitatively: we agreed to

---

reduction of all measurement to quantification: this seriously impoverishes evaluation practice, as it makes invisible so much that is of value to stakeholders (see Weiss, 1972, 1973, 1977; Broskowski, 1976; and Patton, 1978, 1980, 1987 for extended arguments). I therefore have serious reservations about purely quantitative methods, like Data Envelopment Analysis (e.g., Norman and Stoker, 1991), unless they are used as part of a wider systemic intervention practice such as the one described in this book.

circulate the figures at the end of the year to a wide range of stakeholders and collate their feedback for use in writing a commentary on their meaning. We also suggested (and this was accepted by the staff team and management group) that we should 'triangulate' the quantitative data with qualitative information collected through interviews. Triangulation means comparing, cross-checking and contrasting data produced through the use of a variety of different methods (see, for example, Brewer and Hunter, 1989).

Another problem in measuring success was in assessing the quality of outcomes. For example, a person may be successfully diverted from custody, but at the end of the day were they any better off for it? Did they receive appropriate treatment? And in whose view? Did they end up back in custody a month later, or did the diversion have a lasting effect in keeping them out of the criminal justice system? Did the diversion have effects in other areas of their lives (e.g., in terms of employment or family relationships)? These would not be easy issues to quantify, either because they are inherently unquantifiable (e.g., if there are different views of success in play) or because logistical problems would prevent the collection of robust data (e.g., we anticipated that the follow-up of users would be difficult, and in some cases impossible). It was therefore decided that the main *quantitative* outcome to be assessed had to be the simple fact of whether or not a diversion had been achieved. However, by talking with a small selection of mentally disordered offenders, their diversion keyworkers and other agency representatives involved in individual case-work, we could build up qualitative case studies to triangulate with the quantitative data in order to give more depth to the analysis.

The staff and management teams also wanted us to record and use other information about referral agency; gender; age; race; previous contacts with mental health services; accusation of offence; previous criminal record; other major unaddressed psycho-social problems (e.g., homelessness, drug addiction); and an assessment (by the key-worker) of whether the person could be considered as a danger to him/herself or others. This information was considered useful by the Project for three reasons: (i) in their day-to-day operations, the staff team would need to have it to hand; (ii) in the longer term, there were plans to highlight any major differences between social groups, both for research purposes and to monitor equal opportunities[205]; and (iii) it was considered important for a quantitative profile of the user group to be compiled.

[205] It was not anticipated that the data set collected during the evaluation period would be large enough to draw robust conclusions about these issues, but the system should be set up so that this kind of monitoring and research could be conducted over the longer term.

The latter was considered to be necessary because, if the users were shown to be particularly 'difficult', even moderate success in diverting them from custody might be seen as a major achievement. In contrast, if the user group turned out to be mostly first-time offenders with few other social problems, then a higher rate of diversion might be expected—and indeed, questions could be asked about whether the service was actually reaching the right sort of people.

16.9 Qualitative Methods

A variety of qualitative methods were used to answer other questions, or were triangulated with the quantitative data, as follows.

16.9.1 Stakeholder Analysis

First, we asked what the best way would be to evaluate aspects of the Project other than 'success' as defined in terms of the number of diversions achieved. To begin to answer this, we conducted a stakeholder analysis to identify key people whose views should be accounted for in the intervention. Our method was to take the objectives of the Project (as defined in the earlier team-building and planning workshops) to meetings of the staff team and steering group, and ask what categories of people would be involved in, and affected by, meeting those objectives. This gave us a list of stakeholder groups.

16.9.2 Interviews

This list was then used to guide the selection of interviewees. Interviews with those directly involved in the Project were conducted periodically throughout the life-time of the evaluation. Interviews with other stakeholders (e.g., staff of the Courts, Custody Sergeants and lawyers) were conducted in the second half of the evaluation, once people had become accustomed to working with the Project. The interviews themselves were semi-structured: we had some key areas that we wanted to cover, primarily derived from the objectives of the Project (e.g., multi-agency working, perceptions of relationships with the staff, the value of diversion, and problems of achieving diversion), but we also allowed interviewees plenty of space to surface issues in their own terms before we introduced our topics. This technique has the advantage over traditional semi-structured interviewing of giving stakeholders the chance to introduce evaluation criteria that the interviewees, and those involved in setting the objectives of the Project, might not have thought of themselves. In this way, the evaluation criteria were evolved in a partially-directed and partially-participative manner, although even those criteria that were introduced by ourselves following reflection on the objectives were derived participatively in the sense that the objectives had been consensually defined in the earlier workshops with the staff team and management group. As a result of this approach, a number of issues to do with the detail of the way the Project worked, and to do with practices within various agencies, were surfaced that I am sure we would not have thought of had we taken 'expert' control of setting the evaluation agenda (the various stakeholders were the experts in this situation, not us).

Of course, one stakeholder category was service users. While we intended to interview twenty users in the process of shadowing the staff's work (see later), this did not prove possible. The staff team said that it would be unethical to interview someone while in a distressed state (and in many cases it would be impossible anyway because of the extent of emotional and/or mental distress), so they wanted a veto on who we would interview and in what circumstances. Most of the users we saw when shadowing the staff team were indeed in a distressed state, either because of the effects of their mental illness or because of the situation they found themselves in. It was therefore only possible to interview two users in this direct manner.

To make up for this obvious deficiency in our gathering of viewpoints we asked ourselves, how else can we access user views? The answer we came up with, in discussion with the staff team, was to hold a workshop which all previous users (except those in prison[206]) would be invited to participate in. At that time, about fifty people fell into this category. Letters were sent out, and we offered to pay money to cover expenses. We expected a low response given that almost all the people involved were facing serious life crises, but we did not anticipate what actually happened: not one person came!

Once again we had to ask ourselves, how do we gather user perspectives? We came up with two answers, neither of which was as satisfactory as interviews, but they had to suffice in the circumstances. First, whenever we interviewed other stakeholders, we determined to ask them what users had said to them about the Project. Second, we decided to set aside some time and confidential space at the strategic

---

[206] Although those in prison might have had some important stories to tell, we needed to hold the workshop at a venue that most people with mental health problems would consider 'safe'. We therefore selected the offices of the mental health voluntary organisation which was hosting the Project.

planning event to be held at the end of the evaluation, and to which users would be invited, to gather user views.[207]

16.9.3 *Shadowing*

Although we believed that we were going to get some useful material from the interviews, we were also aware that interviews are all about *verbal* descriptions of events. We felt that we needed to be able to separate the 'rhetoric' of the Project from what the staff team actually did on a day-to-day basis. To enable this, we decided to shadow the staff members for several days each: i.e., we accompanied them in their work, observing what went on throughout the time we were with them. Of course, shadowing does not produce an absolutely 'objective' picture of the activities being shadowed because, amongst other things, people are aware of the presence of the evaluators: they might do things that they would normally let slip, and other stakeholders might take the opportunity to raise or suppress issues that they would otherwise have handled differently. Also, even if it were possible for shadowers to avoid having an effect on those being shadowed, their observations still cannot be 'objective' or 'neutral': what is actually seen, and what slips the eye, is in part determined by the framework of concepts and values used to interpret events (see Chapter 6 for a further argument relating to this point). Nevertheless, if triangulated with data from the interviews (and placed in the context of explicit questions guiding observation), we felt that shadowing could produce some useful insights.

16.9.4 *Participant Observation*

In addition to the shadowing, we also engaged in participant observation to a limited degree. We attended all the steering group meetings, and a number of the staff team and management group meetings. Here we were active participants, not observers. We contributed to discussions, and made regular presentations of results to date. We also sought approval from each of the three levels of management for all the methods we proposed using. Participant observation has the same limitations as shadowing, in the sense that no observation can be absolutely objective, but proved a useful extra source of information, especially as it allowed us to see at first hand some of

---

[207] Of course, this event could have fallen prey to the same problem of inadequate user involvement, but we put a great deal of effort into making sure this would not happen. See later in this chapter for details.

the different issues that were preoccupying people at the different levels of management.

16.9.5 Documentary Research

Also, as the Project had been established before either the staff team or ourselves had been recruited, we needed to find out about its history. Conversations with steering group members helped here, but we also accessed documents relating to its foundation. Documentary evidence was also useful for analysing issues surrounding multi-agency working: while the interviews provided the bulk of the material, some claims made in them could be checked by reference to policy documents, minutes of meetings, etc., shown to us by the various participating agencies.

16.9.6 Case Studies

Finally, there was a question about how the qualitative material could be most effectively used to promote worthwhile change. We agreed with the steering group that we would provide quarterly progress reports which would draw upon the data as it was collected. These would be used to inform an on-going review of policy and practice. We also agreed to write a final report which would be distributed to steering group members in first draft form prior to amendment and publication (this would be used to inform a decision on whether or not the Project would receive continuation funding, and would also be passed to the Home Office). In writing the reports, we drew upon the data to construct a narrative. We also developed illustrative case studies to highlight key issues (the choice of user was driven by the issue): the stories of users' experiences punctuated our final report (Cohen and Midgley, 1994), adding depth and human interest to the relatively dry quantitative data.

16.10 Key Issues

The key issues to be examined in this evaluation, identified in an on-going process of debate with stakeholders, personal reflection and data collection were as follows:

- Outcomes for users, and their significance in terms of changes that might need to be made in the diversion system;
- The management of the Project;

- The recording and accessing of information on users;
- The reporting systems, where team members had to report to both the Project and their seconding agencies;
- The resources provided;
- Working practices within the various agencies; and
- The co-ordination of multi-agency working.

A chapter in our final report was devoted to each of these subjects.

16.11 The Strategic Planning

Having conducted the bulk of the empirical work on our evaluation, we came to the strategic planning that we had negotiated right at the beginning. By this time, we had a tentative answer to one of my initial questions: was the Project's focus on direct work with users successful in achieving change in the agencies, or were wider changes in the mental health and criminal justice systems needed that the Project had not managed to facilitate? Our judgement, and the judgement of most stakeholders, was that the focus on work with users had been remarkably effective in getting the agencies to review their practices *in individual cases*, but the fact that the rate of referrals to the Project did not lessen over the year suggested that, on the whole, the Project was being used reactively (to deal with individuals in trouble) rather than proactively (to facilitate change so that these individuals did not get into trouble in the first place). Nevertheless, there were some significant successes in terms of policy changes in the agencies, such as a decision by one Health Authority to support a rota of psychiatrists to be based in the Courts ready to take referrals from the cells.

Some major gaps in the wider system were identified, however: in particular, there was no secure facility for mentally disordered offenders, other than a forensic ward in the local psychiatric hospital which was heavily over-subscribed (and people had to be responsive to medication to be admitted). As a result of this gap, people who had committed serious offences that would normally receive an automatic prison sentence were much less likely to be diverted than those who had committed relatively minor crimes: judges were reluctant to allow a diversion if they felt that the person was not going to be held in secure accommodation. In other words (and this is quite understandable), public safety was given priority over the rehabilitative needs of the offender. Unfortunately, a new secure unit for mentally disordered offenders was beyond the immediately available resources of any of the agencies in the locality, although everybody (including the users) thought that one was needed.

Other gaps in service provision were revealed by the fact that a 'hard core' of mentally disordered offenders were being referred to the Project again and again as they were committing petty crimes on a regular (sometimes daily) basis: none of the current services were meeting their needs. Therefore, the answer to my question about whether the Project had been right to focus on work with users rather than strategic planning is probably both yes *and* no: the staff did a great deal of good in individual cases, but the wider problems were only partially addressed. Given this scenario, our proposal to do some strategic planning (which had only been accepted reluctantly when we first negotiated the remit) was now welcomed enthusiastically.

We needed to make a short sharp intervention, as we had very little time left. We also wanted to make sure that users participated properly in this phase of our work, as one of the problems we had previously faced in the evaluation was a lack of effective user involvement. We could at least ensure that they had a meaningful voice in drawing up plans for future change (and we also planned to take time out to ask them their views on the Project).

In terms of the creative design of methods, we had a number of questions which needed to be answered: how could we ensure that users actually participated? (We didn't want a repeat of our previous 'non-event'). How could we give users space to speak openly and honestly about the issues of concern to them? How could we deal with the emotions that the planning might surface in them? How could we ensure that the planning was widely-focused and solution-orientated rather than narrow and problem-focused? How could we gain the commitment of professionals as well as users to proposals for change? How could we ensure that contentious issues, such as the role of enforced treatment and the rehabilitation/punishment balance, were dealt with openly? How could we ensure that ethical issues were addressed in detail? (Most of the issues surrounding diversion can be considered ethical in nature). And how could we address any disagreements between users and professionals that were surfaced?

It seemed to us that no one method could provide all the answers— but a synergy of two methods, operated in an appropriately tailored manner, might just work. These were the methods from Critical Systems Heuristics (CSH) (Ulrich, 1983) and Interactive Planning (IP) (Ackoff, 1981). I also used this synergy in my work on planning housing services for older people (Chapter 14), but actually the Diversion from Custody Project was the first time the synergy was tried (and since then I have used it on numerous occasions, as have several of my colleagues). Very brief reminders of the key methods and principles from CSH and IP are provided below (I have included the principles because, in this particular case, it was reflection on these that was instrumental in producing the synergy).

16.11.1 *Critical Systems Heuristics*

CSH gives a list of twelve questions that can be used to generate debate during planning. These focus on various issues such as whose interests ought to be served by the development of a system, whose 'expertise' should be accepted, what criteria of evaluation should be used, and who should participate in planning and management. In terms of its principles, Ulrich claims that there is a need to challenge 'experts' and those in positions of authority when they do not take account of others affected by their activities. He suggests that CSH can have a useful role in confronting 'pseudo-dialogue' (insincere communication), but he also suggests that his twelve questions, if answered in meaningful dialogue with stakeholders, can help establish boundaries within which further systems interventions can take place that allow for the transcendence of narrow self-interests so that everybody can benefit. There are therefore two principles lying behind CSH: the 'emancipatory principle', which assumes that there is sometimes a need to challenge those with power because they pursue their own interests with little regard for the interests of others, and the 'participative principle', which (in Ulrich's view) assumes that people can be supported by the use of boundary questions in gaining the competence needed to enter rational debate with others, using a common language, and reach accommodations so as to transcend narrowly-defined interests.

16.11.2 *Interactive Planning*

IP has several aspects to it (see Ackoff, 1981, for full details), but a central concept is 'idealised design'. Idealised design involves the generation through participative debate of a list of 'desired properties' of a system, followed by the production of a design that, if implemented, should make those desired properties a reality. While implementation in its complete form might not be immediately possible, the idealised design nevertheless offers a vision of the future to work towards. As with CSH, it is the participative principle that lies behind IP—although Ackoff supports competent participation through the use of a 'democratic' planning process rather than questions about boundary judgements. Ackoff (1981) actually claims that *any* issue, however large, can be addressed through participative planning, *if* everybody involved is willing to open themselves to dialogue and is also prepared to transcend narrowly defined interests. Here it is important to recognise Ackoff's acknowledgement that pseudo-dialogue will obstruct IP. Indeed, the methods Ackoff offers are not designed to cope with it. The potential for debate and accommodation is essential if IP is to be used effectively.

16.11.3 The Workshops

We decided to work with both service users (people with mental health problems caught up in the criminal justice system) and the staff team and management group[208], whom I will henceforth call the 'professionals' (their role in the strategic planning was not to represent the Project, but to give their professional opinions about the wider changes needed in the mental health and criminal justice systems). However, we were concerned about potential problems of open communication—not because we expected pseudo-dialogue on the part of the professionals, but because we feared that people with mental health problems who depend on professionals for decent treatment (indeed, for their liberty) might be unwilling to contradict professional views.[209] In anticipation of such problems, we decided to conduct separate workshops with the service users and professionals so that both groups could discuss relevant issues in confidence.

We obviously wanted to avoid the fiasco that we had previously experienced when we tried to set up a users' workshop and nobody attended. Therefore, we consulted a local user involvement worker (whose role was to facilitate the participation of users in decision making in mental health services), and she said that she would offer us her support. In addition to mailing all the previous users of the Project, we decided to open the doors of the workshop to anyone with a mental health problem who had recently been in prison. We did this partly as an insurance policy against a low turnout, and partly because, as our main focus was going to be reforming the wider mental health and criminal justice systems (not specifically the Project), mentally disordered offenders other than users of the Project would have an interest in this. To reach these people, the user involvement worker introduced me to a drop-in where I talked with a number of users who

[208] We wanted to work with the steering group too, but they delegated this to the management group.

[209] For more detailed discussions of communication problems during systems interventions in mental health services, see Midgley and Milne (1995) and Thompson (1995).

said that they would spread the word. She also sent letters to a range of services in the locality, asking for appropriate people to be identified and informed of the workshop. Finally, she made personal contact with a handful of people that she knew had been in prison. A total of twelve people turned up for the workshop, about half of whom had been Project users. Interestingly, none of these people were already aware that there were many people with mental health problems in prison without access to appropriate treatment: they all came in feeling that the problem was unique to them. The workshop was therefore a powerful experience for them, allowing them to share their experiences with others who had been there too, and for this reason alone it was worth running.

Each workshop (one for the users and one for the professionals) lasted a full day, with breaks for coffee and lunch. We paid the users £5 to cover their lunch and travelling expenses. I facilitated them alone: in the users' workshop, a user volunteered to act as scribe (recording comments on a flip-chart); in the professionals' workshop I both facilitated and scribed.

I used the twelve questions from CSH in the 'ought' mode to elicit the 'desired properties' of an ideal mental health and criminal justice system that would respect the needs of mentally disordered offenders. However, I rephrased Ulrich's questions in plain English and made them specifically relevant to the context. I then embarked on the second part of IP: production of a creative design that embodies the desired properties. The participants were asked to imagine that both the mental health and criminal justice systems had ceased to exist, and it was their job to design the skeleton of an ideal system—one that would exhibit all the desired properties already listed. In developing their designs, participants were allowed to set up, or propose changes in, any agency they wished, as long as their plans were technologically feasible, viable and adaptable (refer back to Chapter 14 for specific definitions of these terms, which originally came from Ackoff, 1981).

Finally, participants were asked to produce a further design, this time working with current resources only (a team of three people, an office, etc.). In this more restricted design, participants had to assume that the only changes they could make in the system were ones that could be brought about by the people already involved. Generating this more restricted design enabled us to compare the distance between what was immediately achievable and what needed to be done in the longer term (from the various stakeholders' viewpoints). It has to be said, however, that both sets of participants were reluctant to dwell on the constrained designs once they had seen what might be achieved by looking more widely, and therefore these 'immediately achievable' designs were perfunctory at best.

The main output, then, was the two skeletal designs of what the situation ought to be like, one produced by service users and the other by professionals. We reproduced the designs in the form of a report which highlighted the substantial similarities between them, but also indicated where differences lay that would require further discussion and accommodation (Cohen and Midgley, 1994). We would have liked to have facilitated further discussion ourselves, but as we had been forced into doing this strategic planning right at the end of the evaluation, we had to leave it in the hands of those involved in, and affected by, the Project to take up the debate once we had left. Usually, I would not hesitate to continue working on an intervention after its official end, if this is what is needed. However, in this case Claire Cohen and I had to move on to writing up the final report, and we both had other commitments outside this intervention that we had to begin to prioritise.

My experience of conducting these workshops was very positive, especially in the case of the people with mental health problems. Often there is a tendency for such people, when involved in group work, to dwell on problems in a manner that makes them appear unresolvable, but the CSH questions enabled the participants to concentrate on possible *solutions*. This created an uplifting rather than a depressing atmosphere. A down-side, however, was that one individual was overcome by emotion, and left the room: the group tried to provide support within the workshop, but to no avail. Someone went with him to make sure that he was OK, and she returned after about ten minutes on her own. I tried to follow up this person via the mental health voluntary organisation hosting the Project, but unfortunately nobody had met him before and didn't know who he was. In retrospect, I should perhaps have allowed more time in the workshop format to deal with the personal issues that would inevitably be raised.

Now, I have described my method as a *synergy* of CSH and IP. Let me explain why. To recap, the creative design of methods involves understanding the problem situation in terms of a series of systemically interrelated questions expressing the purposes of agents, each of which might need to be addressed using a different method, or part of a method. A *synergy* is generated that allows each question to be addressed as part of a whole *system* of questions. The strategic planning was conceived in just such a manner. Relevant questions to be addressed in the design included, how can we deal with the power relationship between professionals and users? How can we create a clear focus on what *ought* to be done? And how can we liberate people's minds from the constraints of what currently exists so that they can see what might be possible if they were to consider change? Taken as a whole, the method that I designed addressed all of these questions: it reflected a synergy of the emancipatory principle from CSH (concentrating on the identification of power issues) with the participative principles from both CSH (supporting competence in participation through the use of boundary questions) and IP (supporting competence in participation through the constitution of planning groups reflecting different needs and expertise).

16.12 Findings

In our final report (Cohen and Midgley, 1994), details are provided of the characteristics of Project users; numbers of diversions achieved at different stages of the criminal justice system; and the meaning of these figures in the eyes of stakeholders, supported by case study evidence. Key issues were also discussed, and recommendations made, on resourcing for the Project; working practices within the individual agencies; difficulties faced by the Police; the role of psychiatrists; the relationship of the Project with the Crown Prosecution Service; and the co-ordination of inter-agency working. We also presented the full outputs from the strategic planning exercise so that these could be used as blueprints for future change.

I will not go into too much detail here, but suffice it to say that the quantitative analysis revealed a much higher success rate than anyone had expected—even the staff team, who were working directly with users, were genuinely surprised by the figures. The findings of our evaluation can be summarised as follows:

16.12.1 User Characteristics

The user group was predominantly made up of men in their twenties. Many had serious difficulties that were unaddressed at the time of referral (e.g., concerning mental health, drugs, alcohol, family breakdown and/or money), and the Project had to bring support services on line to deal with these. The majority had both previous convictions and a history of mental health problems. A large minority were also considered a danger to themselves or others. In short, the Project aimed its service at those with the most difficult issues to address: many of its users were trapped into cycles of offending and recurring mental health problems. We concluded that any intervention which can be shown to help these people should be looked upon as an important achievement.

16.12.2 Results of Final Interventions

The information above gives some context to the activities of the Project. These activities were described quantitatively through an analysis of the final intervention undertaken with each user (recognising, of course, that the *final* intervention was actually the tip of the iceberg in many cases—27% involved multiple interventions).

11% of final interventions took place in Police Stations. Here the staff team worked with the Police to ensure that people who had been accused of minor offences were diverted before charge—or, when charges were brought, to help secure Police bail. 44% of final interventions took place at bail application in the Courts. Here the team tried to ensure that users were not remanded into custody. These were cases where the crime a person was accused of was unlikely to result in a custodial sentence, but the person was still at risk of being held on remand. The final 45% went all the way through to trial and/or sentence. These were users whose offences were serious enough to put them at risk of a custodial sentence. Here, the Project worked to secure a non-custodial alternative to prison.

Altogether, 85% of final interventions resulted in diversions (a more detailed breakdown of outcomes is provided in Cohen and Midgley, 1994). It is reasonably safe to conclude that many of these diversions would not have happened without the team's interventions: the staff acted in an assessing and co-ordinating capacity, bringing services on line to make non-custodial outcomes possible for mentally disordered offenders. This was something that people in the criminal justice system said was not being done systematically prior to the recruitment of the team.

A simple figure like the above does not say much about the complexities of many of the cases that the staff team dealt with. Sometimes a number of setbacks were experienced on the way to a diversion. Nevertheless, in our final report we argued that, given the nature of the user group, the figure of 85% was something that the Project should be proud of.

The quantitative data, triangulated with the case studies and interview materials, resulted in us recommending that the agencies financing the Diversion from Custody Project should renew their funding for a minimum of five years to give the Project a stable future. Also, we suggested that the steering group should consider how a new service could be offered (to complement the work of the Project) that would provide secure accommodation; would not focus on psychiatric treatment; but would nevertheless have a therapeutic orientation. This would be an alternative to prison for mentally disordered offenders who do not respond to traditional psychiatric treatment and who have committed serious crimes.

16.12.3 The Management Structure

We looked at the management structure and the effectiveness of multi-agency working on the Project. We concluded that decisions made by the steering group appeared to be well informed, and communications between the management group and the steering group were good. However, some problems were experienced in communications between the staff team and management group. These were of a kind that are to be expected in a hierarchical organisation: the staff team had a perception of being excluded from decision making (they were not invited to steering group meetings) and as a consequence found it difficult to broach issues with the management team despite the latter's clearly expressed openness. The hierarchical structure had advantages in a multi-agency project such as this, because senior representatives of all the key agencies could be involved in the steering group, making it easier for the Project to influence the practices of these agencies. However, its disadvantages led us to recommend that a review of the management structure be undertaken, involving collaboration between the steering group, the management team and the staff.

16.12.4 Multi-Agency Working

Multi-agency working took three forms: participation on the steering group; joint working in the staff team; and co-operation between the staff team and other agencies. All the key agencies seemed to participate effectively on the steering group. Also, good working relationships were maintained within the staff team. However, the Project's relationships with other individuals and agencies on the ground were mixed. This was only to be expected given the complexities of multi-agency working, but it has to be said that a minority of stakeholders were actively hostile to the staff team. Our final report details successes and difficulties experienced by the Project in its efforts to work with three key agencies: psychiatrists, the Police and the Crown Prosecution Service.

16.12.5 Results of the Strategic Planning

The outputs from the strategic planning workshops revealed a remarkable similarity between professionals and users in their attitudes to diversion; in their identification of problems; and in their proposed solutions. There was a common focus on the importance of multi-agency co-operation, and an agreement that *nobody* with a mental health problem or a learning disability should spend time in custody—alternatives should *always* be found. New facilities were proposed, such as secure accommodation with a therapeutic orientation for people who are considered a danger to others. Also, the importance of training for the Police in working with mentally disordered offenders was discussed, and an ideal model of multi-agency working was developed.

The only major difference between the users and professionals was in respect of the nature of the proposed 'secure accommodation': the professionals gave Rampton as an ideal model, while the users said that replicating Rampton is precisely what should be avoided (in their view, Rampton used an oppressive regime).[210]

These workshops led us to recommend that the Project should use the ideas for change produced by the users and professionals as the basis for future planning within the locality to guide the co-ordination and development of the mental health and criminal justice systems as a whole. In line with this, we also suggested that, once the Project had been made financially secure, the steering group should place less emphasis on project management and more on their wider planning role (also paying attention to the identified need to contribute to the development of diversion nationally). Finally, we recommended that the Project should set up a user planning forum which has flexible representation on the multi-agency group. This recommendation was made, partly due to a general recognition by stakeholders that user involvement in mental health services is important, but also because the user group wanted to meet again and maintain a continued identity with a role in the Project.

16.13 Outcomes

When we circulated a first draft of the report to stakeholders, there was a great deal of written feedback—especially from some of the agencies which had been criticised for failing to work effectively with the Project. What was interesting, however, was the speed at which the problems we had identified were rectified. One particular

[210] Rampton is a 'famous' or 'notorious' (depending on your point of view) psychiatric hospital in the UK which specialises in the treatment of dangerous criminals. Unlike most psychiatric facilities, which offer only short-stay accommodation while a person is in crisis, people are often kept in Rampton for many years.

problematic relationship, which had caused protracted difficulties for the Project for the whole year, was turned around in a matter of days: the staff team said that the change was amazing, and completely unexpected. Having taken speedy action, the senior agency representatives on the steering group then called us to a special meeting at which they asked us to amend the report, making it clear that these problems were now in the past and that effective action had been taken. We checked with the staff team to see whether the changes appeared to be lasting, and when they confirmed that this was the case we agreed to make the amendments.

It is interesting to note that this experience runs contrary to the usual reaction reported in the literature to summative evaluations which deliver mixed or bad news: it is common for reports to be shelved, and then to gather dust, with no action being taken to remedy the problems identified (Cronbach, 1982). I attribute the prompt action taken by the agencies in this case to the fact that the Home Office (which they were accountable to) was going to receive the report, and they wanted to avoid public criticism. Therefore, it would appear that the choice of a summative method was the right one in this instance.

I should also note that the Project did indeed receive continued funding, and eventually new premises of its own. Feedback from the Home Office was also very positive: the Chair of the steering group reported that the Home Office was encouraged by the successes demonstrated through this work, and made particular mention of the synergy of Critical Systems Heuristics and Interactive Planning which produced clear and compelling recommendations for the future of the wider mental health and criminal justice systems.

16.14 Conclusion

In this chapter, I have provided an example of the practice of the creative design of methods (which, of course, incorporates boundary critique—refer back to Chapter 14 for a detailed example of this). Over the course of the intervention we supported team-building (using our own method focusing on objectives and language); operational planning (using the methods from Soft Systems Methodology); and IS planning (linking Soft Systems Methodology with Jackson System Development). We also evaluated the Project using a mixture of quantitative and qualitative methods (quantitative data gathering, stakeholder analysis, interviewing, shadowing, participant observation, documentary research and the compilation of case studies). This evaluation was essentially summative (based on the presentation of a final report), but had formative elements too, in that all levels of management in the Project received regular up-dates. There was also an emphasis on participation: again, all levels of management were involved in the decision making about which methods to use (although user involvement proved problematic), and we were always able to proceed on the basis of a consensus. Finally, we engaged in a short strategic planning exercise, synergising aspects of Critical Systems Heuristics and Interactive Planning into a single method which produced compelling results in terms of re-visioning the mental health and criminal justice systems in the locality so that the needs of mentally disordered offenders could be properly accounted for in future years.

17

Developing Services with Young People (Under 16) Missing from Home or Care

The intervention reported in this chapter involved the mixing of methods through creative design (like the Diversion from Custody evaluation reported in Chapter 16), but was substantially different in its orientation. There was no request for a summative judgement, and indeed extensive discussions about methodology with some of the stakeholders revealed that they shared my own misgivings about summative evaluation—that it tends to make people defensive and resistant to change (see also Cronbach, 1982). While I would never say 'never' to conducting a summative evaluation (I have actually engaged in several, such as the one described in Chapter 16), I nevertheless argue that the circumstances have to be right for it to be a valuable form of intervention (as they were, in my view, in the case of the Diversion from Custody evaluation—but were not in this case).

17.1 Young People on the Streets

The intervention reported in this chapter was commissioned by a consortium of voluntary organisations (the Children's Society, NCH Action for Children, and the Manchester & Salford Methodist Mission) who wanted some action research to (i) investigate the situation faced in Central Manchester by 'detached' under-16 year olds (young people living on the streets[211]); (ii) identify appropriate responses and services

[211] Many young people in their teens resent being called 'children' and prefer the label 'young people'. As a mark of respect for this, I have called them 'young people' in this chapter, even though 'young people on the streets' doesn't have quite the same emotional impact as '*children* on the streets'.

to meet the needs of this group; and (iii) assess the feasibility of potential projects.

Stein *et al* (1999) conducted a nation-wide survey of young people under 16 in the UK and concluded the following:

> "Overall, we estimate that 11% of young people run away for one night or more on one or more occasions before the age of 16, amounting to around 77,000 young people running away for the first time each year.... At first glance, these findings may seem rather bland. However, they are of considerable importance in that they represent, for the first time, firm and reliable evidence that there is a significant prevalence of running away in all parts of the UK. The implication of this is clearly that services must be developed in all kinds of areas.... if there is to be an effective and inclusive response to the needs of young people who run away. Whilst many young people only run away once, others go on to run away repeatedly. We estimate that there are a total of around 129,000 incidents of running away over night per year in the UK" (Stein et al, 1999, p.38).

Some of these young people stay missing for a considerable time and become completely 'detached' from the parents, carers and agencies that should be supporting them (English, 1973; Barter, 1996). As they have no source of income, they have to develop survival strategies that can expose them to risks of violence, involvement in crime, drug or alcohol misuse and sexual exploitation (Morgan-Klein, 1990). They do not have access to educational opportunities, primary health care or safe accommodation (Stein *et al*, 1994; Power *et al*, 1995), and their mental and physical health can be compromised (van der Ploeg and Scholte, 1997). In the UK, work with these young people is generally underfunded, unpopular and poorly co-ordinated: responsibilities tend to fall between different agencies, and service provision is piecemeal at best, failing to meet the full range of needs of young people on the streets (Boyd *et al*, 1999a).

During our intervention, the Police in Manchester intuitively estimated that they dealt with about 2,000 young people living on the streets of that city in the previous year alone. When I first heard this I was shocked, and most of the people I have talked with about it have the same reaction. It is largely a 'hidden' problem—something we expect in Latin America, where absolute poverty can be found in just about every major city, but not in the UK with its supposedly advanced welfare state. It seems that, for the most part, our society manages to keep absolute poverty (including that faced by young people under 16) invisible.[212]

---

[212] I am aware, from living in London until 1990, that some shops in the centre of London pay a 'service' to go onto the streets at 5.00am and hose down homeless people sleeping in the shop doorways. By the time the tourists arrive the homeless people have disappeared,

17.2 First Contacts

I first found out about the possibility of conducting this intervention when Mandy Brown, one of my colleagues, was contacted by the Children's Society, NCH Action for Children, and the Manchester & Salford Methodist Mission. They had heard about our work at the Centre for Systems Studies from Roger Adams, an internal consultant in the Children's Society. The consortium had short-listed ourselves along with nine other research groups whom they thought might be appropriate to undertake the intervention. We were invited to tender for the work.

The money on offer was sufficient to employ a half-time Research Assistant for nine months, and to pay for a couple of months of Mandy Brown's and my own time. We decided to invite Alan Boyd, who had recently started a part-time Ph.D. under my supervision and who lived in Manchester, to work with us on developing a tender, with the idea that he would be the half-time Research Assistant if our bid was successful.

The request for tenders was quite explicit about the fact that the voluntary organisations wanted the work to be undertaken in an *action research* mode (i.e., it was quite clearly an intervention, not a pretence at observation); they believed that the exploration of values should be at the heart of the intervention; the interveners had to have experience of working with young people[213]; and they had to demonstrate commitments to the participation of young people, equal opportunities and anti-oppressive practice. Given everything I argued about methodology in Section Two of this book, it should be quite clear that the values of the commissioning organisations and our own were very close (at least it seemed that way from the documentation, and this was later confirmed by our experience of working with the agency representatives). In addition, the three of us were in agreement that this was an important project to put our time into, in the sense that

---

and London looks prosperous and clean. Of course, as anybody who has regularly walked through London knows, they re-appear in the course of the day, many begging on the streets. In my experience, in the 1970s, homeless people were few and far between in London, and individuals were mostly known to their local communities, but in the 1980s the numbers began to increase enormously and they mostly became anonymous bodies on the streets. In addition to the people with serious alcohol and/or mental health problems who have always been on the streets, there are now thousands of ordinary (mostly young) people who simply do not have an income or a home to go to.

213 Although none of us had previously done systemic intervention work with young people, Mandy Brown had worked as a youth worker in a children's home, Alan Boyd had run a church youth group, and I had worked in a residential home for adolescents with mental health problems.

anything we could contribute to improving the lives of young people on the streets, many of whom are in crisis and extremely vulnerable, would be very much worthwhile. We felt that we could make a positive difference. The obvious 'fit' between the consortium's and our own expressed purposes and values encouraged us to put some considerable work into producing our bid, and we were rewarded with the contract.

17.3 Our Initial Proposal

Our specification of methods in the tender document was quite general, proposing an initial period of interviews and workshops with young people on the streets, feeding into a series of multi-agency workshops (with participation from young people) to develop ideas for change. We proposed interviews followed by planning workshops for three reasons. First, it is important for agencies to listen to the voices of young people so that the services they design make a genuine difference to their lives, in the young people's own terms rather than in terms set by a purely 'adult' agenda. A young person rarely runs away 'for fun': there is usually a serious problem that s/he is trying to escape, and for this to be addressed the young person must first be heard.[214]

Second, young people still occupy a marginal position in British society (albeit less so than was once the case), in the sense that in many families they are seen as subordinates; they are not regarded in law as capable of making 'adult' decisions; and they are often looked upon as less 'rational' than people over 16. Indeed, young people have no legal right to control over some aspects of their lives that I personally regard as essential.[215] We therefore wanted to be sure that young people's views were included in a meaningful manner, and we were conscious of the need to avoid replicating their marginalisation (as far as possible) in the design of our intervention.

---

[214] There are, of course, exceptions to this generalisation, like a group of regular runaways we interviewed in a children's home who claimed that they absconded in order to make money from illegal activities. However, all the other young people we talked with said that they were running away from abuse or entrenched problems at home.

[215] We came up against a good example of this in our intervention. When children are in care, it appears that parents have an automatic right to visit them—regardless of whether the child wants this or not. We met several young people who said that they did not want their weekly visit from a parent whom they regarded as an abuser, even though the meeting was supervised. One young woman said that she wanted to break off all contact with her mother, yet she was being prevented from doing so by the care home staff. However painful this is for the parent, I feel that young people should be able to choose not to have enforced social contact of this kind.

Our third reason for interviewing young people before starting a series of planning workshops was strategic. We were aware that most workers in the 'caring' professions are graduates who have been socialised into an 'observational' mode of research: even if qualitative methods are accepted, there is usually still an assumption that these will be used to *find out about* the situation rather than change it. As the commissioning organisations had asked explicitly for *action* research, we knew that we would not encounter major resistance to intervention, but we were still aware that some interpretations of action research are less explicitly interventionary than our own (applied social science, where researchers conduct a study and then write it up for others to use to make changes, is occasionally confused with action research). Therefore, we thought it would be wise to talk about finding out about the situation before launching into planning workshops.[216]

Although we set out this general direction for the intervention, we made it clear that local contingencies may require a change of direction, and if this turned out to be the case it would be negotiated with our steering group (made up of representatives of the three funding organisations). We did indeed find it necessary to depart from our original proposal, and many things happened that we had not anticipated—so, rather than dwell on the original proposal, I will provide an overview of our approach (next), and then a narrative of the course of events, the questions they raised, and how we answered them through the creative design of methods.

17.4 An Overview of the Intervention

Altogether, there were four phases in the intervention. Phase One was about identifying key issues (asking 'what *is* the case?') through interviews with young people on the streets, thereby building up a picture of their situation from their own point of view. Phase Two was about identifying important goals for different stakeholders (specifying 'what *ought* to be the case?'), including young people aged under 16; young people over 16 with previous experiences of running away; carers; and all the various agencies involved in working with young people.

[216] In an evaluation session, conducted near the end of our intervention, Roger Adams (a Children's Society action researcher) expressed surprise that we had not gone straight for planning workshops involving a partnership of young people and professionals. Interestingly though, the consensus amongst the agency representatives was that our initial investigative research based on interviews with young people was an essential aspect of our intervention: it produced a powerful picture of the needs of young people which agency representatives felt that they had to respond to by making positive changes.

This was done through a series of workshops. Phase Three involved workshops with individual agencies, and multi-agency groups, to plan how to realise the goals identified. Phase Four involved communicating about the project through a variety of publications aimed at different audiences.

Phases One to Three took nine months to complete, and then Phase Four (dissemination) was completed in the six months after the official end of the work. A narrative of our progress through the intervention is provided below.

17.5 Establishing the Ground Rules

We started the intervention with a series of meetings with our steering group to work out the details of how we were going to proceed. Because we were going to talk with vulnerable young people under 16, we were asked to produce written documentation about our ethics, principles and practices; design a complaints procedure in case any young person had a grievance against us; and agree a confidentiality statement.[217] We also agreed that any quotations by young people reproduced in reports would be anonymised. It is a legal requirement in the UK that, whenever someone under 16 is interviewed, an 'appropriate adult' other than the interviewer is present to represent the young person's interests (see Sieber, 1992, for a discussion of the ethical issues associated with involving potentially vulnerable people in research). The Children's Society volunteered two of their street workers, a man and a woman, to act as appropriate adults and accompany us during interviews and workshops.

Another feature of these meetings was a series of discussions about how we were going to get access to young people and involve them in the intervention. In Chapter 16 I described the almost insurmountable difficulties I faced when trying to get 'mentally disordered offenders' to participate in interviews and a workshop—yet in this case we were talking about young people with no fixed abode, many of whom were actively avoiding any contact with the 'authorities' for fear of being returned to the abuse they had fled from. The situation was therefore

[217] The confidentiality statement we ended up with was adapted from one used by the Children's Society, whose street workers go out at night to make contact with young people on the streets. It basically states that all discussions remain confidential unless a young person discloses harm being done to themselves or others. This caveat was necessary because of the possibility that a young person might allege abuse. Indeed, an allegation of physical abuse at a children's home was made during the course of our intervention, which the Children's Society followed up on our behalf.

even more difficult than the one detailed in Chapter 16. I wanted to learn from the problems I had encountered in that earlier intervention, and make sure these issues were addressed from the start. Again, for the interviews, the Children's Society street workers were the answer: they knew where the young people hung out, and volunteered to take us around in the evenings and at night to make contacts and conduct interviews. However, we were told that it would most likely be impossible to arrange a participative workshop with young people: they tend to live hand-to-mouth, one day at a time, and would not be able to plan for attending such an event. An alternative, however, would be to hold a workshop with young people who had previously run away from children's homes, so this is what we agreed to do.

17.6 Phase One: Identifying Key Issues

We considered a variety of options for identifying the key issues from the perspectives of young people—we did not automatically go for interviews. One idea was to produce a video or photographic exhibition in partnership with young people on the streets, as suggested by Dewdney *et al* (1994).[218] Although this seemed like an attractive option because it would present the issues in an emotionally powerful manner, making it hard for the agency representatives planning new services not to respond positively (and it would also be fun for the young people), we ended up rejecting it on the grounds of lack of time, the need for confidentiality, and lack of sufficient expertise to guarantee a polished product.

We therefore accepted that interviews were going to be the most effective option, but what kind would be best? I had recently co-facilitated a training course with senior managers in health services, and the lead facilitator was Susan Weil (from the SOLAR Centre at University College Northampton). She had introduced me to a technique for surfacing issues that involved the use of photographs to stimulate debate. The idea was that people pick photographs that have some meaning for their lives, and then discuss the issues they raise (Prosser, 1998). Alan Boyd and Mandy Brown had also experienced

[218] Maturana's (1988a,b) theory of autopoiesis, and Bilson's (1996, 1997) application of it in interventions, informed my thinking about this. Maturana (1988a,b) argues that people are able to switch to a new 'rational domain' (they can think in new ways) if an emotional reaction is triggered that makes the switch compelling. Bilson (1996, 1997) explains how emotions can be stirred through the use of personal anecdotes when presenting research data. A video, which allows personal appeals to be made by young people, would be even more compelling.

the use of this technique through their participation in one of Susan Weil's workshops (Weil, 1998b), had found it useful, and were keen to try it out.

However, because the situation we were entering was very different (the interviews would be on the streets at night, not in a comfortable, centrally-heated office), we were unsure exactly how the use of this method might pan out. We therefore decided to pilot it. We gathered together two sets of images: one was photographs, culled from magazines, that seemed (in our eyes) to be relevant to the situation of homeless young people; and the other was a set of cards with evocative paintings which had been specially designed for raising issues in the context of counselling.[219] We went out for a couple of nights, using the cards with two young people; the photographs with another two; and ordinary questioning (without visual images) with a final two. We compared the outputs from the interviews and asked the young people themselves what they thought of the pictures, and the consensus seemed to be that the visual images did not add much value to the interviews. There was one exception to this, however, which was a young man with a learning disability who seemed to find the cards helpful. We therefore decided to go with ordinary semi-structured interviews, and hold the cards in reserve in case they were needed in individual cases.

17.6.1 Designing the Interviews

In planning the interviews, consideration was given to producing an account of the situation of 'detached' young people that would engage and interest other stakeholders. We therefore wanted to elicit compelling stories about the young people's lives. We also wanted to strike a balance between our own wish to get these stories (which would require some imposed structuring from us), and the wishes of young people to use the interviews in other ways (if indeed they did): for example, we foresaw the possibility that some young people might want to use the interviews therapeutically (which might happen directly through the discussion of issues, and indirectly on account of information about their individual situation being heard by the appropriate adult who would be in a position to offer longer term support).

---

[219] A couple of years ago I visited a counsellor when I found that personal and work issues were becoming entangled in an uncomfortable and unpredictable manner. She helped me work through the issues, and one of the techniques she used was to have me explore the meanings of these picture cards.

We used some of the literature on interview methods[220] to inform the drawing up of guidelines (which were approved by our steering group), and these can be found in our final report on the intervention (Boyd *et al*, 1999a). We also used a pre-prepared list of questions to guide the interviews, but this was not intended to be overly prescriptive, and generally questions arose naturally out of a fairly open-ended discussion (see also Foddy, 1993). An expenses payment of £5 was made at the end of each interview.

There were, of course, inherent limitations regarding what could be achieved by way of involving 'detached' young people through interviews, and we decided that we simply had to accept these. Young people who had been away from home for a relatively brief length of time might either be in a state of crisis or on a high with the novelty of being away, and participation in our intervention might simply be seen as an unwelcome imposition. Also, these young people would quite likely move on within a few days, making them unavailable for more than one-off involvement. Furthermore, we envisaged that involving black young people would be more difficult than involving their white counterparts, since the former tend to make more use of networks of friends and family, and generally adopt a lower profile to avoid racism (Julienne, 1998).

We recognised, however, that it would be possible to contact other young people with relevant views and experiences, including older young homeless people who had been runaways in the past (van der Ploeg and Scholte's, 1997, research suggests that a relatively high proportion of homeless adults started living on the streets as teenagers). It would also be possible to involve young people who had been missing in the past, but who were currently living at home or in care (as suggested by our steering group). More generally, we felt that young people in care could make a useful contribution, even if they had not gone missing themselves: this is because they are at a high risk of running, and most would be living with other young people who had gone missing in the past.

Our steering group suggested that some young people might prefer to be interviewed with just ourselves and the appropriate adult present, while others might prefer to be seen together with a group of friends (for support). They argued that we should respect the young people's wishes on this, not try to make them talk individually if they did not want to. When it actually came to it, most were happy to talk on their

---

220 See, for example, Rich (1968); Bradburn and Sudman (1980); Powney and Watts (1987); Fine and Sandstrom (1988); Freire and Faundez (1989); Morrissette (1992); Weber *et al* (1994); Holstein and Gubrium (1995); Drever (1995); Kvale (1996); and Graue and Walsh (1998).

own, but four group interviews were undertaken. In these, we tried to encourage open, respectful but critical discussion amongst the participants.

At this early stage we also asked ourselves, should we involve other stakeholders? We thought of various 'pros' and 'cons'. An important 'pro' was the fact that involving others would enable us to build a richer picture of the situation, including a diversity of perspectives. A 'con', however, was the desire to ensure that young people were really heard by the agencies: there was a possibility that their voices could be diluted by mixing in other viewpoints. We therefore decided not to involve any 'professionals' at this stage, but thought it worthwhile to consider the involvement of other non-professionals who might find it difficult to get their voices heard. We had parents and foster carers in mind. Our steering group advised us that it would not be easy to involve parents, partly because they would be difficult to contact (none of the organisations we were in touch with at that time kept records of their names and addresses), and it was also unlikely that they would want to participate because of the stigma, and because of their emotional state if their child was missing at the time. Nevertheless, we thought we should try. We got the support of the local newspaper and local radio, and asked for members of the public with an interest in child homelessness, including parents of young people who had run away, to come forward. Unfortunately, none did. However, involving foster carers (whom we contacted via Social Services) was much easier, and several were interviewed.

17.6.2 Conducting the Interviews

When it came to interviewing the young people, we felt that it was important to have some preliminary discussion with potential interviewees to assess their current situation, and for an informed decision to be made about whether the interview should go ahead. Introductions were led by the appropriate adult, who in most cases was already known to the young person. We always tried to maintain a gender mix among the adults: if the interviewer was male, we asked for a female appropriate adult, and sometimes Alan Boyd and Mandy Brown would conduct interviews together. Before beginning, we ran through the contents of a leaflet which we would give to the young person, explaining the main aims of the intervention. We then gave the young person a copy of the confidentiality statement, explained it, and asked if it would be OK to record the interview and take some notes (saying that these would help in putting across to the agencies what young people had to say). Potential interviewees were encouraged to ask questions, expressing any concerns that they might have about the interview, and to decide whether they wanted to go ahead with it, and on what terms.

In addition to participating in the discussions prior to the interview, the appropriate adult kept time and observed what appeared to make the process effective or ineffective for the young person. By and large, the roles of the interviewer and the appropriate adult were kept separate, but occasionally the appropriate adult did ask questions during the interview, either out of interest, or sometimes because they had encountered the young person previously and felt that they could help the interviewer to reveal relevant information.

For each interview, the date, time and location were noted, together with the age, gender and ethnic origin of the interviewee. Young people were not asked about their ethnic origin or sexual orientation, as it was felt that this might inhibit the establishment of a rapport, so ethnicity was a judgement by the interviewer. Some young people did, however, make their sexuality known.

The series of interviews ceased when we felt that no new information was being revealed. 23 young people were involved in total. Specific details of the characteristics of these interviewees can be found in our final report. No young person refused to be interviewed or to have their interview tape recorded, and all except two of the recordings could be fully deciphered. Because the young people volunteered to be interviewed, and were keen to tell their stories, the presence of an appropriate adult and a tape recorder appeared to be relatively minor barriers. The 'externality' of ourselves as interveners, being from a University and (in the case of two of us) from another city, may also have enabled interviewees to be open about their experiences.

At the end of the interviews, interviewees were asked how they had felt during the interview and how it could have been improved. Although they inevitably found aspects stressful, as confirmed on occasion by their body language, they did not find the experience negative overall, and some said that it had definitely been worthwhile from their point of view. On a number of occasions, the appropriate adults who sat in on the interviews said afterwards that they had learned new information about the young person, some of which was relevant to their future work with him or her. The value of the process to the young people was indicated by the fact that, although the majority of them were happy to end their involvement with the intervention at this stage, four subsequently became involved in later phases of it.

The expenses payment proved to be an incentive for some young people to participate who would not have done so otherwise, this being particularly apparent for young people at a children's home who regularly went missing and claimed that they did so in order to acquire money through illegal activities. For most, however, the chief motivation appeared to be the opportunity to tell their story to someone who would listen, and might bring about change.

This first phase of the intervention lasted two months, and we worked a couple of nights per week on it. Although only 23 young people were involved, it was generally only possible to conduct a couple of interviews per night, as most of the time was spent just looking for new interviewees to talk with.

17.7 Phase Two: Evolving Stakeholder Goals

Having explored what *is* the case (from the perspectives of young people, plus several foster carers) in the first two months of the intervention, we then wanted to move on to look at what *ought* to happen (from the points of view of a wide range of stakeholders).[221] In putting together our original tender document, we had in mind that the interviews would be followed by a series of planning workshops, and that these would be similar to the planning workshops I had run in previous interventions with the Diversion from Custody Project in 1992-3 (Chapter 16), and with agencies involved in providing housing services for older people in 1994-6 (Chapter 14). In both those interventions, which we judged to be successful, we created a synergy of methods from Critical Systems Heuristics (CSH) (Ulrich, 1983) and Interactive Planning (IP) (Ackoff, 1981). See those chapters for details.

17.7.1 Designing the Workshops

As we entered the second phase of our intervention, we reviewed the idea of a CSH/IP synergy in discussion with our steering group. First of all we asked the question, who are the key stakeholders? In addition

[221] This division between 'is' and 'ought' was derived from Ulrich's (1983) methodology of Critical Systems Heuristics (CSH), although his specific method (use of the twelve questions detailed in Chapter 7) played a fairly small role in the intervention (see later). Of course, the distinction between 'is' and 'ought' is not unique to CSH: it derives from Kant (1787, 1788), and has been advocated by a variety of critical theorists (e.g., Horkheimer, 1937; Wellmer, 1970; Habermas, 1971; Fay, 1975).

to young people and carers, a wide range of agencies were identified, including Social Services (which runs several children's homes, and has responsibility for child protection more generally), the Housing Department, the Police, the Education Department, a Methodist Housing Association, the local TEC (a training facility for unemployed young people), the Careers Service (which provides careers guidance, mostly in schools), a drop-in for homeless people, Safe in the City (a Children's Society project working on the streets with homeless young people), a mental health crisis service, a mental health drop-in centre, the Health Authority, NCH Action for Children (a voluntary organisation running children's homes) and the Manchester & Salford Methodist Mission.

We discussed the fact that, at this stage, there was no need for a great deal of detail: the agencies did not already have a clear idea how to move forward, so it would be more important to communicate what the young people had to say, and to surface the key values which should inform the development of services. In addition, it was anticipated that time would be an issue for the agency representatives: it would be unlikely that they would be able to put aside more than half a day at a time, and then only on an occasional basis (none of the agencies other than the Children's Society had staff whose sole remit was homeless young people). We were told that there was some scepticism about the possibilities for effective multi-agency working in the locality (there was a history of conflict between a couple of the agencies, and some previous attempts at co-operation had been seen as 'all talk and no action'), so it would be important to make a positive impact first time, otherwise there was a risk that participation could fall away sharply. All of this confirmed our belief that the CSH/IP synergy was appropriate: it can indeed be carried out in half a day (unlike Soft Systems Methodology, for example, which I have found has to be conducted over a period of several days at least if meaningful results are to be obtained); it focuses on values; it does not assume that people already have a strategic direction in mind; and it tends to produce a striking consensus (in my experience the areas of disagreement that are surfaced are minor in comparison with the areas of agreement), so people almost always get a sense of positive achievement from its use.

We also discussed with our steering group the possible participation of young people in the workshops. We had already agreed that we should work with groups of ex-runaways in children's homes rather than young people who were living on the streets because of the logistical problems of getting the latter to attend workshops (see earlier). So the question was not 'who should be involved?', but 'how should the involvement be handled?' We determined that it would not be advisable for young people and other stakeholders to attend the same workshops, at least initially: some young people may be embarrassed in front of professionals by poor reading and writing skills; they would be less used to participating in relatively large meetings; and they might prefer a more 'playful', creative approach. In addition, young people might be inhibited by the presence of staff from Social Services and the Police, as many regard these agencies as part of the problem they are trying to deal with (both agencies are required by law to return young people to where they have run from unless there is clear evidence of abuse[222]). Also, more generally, lower status tends to be accorded to children's views compared with the views of adults (see the earlier discussion of the marginalisation of young people), and we did not want to replicate this dynamic in our workshops. For all these reasons, we decided to work separately with professionals and young people. Indeed, working with young people in a separate group had the added advantage of enabling their viewpoints to be built up and thought through without being 'squashed' at an embryonic stage by adults.

The next question was, would it be better for all the agencies to meet together in a multi-agency forum, or would it be preferable to keep them separate? Again, there were both 'pros' and 'cons' to consider. In favour of working with the agencies separately was the history of conflict that had dogged some inter-agency relationships—and the Children's Society, one of the funders of our intervention, was an agency that had been a participant in this conflict (so we had to consider the possibility that we might be perceived as having a vested interest[223]). The last thing we wanted at our crucial first meeting was pseudo-dialogue, with agency representatives skating around difficult issues, knowing that they might reach a verbal agreement but actions would not follow from their words. However, in favour of bringing the agencies together was the fact that they were mostly at a very early stage in their thinking about how to offer support to homeless young people (some had policy statements, but only the Children's Society was actually offering a service to young people on the streets), suggesting that the agency representatives might be open to setting aside previous

---

[222] The problem is, when there is abuse, young people are generally reluctant to disclose it to the Police or Social Services, so they are nearly always returned to the abusive situation.

[223] As it happened, the fact that we were employees of a University from another city meant more to the key stakeholders in the conflict than where the money for the intervention was coming from. We were therefore viewed positively, as external facilitators able to broker change, rather than people with a vested interest in taking sides.

conflicts in order to forge a new, consensual direction. In addition, time was short[224], and working with all the agencies together would mean one or two workshops rather than eight or ten. We therefore went with the idea of a multi-agency workshop, knowing that we would have to deal with the history of conflict along the way.

In making ready for this, we prepared the following set of goals which we sent through the post and asked the participants to commit to from the start:

Overall goals:

- Produce an innovative vision to guide the genuine improvement of services for young people who go missing from home or care;
- Say who should be involved in the more detailed planning of these services, and in what ways;
- Identify what can realistically be achieved;
- Identify barriers to be overcome and possible constraints to be tested; and
- Build commitment to bringing about useful change.

Goals for individual participants (including the workshop facilitators):

- Identify changes that might usefully be made to your situation and your role in it;
- Develop productive relationships with other participants; and
- Learn new approaches for developing services and working in groups.

We also decided to build in some formal evaluation of our activities to inform both our future decision making in this intervention and our learning more generally. Alan Boyd therefore designed a questionnaire to be filled in by participants after their involvement in a workshop. This asked questions about people's perceptions of what happened, and their commitment to change (or lack of it) based on the results.

---

[224] Although we had nine months to complete the intervention, in our judgement we were trying to cram eighteen months worth of work into this time. We were explicit about this from the start in discussions with our steering group, as we knew that negotiating what we did in relation to the time available was always going to be an issue.

17.7.2 Conducting the Workshops

We set out to conduct three workshops at this point: one with young people under 16 (from children's homes); one with representatives from the various agencies; and one with homeless young people over 16 who had previously been runaways. Only one homeless person over 16 turned up to the latter workshop, so we talked with her individually. However, the other two workshops were well attended: five young people aged 12-15 came to the young people's workshop (all of whom had previously lived on the streets), and twelve agency representatives came to the multi-agency workshop.

17.7.3 The Multi-Agency Workshop

The first one was the multi-agency workshop, which was attended by representatives of all the organisations listed earlier (except the Education Department). Interestingly, the Police and Children's Society representatives came early, and began to chat informally about some of the inter-agency conflicts that they had both previously been involved in. The two agencies were concerned about how the media were to be handled in this intervention because it turned out that, in their view, a great deal of the conflict resulted from the media misrepresenting the Children's Society's point of view on a sensitive local issue, thereby causing a problem for the Police. Therefore, before the workshops had even begun, work was already in hand to set the conflict aside—and without the need for our own intervention, other than setting up a forum where contact between people was made possible (these discussions continued in the break, and then many jokes about the conflict were cracked after the workshop was over, suggesting that it was well on the way to being overcome).

The workshop started with a presentation of the key issues that had been revealed by the interviews with young people. Prior to the workshop we asked ourselves, how can we present the issues in a manner that will engage the agencies, but without breaking the confidentiality of any of the young people by revealing too much from one person's story? Also, how can we demonstrate the systemic nature of the problems? They were highly interactive, and young people often described vicious cycles that they got into. Finally we asked ourselves, how can we show that each of the agencies can play a part in solving these problems?

We were aware of several techniques for presenting systemic problems. First, there is Checkland's (1981) idea of producing 'rich pictures' (visual representations of issues linked with arrows, as described in Chapter 15 of this book). The only drawback of this, in my experience, is that the method works best when people can produce their own rich picture: if a picture is presented to an audience by an 'outside' analyst, it is only meaningful if it is fairly simple. Another possibility was the use of 'signed digraphs', 'influence diagrams' or 'qualitative system dynamics models' (Coyle, 1977; Roberts *et al*, 1983; Flood and Carson, 1993; Morecroft and Sternman, 1994): these are three different terms for a basic kind of system model representing causal links between phenomena (i.e., phenomenon *X* makes a quantifiable change in phenomenon *Y*). The advantage of these modelling techniques is that they can make feedback loops (vicious or virtuous cycles) very visible—but a significant disadvantage is that they use symbolic languages which, while fairly basic, are not immediately transparent to an untrained audience. They are arguably more self-explanatory than a complex rich picture, but they lack the appeal of visual imagery. Another alternative is 'problem mapping' (Midgley *et al*, 1997), which I used in the intervention reported in Chapter 14, and which shows the links between stakeholders' verbal problem statements—but like influence diagramming, it does not have the immediate visual appeal of pictures.[225] Eventually we decided to go for rich pictures, but endeavoured to keep them simple, using cartoons (clip-art from a computer package) linked by arrows to make the vicious cycles as immediately apparent as possible.

We used the rich pictures as a visual aid to describe the systemic problems that young people found themselves caught up in, and added quotations for emotional impact. When our presentation was finished, there was a general discussion in which agreement was expressed that the issues were accurately represented and something definitely needed to be done about them. There was a strong feeling that the words of the young people were powerful and should not be ignored. We therefore moved on to the CSH/IP synergy, as planned. As in previous interventions (Chapters 14 and 16), we asked the participants to imagine that all services for young homeless people had disappeared from the face of the Earth, and their job was to plan a new system from scratch. We presented the three constraints which were designed to stop the plans from becoming utopian (again, see Chapters 14 and 16 for details), and used the CSH questions to facilitate debate. However, in

[225] See Flood and Carson (1993) for a good overview of diagramming techniques (although 'problem mapping' is not included, as it was invented later).

contrast with the Diversion from Custody evaluation (Chapter 16) and the redesign of housing services for older people (Chapter 14), we did not work through the CSH questions systematically: we simply held them in mind and went with the flow of discussion.

There was some unwillingness to get into a great deal of detail about the desired properties of services, and group reflection on this brought up the issue of scepticism about multi-agency working. Several agency representatives said that there was no point getting into the detail when they were unsure about whether there was even agreement on basic values—especially as some of the agencies' statutory roles were part of the problem (e.g., the obligation on the Police and Social Services to return young people to the places they run from). We therefore spent most of the rest of the workshop discussing values before producing a very general ideal for the production of new services. Encouragingly, there was indeed agreement—and the Police and Social Services representatives made it clear that they shared the values being expressed by others, and were aware of (and wanted to do something about) their roles in the vicious cycles.

At the end of the workshop we had a list of core values and a skeletal ideal service consisting of three general elements:

- *Prevention* of young people going missing;
- *Support* for young people who are missing from home or care; and
- *Information* to underpin the prevention and support services.

The agency representatives were not at all discouraged by the lack of detail. On the contrary, they were very pleased to have experienced a consensus on values, and wanted to hold a second workshop as soon as possible. We agreed to this, and in between the two workshops we produced a diagram showing how all the group's values could be related together. This is reproduced in Figure 17.1, and we sent this out by post before the second workshop was convened.

The evaluation questionnaires returned by the agency representatives revealed that, while people were indeed pleased that they were moving towards a consensus, by and large they did not yet feel that sufficient progress had been made to enable them to commit to any action. Barriers they mentioned included the lack of involvement from strategic policy makers who controlled access to resources (some agency representatives were in a position to make decisions themselves,

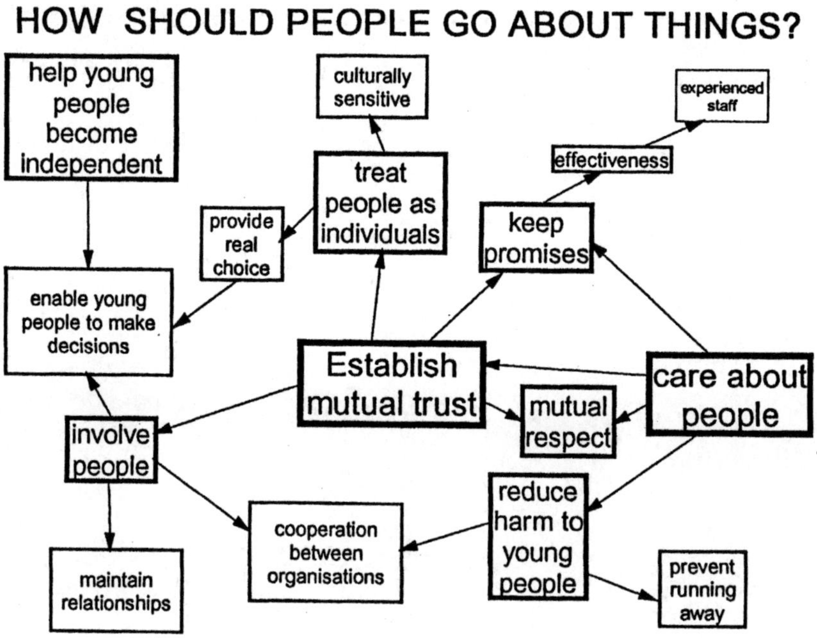

FIGURE 17.1: Principles and values to underpin work with young people

while others had to report back to senior management who would make the decisions); a lack of involvement by the education sector (while the Careers Service had become involved, the local Education Authority had not); lack of time due to workload pressures; and a feeling on the part of some that the issue was not a top priority for their organisations.

17.7.4 The Young People's Workshop

A few days after the multi-agency workshop, we held our workshop for young people under 16.[226] At first I was concerned that

[226] In-between the two workshops, we were also given an unexpected opportunity to visit a Social Services children's home and meet with twelve of its staff, including the manager. However, rather than schedule a planning workshop with these staff, we decided to make this a group interview and explore their perspectives on the issues raised by the young people in the first phase of our intervention. This material added to the insights that we had already gathered from the young people and foster parents, and enabled us to enrich the narrative about the young people's problems that we wrote for our final report (Boyd *et al*, 1999a).

there were going to be almost as many adults as young people (there were five young people, three facilitators and one appropriate adult), and that the presence of so many adults would be overpowering.[227] However, in the event, this did not seem to be even the slightest barrier to participation.

Prior to the workshop, we discussed what we wanted to do. We anticipated that we would have to moderate our language (e.g., stop talking about 'agencies', 'stakeholders', etc.)[228] and make sure that the participants had plenty of opportunities to question us. Also, we agreed that we would use the synergy of CSH and IP methods, despite an earlier thought that we might try something more 'playful' with the young people. We made this decision on the grounds that we wanted the results of the workshop to be comparable with what the multi-agency workshops produced, and we didn't want the output from the young people to appear less 'rational'. Again, this would have replicated the marginalisation of young people—especially *these* young people, from children's homes, who could easily be stereotyped as 'disturbed adolescents'.

Another issue we were not sure of was the degree to which the young people would be able to think in the 'ideal' world. Would they plan with a clean sheet, or would they be unable to think beyond what currently exists? Ackoff (1981) describes how many adults, let alone children, have difficulty keeping focused on an ideal, and there was a difference of opinion between the facilitators about our expectations of the young people: one view was that they would be very concrete thinkers, and another was that they would find it *easier* than adults to think about ideals because they would not be operating with the same history of preconceptions about what is and is not possible. We discussed how we would handle the situation if indeed they could not conceptualise an ideal: we agreed that we would make comparisons between proposals for new services and their children's homes; ask them what they would change at their homes, and if the improvements they wanted would be relevant for the new services.

---

[227] I guess there was an element of selfishness in the decision for all three of us to facilitate. We rationalised it, saying that we could all play different roles in the process, but the bottom line (at least for me) was that this was undoubtedly going to be the most interesting and rewarding workshop, where we would see young people creating plans for services of direct relevance to them, and I didn't want to miss it!

[228] This was important. We quickly found out that even words like 'organisation', which we used without even thinking about it, were not meaningful to a couple of the participants. However, they had no trouble challenging us when we were unclear, and we quickly adapted to each other's language.

In the event, we need not have worried so much about this issue. The young people focused most of their attention on two issues: the need for some form of 'refuge' in the city which runaways could go to for food, shelter, a shower, confidential advice, etc.; and the need to improve children's homes so that young people would be less likely to run from them in the first place. What we found particularly striking about this workshop was the participants' ability to take an overview of the issues and still produce a great deal of detail about how these services should be run (this detail was in marked contrast to the agency representatives' ideal design): they handled a variety of difficult and/or sensitive issues, such as producing a drugs and alcohol policy; rights to confidentiality; informing parents; the attitudes and values of care staff; managing violence; the size of children's homes; etc. In my view, they demonstrated a great deal of sense, compassion and responsibility in the production of their ideal design, and thereby produced a compelling argument for change.[229]

The evaluation questionnaires filled in by the young people revealed a much higher level of personal commitment to action than expressed by the agency representatives in the previous workshop. Most of the young people said that they would be keen to be involved in the future design of actual services.

17.7.5 The Second Multi-Agency Workshop

After the young people's workshop, we set up the second multi-agency event. The first thing that struck us about this was the fact that all the agency representatives returned, except those from the Police and Social Services—the two agencies which had been criticised by young people for being part of the problem. The Police representative was scheduled to participate, but did not turn up on the day (later, it turned out that he had been unavoidably detained—no snub was intended). The Social Services representative had been uncontactable, and therefore we had to go ahead without him (again it turned out that he was simply away from his office on business, and no snub was intended). There was some disquiet in the group about the fact that

[229] A striking example of the exercise of responsibility was when they discussed the possession and use of drugs in a refuge. In their experience, drugs and alcohol were often closely linked to violent behaviour, so while they wanted to offer support to young people with drug and alcohol problems, they also thought that some basic rules about possession and use of drugs on the premises should be put in place to protect others. When one girl was challenged by another participant for her support for this, on the grounds that she often took drugs herself, she laughed and said "just because I do it doesn't make it right". There were many instances of the young people being able to separate their own individual desire to do something from what would be necessary in a refuge for the wider good.

these two agencies were absent, especially as (at that time) no reasons were given. A view was expressed that the group was obviously not as far advanced in terms of building bridges and constructing a consensus as they had thought, and that it would not be possible to make any concrete decisions without the involvement of the Police and Social Services.

However, the participants were keen to make good use of their time together. They particularly wanted to see what the young people had proposed as their ideal, so we made a presentation of this. There was a strong feeling that the young people and professionals had similar values, and participants were also reassured to discover that the young people's desired services were similar to their own. The participants could see no reason to object to anything that the young people had asked for, and welcomed the level of detail in their ideal design—but they were concerned that Social Services in particular were not present to hear the presentation and be influenced by it.

The only difference between the young people and agency representatives that was noted by the latter was that the professionals seemed to put greater emphasis on 'prevention', which they said needed more consideration. They therefore spent most of the rest of the workshop discussing what preventative services might look like.

Finally, we brought the workshop round to the issue of how we were to use the rest of our time in this intervention, given the fact that no multi-agency agreement had yet been secured. We wanted some guidance about how to proceed. There were two reasons for asking for this guidance: first, we assumed that the agency representatives would know better than us what would overcome the obstacles we were experiencing (in terms of the absence of key agencies); and second, by giving them some control over how we used our time (by making us their instrument), we anticipated that they would be more motivated to continue with their own involvement. The response we got was that we should endeavour to work with some of the agencies individually (especially the Police, Social Services and the Education Department whom the group thought might feel threatened) to discuss possible routes for action before convening one final multi-agency workshop. The participants argued that top managers needed to be engaged, and agreement reached on principles and mechanisms for stakeholders to work together to develop and implement plans (each agency representative undertook to do this within his/her own organisation). The aim should be to develop a genuinely shared vision prior to the final multi-agency event. The participants also felt that the evidence from young people was powerful enough to persuade the other organisations to get involved once again.

Here, in offering to be the instrument of the group, we deliberately blurred the boundary between our own identity as a team, and the identity of the group as an agent of change. In terms of the process philosophy outlined in Chapter 4, and the corresponding theory of agency described in Chapter 6, our actions following this workshop could then be seen either as *our* actions alone, or they could be seen as actions motivated by the whole group's desire for change. As I see it, this blurring of the boundaries of identity had a doubly empowering effect: it gave the group a new resource (ourselves), and it granted a new credibility to our actions, stemming from the group's consensus (which was added to the credibility stemming from the voices of the young people). I anticipated that the combined pressure from the young people, the multi-agency group and ourselves would be difficult for the non-participating organisations to ignore.

When people completed our evaluation questionnaires after the workshop, we found that the agency representatives were now much more positive, and most had taken ownership of the change process. This therefore provided a solid platform upon which to build the next phase of the intervention.

17.8 Phase Three: Action Planning

Going into the next phase, we decided not to choose or design any *general* methods: we knew that we were going to have to work individually with a variety of agencies, and that each one was likely to present different challenges. Therefore, we decided to plan each engagement individually as it became apparent what it would entail.

First, we had a meeting with our steering group. There, we discussed our proposed strategy. Our funding organisations were in agreement that we should concentrate in particular on the Police, Social Services and the Education Department. However, I also raised another issue that had occurred to me outside the context of the workshops: there was a possibility of public resistance to the idea of a refuge for homeless young people (if this was indeed going to be one of the proposals to emerge from our intervention), as there could be a perception that it would attract 'undesirables' to the city.[230] To check

[230] A similar objection to a refuge had been raised by a member of staff at a children's home who talked with us as part of a group (see footnote 226, earlier in this chapter, for details of the group interview): he said that he would not be in favour of such an initiative in case it

this out, we proposed that we should contact the Chamber of Commerce to discuss the matter with owners and managers of retail outlets in the city (many homeless young people get involved in shoplifting just to survive). The steering group agreed that it would be worthwhile to identify any such potential resistance so that appropriate preventative action could be taken if necessary.

We held several interviews with retailers, and were encouraged to find that everyone we spoke with had a concern for the young people, and did indeed want something to be done to help them. They thought that a refuge would be a good idea, as it could put young people in touch with agencies who could help them, thereby taking them away from a hand-to-mouth existence where shoplifting became a necessity. They therefore saw it as in their interests to support such a proposal rather than resist it.

We made contact with Social Services, and they agreed that two of their officers would meet with Alan Boyd. This was a productive meeting for a variety of reasons. First, it emerged that their representative had not deliberately pulled out of the earlier multi-agency workshop, and they did indeed want to discuss the possibilities for change. Second, they were keen to discuss the views expressed by young people, and were encouraging about the possibility of a positive response: in particular, they were in the process of reviewing the running of their children's homes, and the young people's ideas were broadly in line with the changes they hoped to be making. Third, it became evident that the officers we were in touch with did not have the direct authority to implement proposals for change. Alan Boyd therefore worked with them to identify the appropriate route into the organisation's planning process that would have to be taken, without which change could not be achieved.

We also made contact with the Education Department who agreed to a workshop. This was a breakthrough because they had previously not been able to allocate personnel to work with the multi-agency group. We discussed how we were going to facilitate this workshop, and agreed to go with the CSH/IP synergy as used previously, primarily because it would produce results that would be comparable with the material already generated, and would bring the Education Department 'up to speed' with what we had done. The Education workshop resulted in a good number of suggestions for service developments, but was not

---

encouraged young people to run away from their home. However, in the young people's workshop, they had anticipated this objection and had been quite derisive of this viewpoint, saying that no young person runs away from a children's home just for a shower and a bed for the night. There is almost always a problem which, for one reason or another, cannot be addressed at the home itself.

able to identify concrete ways of moving forward on them in the short-to-medium term. The primary barrier was perceived to be resource constraints which, they said, were so severe that they had little staff time and money available for anything other than fulfilling their statutory responsibilities. This was disappointing, but we resolved to take the issue to the multi-agency group for discussion to see what support might be forthcoming.

Arguably the most significant breakthrough came in a workshop with the Police, set up by the two Police representatives who attended the first multi-agency workshop. Eight officers with different planning and managerial responsibilities connected with the issue of homeless young people participated in this workshop, most of whom had never met previously and were glad of the opportunity to work together. Again, Alan Boyd, Mandy Brown and I spent some time prior to the workshop discussing how we would facilitate it. It had been made clear by the representatives setting up the workshop that they wanted to hear about the results obtained so far. We were therefore asked to present these, and were happy to oblige: we used the various rich pictures we had developed in order to give an overview of the issues, and also introduced the idea of a refuge that had come from the young people.

We were also aware that many of the young people whom we had interviewed in the first phase of the intervention had made comments about the role of the Police, and even related specific incidents that they were unhappy about. We were keen that these voices should be heard in the workshop, and that the Police should have the opportunity to think about how the problems could be addressed. The question was, how should this be achieved? We considered the production of a written narrative, containing quotations from the young people, which could be circulated before the meeting. However, we were concerned that we might be perceived as constructing the issues: it would be much more powerful if our words could be excluded. We therefore prepared a hand-out that simply contained a list of quotations, taken from the interview transcripts and anonymised, all of which were of direct relevance to the Police. We were not selective: we simply listed every mention of the Police, thereby letting the voices of the young people speak for themselves. We gave the hand-out to the participants after our initial presentation, and as they began to read it, the room went very quiet. Everybody took the time to digest it thoroughly before speaking.

The first comment was striking. The man who spoke said that he recognised the validity of everything that had been said, and could even take an educated guess at the identities of some of the officers who had been involved in the incidents related in the quotations. All the participants were in agreement, and many similar comments followed. Our overview of the issues, together with the quotations, had presented a picture of an unproductive cycle of young people running away from home or care, being returned by the Police, and then running away again, which had led to a negative attitude among many officers that 'missing from homes' are a nuisance. All the participants agreed that this cycle needed to be broken.

Our plan at this point was to enter the CSH/IP synergy once again, to produce results that would be comparable with the ones from the other workshops. However, there hardly seemed any need for facilitation: the participants immediately started coming up with ideas for improving the situation. There was an atmosphere of creative excitement in the room, and one after another the ideas went down on paper. Having generated a whole set of possibilities, the participants asked themselves which ones could be pursued straight away. They identified two, both of which would need the co-operation of Social Services and the Children's Society. The main innovation was that, while the Police and Social Services both have a statutory obligation to return young people to the place which they have run from, the Children's Society does not: this means that, if a Children's Society worker can be involved when the Police and/or Social Services pick up a runaway, s/he can have time alone with the young person before a decision is made about what to do with him or her. While young people are often reluctant to reveal abuse to a police officer or social worker, they are more likely to trust a youth worker from the Children's Society. Thereby, the cycle of abuse, running and return can be broken.

The participants then said that they would have to gain the support of their senior management team before going ahead, but were optimistic of success because (apart from the importance of the issue in itself) there would be a significant resource saving if the vicious cycles could be stopped. They then suggested that the other ideas should be referred to the internal operational research unit in the Police, with a proposal to set up a pilot study (including a controlled experiment) to find out whether the ideas would indeed be worth pursuing. Again, the officers were confident about the possibility of gaining senior management support for this. Shortly after this workshop, the Police got the go-ahead to start negotiations with the Children's Society and Social Services, and made an approach straight away.

17.8.1 *The Final Multi-Agency Workshop*

Following the single-agency workshops, we convened a final multi-agency workshop (as previously agreed with the multi-agency group). As we were aware that this would be the final event, we contacted as many potential participants as possible. The Police, Social Services and the Education Department all came, in addition to the sponsoring organisations, the Housing Department, the Health Authority, foster carers and representatives from a variety of other voluntary organisations.

Again, we needed to design methods for this workshop that would address a variety of questions. These included, how can we make sure that young people's voices are listened to, and remain the foundation for any plans that emerge? How can we create a positive atmosphere that will make people feel that they have achieved something, and will enable them to take ownership of the process of change? Related to this, how can we make sure that the process of change is not dependent on our continued facilitation? And finally, how can we do all of this in half a day!?[231]

We decided to recap the situation as constructed in the rich pictures from the interviews with young people and others, as we were aware that some new participants would be involved. We were also aware that some successes had already been achieved, and these could also be used as the basis for a presentation: there was the agreement with Social Services to participate in making changes; a commitment from a small group of young people to future involvement in the design of specific services; the ideas from the Education Department (although the participants in that workshop were not optimistic about their realisation in the short-to-medium term); the ideas from the Police, which already had internal support and were rapidly progressing through a process of negotiation with the Children's Society and Social Services (importantly, this was now happening independently from our involvement); and the previous commitment from the multi-agency group to pursue change—which they had said was dependent for success on the participation of the Police, Social Services and the Education Department, all of whom were now on board.

As far as we could see, four specific projects had been suggested by various stakeholders (some of these amalgamated ideas from several sources):

[231] The time constraints of the workshops had created difficulties all along, but we had to be realistic and accept that asking agency representatives to give up full days for the workshops, when most of them were doing this planning in addition to their normal jobs, was too much and would kill participation.

- The design and establishment of a refuge;
- Enhanced co-operation between the statutory and voluntary sectors to break the vicious cycle of abuse, running and return;
- The development of a long-term strategy for work in schools to support vulnerable young people; and
- The production and dissemination of information for young people about going missing.

In addition to these specific projects, there was a general commitment to making sure that all the organisations have 'caring about people' as the principle value informing both policies and everyday practice. Also, Social Services had some related initiatives in hand (amongst other things, the review of children's homes) which were being developed independently from this intervention, but would complement the projects identified above. See Boyd *et al* (1999a) for specific details of the projects.

The first half of the workshop would therefore be presentations from us to bring everyone 'up to speed'. The question then was, how to structure the second half? We knew that there would be no time for the use of methods like Checkland's (1981) identification and elaboration of 'relevant systems' (using the CATWOE mnemonic), followed by conceptual modelling, which I have used on previous occasions to move on from the CSH/IP outputs.[232] We therefore needed to be quite directive in the way the second half was structured, to minimise 'drift' in the discussion and ensure that the focus was on how the group could take forward the ideas. We decided to check with the group our own perception that the four specific projects were indeed priorities, and then (if they were accepted) ask people to work in small groups on the one that they had most personal commitment to. The idea was that, if everyone focused on just one thing they were committed to, they would be more likely to take ownership of driving future action.

People did indeed agree on the four projects, and worked in small groups as we had planned. We asked the groups to elaborate the project ideas where necessary, and to make sure that they came out with two things: a clear plan to take the projects forward, and the name of a co-

[232] See Checkland (1981) and Checkland and Scholes (1990) for a discussion of these methods (which come from Soft Systems Methodology), and Chapter 15 (in this book) for an example of their application. I have also applied them successfully in a series of workshops with a Systems Engineering Department in a Colombian University wishing to create a new research direction of relevance to Colombian society. In that instance, Checkland's methods provided an excellent 'bridge' between the values and skeletal plans that came out of the CSH/IP synergy, and the specific action planning that was needed if implementation was to proceed as planned.

ordinator who was prepared to take responsibility for keeping the project on track. All the groups did this. When we came together at the end, the Children's Society representative volunteered to act as 'overall co-ordinator', and we were happy to bow out in the knowledge that there were named individuals willing to accept responsibility for making the changes happen.

17.9 Stage Four: Dissemination

While in many interventions, dissemination is not specifically considered in terms of methods (it is generally taken for granted that a report will be produced as a record of events), this was not the case in Manchester. We were keen (as was our steering group) to ensure that the means of dissemination supported the continuation of the intervention after our departure. We therefore planned to produce 500 copies of a report (Boyd *et al*, 1999a), to be distributed free of charge to people in key agencies. However, in addition to this, we agreed to print 1,000 magazines summarising the results of the intervention (Boyd *et al*, 1999b): these were to be made available to young people as well as professionals. Finally, we produced a set of 1,000 'poster kits' (self-assembly posters) highlighting the main issues. These were for distribution to all services involved with young people, the idea being that they could be pinned on walls and notice-boards to remind people of the concerns of young people and the commitments that the agency representatives had made.

17.10 Reflections

Although only about six months have passed since our involvement in this intervention, we have been kept regularly informed of developments. It appears that the Police, Social Services and the Children's Society are now working closely together in Manchester as planned. In my view, this alone is a significant achievement: prior to our intervention, these agencies were hardly talking to one another. Plans are also in progress to raise money for the refuge. The work on producing information and education strategies is less well advanced, but we expected this as they are longer-term initiatives, and there are significant barriers to be overcome in terms of raising the resources for implementation. The important thing, however, is that these ideas are still 'alive', with people committed to keeping them on the agendas of their organisations. Finally, the collaboration between the statutory and voluntary sectors has spawned a whole new set of ideas which are being pursued, suggesting that the intervention is now firmly in the hands of the local participants. To us, this is the most important indicator of success (in the case of this intervention), as it was always one of our primary purposes to facilitate the engagement of others in change, based on the needs expressed by young people—we were not there to direct change ourselves.

17.11 Conclusion

In this final chapter on the practice of systemic intervention, I have presented a second example of mixing methods which was substantially different from the intervention described in Chapter 16 in two key respects: it followed an action research rather than a summative evaluation design, and it placed a strong emphasis on participation and the exploration of values to inform action planning rather than quantitative data gathering. Nevertheless, the interventions were similar in the sense that they both employed the creative design of methods, and the teams of interveners drew upon the theory of boundary critique to understand many of the relationships between stakeholders.

Although this hasn't been a specific focus of my analysis, I also trust that some of my learning during the past six years is evident: I believe that my thinking and use of methods was more subtle in this final intervention than in the one reported in Chapter 16 (which was undertaken in 1992-3), and I hope that my theory and practice will continue to develop in the years to come.

18

Looking to the Future

In this book I have covered considerable ground: moving from systems philosophy, to the development of a methodology for systemic intervention, to the provision of examples that illustrate how the theory can be put into practice. Below, I offer a brief review of what I have done; reflect on two important areas for further research; and ask you to join me in developing and applying systemic intervention.

In Section One on systems philosophy, I examined the mechanistic worldview with its reductionist methodology, and showed that it is subject/object dualism that underpins this. I then proposed an alternative to this dualism based on a distinction between process and content: the *content* of analyses can be defined through the *process* of making boundary judgements—the 'boundary' idea being fundamental to systems thinking.

In Section Two on methodology, I started by introducing the concept of intervention. I then defined *systemic* intervention as purposeful action by an agent to create change in relation to reflection on boundaries. I then elaborated a theory of *boundary critique*, showing the relationship between boundary and value judgements; how the concept of a boundary judgement can be extended to encompass concerns about how things *ought* to be (as well as what actually exists); the importance of wide-spread stakeholder participation in systemic intervention; and the need for agents to deal with the marginalisation of particular issues and stakeholders within social contexts. I also argued for both theoretical and methodological pluralism to make interventions flexible and responsive to stakeholder concerns. Finally, I discussed how methods could be mixed in practice, and presented an ideal model of interventionist learning. This indicated that there is no need for interveners to have a large amount of methodological knowledge before entering into practice: learning can take place *through and around* practice.

Finally, in Section Three of this book, I focused on practice itself. After providing some background to my own intervention practice in

Community OR, I presented four examples of systemic intervention. These illustrated boundary critique; the assessment of a variety of methods leading to the choice of just one; the mixing of quantitative and qualitative methods from diverse methodological sources to support planning and evaluation; and the mixing of a variety of qualitative methods to facilitate multi-agency working. Most of these interventions had as a principle interest how the voices of marginalised stakeholders and the issues that concerned them could be included in plans for change.

18.1 Further Issues to be Addressed

At this point you may still have unanswered questions, or there may be aspects of my philosophy and/or methodology that you are sceptical about. I hope so, because that is a good basis for further research. Below, I touch on two particular issues that still concern *me*, and remain unexplored in this book. I hope that each of them will be the basis for future research (possibly further books), either by myself or others who decide to pick up on these ideas.

18.1.1 Systemic Intervention and Scientific Research

First, although I have claimed that science could be transformed by seeing it as an aspect of systemic intervention (ethics and research agendas could be the subject of participative debate and development, to the benefit of wider society as well as scientists themselves), I have not yet tried out these ideas in practice. It will be important for me to do so in order to reveal the difficulties and resistances that might be encountered, both within scientific communities and other interest groups. I believe that, if there is a willingness to engage in meaningful dialogue, many entrenched interests can be transcended (certainly the evidence from the application of second wave systems methods in other contexts bears this out—see, for example, Ackoff, 1981). However, even in situations where open debate is not forthcoming, or where dialogue breaks down, it is still meaningful to engage in systemic intervention to test scientific agendas: if consensus or accommodation cannot be reached, at least decisions can be taken in the light of knowledge about how they are seen by others, and some otherwise unanticipated consequences of pursuing particular lines of research can be identified and dealt with in advance. If there are scientists reading this who would like to engage in systemic intervention to support the development of ethically-reflective research agendas in partnership with other stakeholder groups, I would like to hear from you. Together we could test some of these ideas.

18.1.2 The Politics of Systemic Intervention

The other issue of significant interest for me is whether systemic intervention, as I have described it, could inform the construction of a new political theory and practice. At a philosophical level, what I have proposed is quite different from both liberalism and communitarianism (these are two particularly influential political discourses in the USA, and they are now becoming increasingly important in European debates following the collapse of Eastern European Marxism).

Liberalism is said to have its roots in the philosophical treatises of Locke (1689), Mill (1859) and Kant (1787) which, while in many ways substantially different, share a common assumption: that the individual, or subject, is an irreducible moral agent. The individual is usually viewed as having the inherent potential for rational thought—even if, in some cases (such as when people have learning disabilities) this rationality is limited. For Rawls (1971), one of several contemporary liberal writers who have received wide-spread attention, individuals are in a sense 'pre-social': they come into the world as selves, or choosing subjects. When they participate in social interactions and use common understandings they do so as active agents with the capacity for rational choice. This is not to say that human beings are completely autonomous, but they do have partial autonomy within their communities (Rawls, 1971; Caney, 1992), and can choose their community affiliations (Dworkin, 1989).

Communitarians, in contrast, criticise the liberal idea of the autonomous individual with the capacity for rational thought (for two particularly well known critiques, see Sandel, 1982, and MacIntyre, 1985). The basis of their argument is that individuals are socially constructed beings who, as users of shared symbolic languages, simply could not exist independently from communities. Liberals are also accused of over-emphasising rationality, largely to the exclusion of affective commitments to tradition (MacIntyre, 1985). Thus, in the eyes of communitarians, the claim by Rawls (1971) and others that individuals are in some sense 'pre-social' is simply wrong. I would argue that there has been a tendency for communitarians to over-emphasise the extent to which liberals adhere to the idea of the autonomous individual (as Caney, 1992, points out, most stress that autonomy is only partial), but nevertheless there is a substantial difference between liberals and communitarians on whether or not individuals are able to choose their community affiliations. Communitarians stress that communities are held together by implicit, socially shared meanings which take a considerable time to form: they cannot just be chosen. Communities are essentially *traditions* which form the basis for understanding and are not easily opened up to internal critique [although, in MacIntyre's (1985) view, critique is possible when traditions become really problematic and outside influences are sought].

The different understandings of individuals and communities proposed by liberals and communitarians inform opposing normative views on the nature of the good society. For communitarians, the good of the community as a whole comes before any conflicting notions of individual good formed within it (indeed, if the society is properly cohesive, individuals should see personal benefit *in the same terms* as community benefit). In contrast, for liberals, good is a calculation that should take place at an individual level (the only legitimate role for normative action at the community level is in safeguarding individual freedoms—see Nozick, 1974). However, in Chapters 4, 6 and 7 of this book I developed a theory of agency which challenges *both* the liberal and communitarian assumptions. I argued that human agents can be seen as individuals *and/or* collectives (families, groups, organisations, communities, etc.), and that agents act under the influence of wider knowledge generating systems. In this way, it is possible to talk in terms of autonomous moral individuals (when the boundaries of individuals and their knowledge generating systems are regarded as coterminous); individuals acting as part of wider systems; *and* collective agents of various kinds (either acting autonomously or as parts of wider systems). Therefore, no primacy is given to *either* the individual or community: rather, both are considered as valid boundaries for understanding agency and moral decision making, and the emphasis is shifted to how processes of boundary critique and systemic action can be enabled at all relevant levels.

In the longer term, I want to begin to look at if and how systemic intervention can be used to promote multi-layered participative democracy, challenging social exclusion (or marginalisation, whichever term is preferred[233]), and provide a basis for moving towards more ecologically sustainable forms of development. Again, I would like to encourage others with similar interests to get in touch: let us see what we can achieve by working together.

---

[233] In Chapter 1 I argued that the term 'marginalisation' is more appropriate, although 'social exclusion' is already in common use.

18.2 Start from Where You Are

As I made clear in the first chapter of this book, I have laid out an enormous research agenda which will take more than my own lifetime to explore to the full. If you have found what I have had to say in any way meaningful, I invite you to take it further. Most importantly, if you want to apply the ideas in this book in intervention practice, do not feel intimidated by the extent of the literature on methods of intervention that may be drawn upon in support of methodological pluralism: *start from where you are right now,* and build your knowledge about systemic intervention as you go.

Whether you want to support, develop and/or apply my own approach to systemic intervention, or whether you are sceptical and want to critique it, I would encourage you to go ahead—and make the results public. We will all be able to learn from a continuing debate.

References

Abram, D. (1988). The mechanical and the organic: Epistemological consequences of the Gaia hypothesis. In, *Gaia: The Thesis, the Mechanisms and the Implications*. Bunyard, P. and Goldsmith, E. (eds.). Wadebridge Ecological Centre, Camelford.

Ackoff, R.L. (1970). A black ghetto's research on a university. *Operations Research*, **18**, 761-771.

Ackoff, R.L. (1974). *Redesigning the Future: A Systems Approach to Societal Problems*. Wiley, Chichester.

Ackoff, R.L. (1979). The future of operational research is past. *Journal of the Operational Research Society*, **30**, 93-104.

Ackoff, R.L. (1981). *Creating the Corporate Future*. Wiley, New York.

Adorno, T. (1951). *Minima Moralia*. Verso, London.

Adorno, T. and Horkheimer, M. (1944). *Dialectic of Enlightenment*. Verso, London.

Allaby, M. (1989). *Guide to Gaia*. Macdonald & Co., London.

Allen, P.M. (1998). Evolving complexity in social science. In, *Systems: New Paradigms for the Human Sciences*. Altmann, G. and Koch, W.A. (eds.). Walter de Gruyter, Berlin.

Andersen, T. (1987). The reflecting team: Dialogue and meta-dialogue in clinical work. *Family Process*, **26**, 370-379.

Anderson, H. and Goolishian, H. (1988). A view of human systems as linguistic systems: Some preliminary and evolving ideas about the implications for clinical theory. *Family Process*, **27**, 371-393.

Anderson, H. and Goolishian, H. (1992). The client is the expert: A not-knowing approach to therapy. In, *Therapy as Social Construction*. McNamee, S. and Gergen, K. (eds.). Sage, London.

Apel, K.O. (1977). Types of social science in the light of human cognitive interests. In, *Philosophical Disputes in the Social Sciences*. Brown, S. (ed.). Harper and Row, New York.

Arendt, H. (1958). *The Human Condition*. University of Chicago Press, Chicago.

Argyris, C. and Schön, D.A. (1974). *Theory in Practice*. Jossey-Bass, San Francisco.

Argyris, C. and Schön, D.A. (1985). *Strategy, Change and Defensive Routines*. Ballinger, Cambridge, MA.

Aronson, E. (1976). *The Social Animal*. 2nd ed. W.H. Freeman & Co., San Francisco.

Ashby, W.R. (1956). *Introduction to Cybernetics*. Wiley, Chichester.

Ashenden, S. and Owen, D. (eds.) (1999). *Foucault contra Habermas: Recasting the Dialogue between Genealogy and Critical Theory*. Sage, London.

Avison, D. and Wood-Harper, A.T. (1990). *Multiview: An Exploration in Information Systems Development*. Blackwell, Oxford.

Bagby, J.W. (1957). A cross-cultural study of perceptual predominance in binocular rivalry. *Journal of Abnormal and Social Psychology*, **54**, 331-334.

Banathy, B.H. (1984). *Systems Design in the Context of Human Activity Systems*. International Systems Institute, Carmel CA.

Banathy, B.H. (1987). Choosing design methods. *Proceedings of the 31st Annual Meeting of the International Society for General Systems Research*, held in Budapest, Hungary.

Banathy, B.H. (1988). Matching design methods to system type. *Systems Research*, **5**, 27-34.

Barish, N.N. (1963). Operations research and industrial engineering: The applied science and its engineering. *Operations Research*, **11**, 387-398.

Barr, C. and Vangen, S. (1994). Investigating effective collaboration. In, *Community Works: 26 Case Studies showing Community Operational Research in Action*. Ritchie, C., Taket, A. and Bryant, J. (eds.). Pavic Press, Sheffield.

Barter, C. (1996). *Nowhere to Hide: Giving Young Runaways a Voice*. Centrepoint, London.

Bateson, G. (1955). A theory of play and fantasy. In, *Steps to an Ecology of Mind*. Bateson, G. (1972). Jason Aronson, Northvale NJ.

Bateson, G. (1960). Minimal requirements for a theory of schizophrenia. *American Medical Association Archives of General Psychiatry*, **2**, 477-491.

Bateson, G. (1967). Style, grace, and information in primitive art. In, *Steps to an Ecology of Mind*. Bateson, G. (1972). Jason Aronson, Northvale NJ.

Bateson, G. (1968). Effects of conscious purpose on human adaptation. In, *Steps to an Ecology of Mind*. Bateson, G. (1972). Jason Aronson, Northvale NJ.

Bateson, G. (1970). Form, substance, and difference. In, *Steps to an Ecology of Mind*. Bateson, G. (1972). Jason Aronson, Northvale NJ.

Bateson, G. (1971). The science of mind and order. In, *Steps to an Ecology of Mind*. Bateson, G. (1972). Jason Aronson, Northvale NJ.

Bateson, G. (1972). *Steps to an Ecology of Mind*. Jason Aronson, Northvale NJ.

Bateson, G. (1979). *Mind and Nature: A Necessary Unity*. Wildwood House, London.

Beck, U. (1986). *Risk Society: Towards a New Modernity*. Ritter, M. (trans.). 1992 ed. Sage, London.

Beer, S. (1959). *Cybernetics and Management*. English Universities Press, Oxford.

Beer, S. (1966). *Decision and Control*. Wiley, Chichester.

Beer, S. (1975). *Platform for Change*. Wiley, Chichester.

Beer, S. (1979). *The Heart of Enterprise*. Wiley, Chichester.

Beer, S. (1981). *Brain of the Firm*. 2nd ed. Wiley, Chichester.

Beer S (1985). *Diagnosing the System for Organisations*. Wiley, Chichester.

Beer, S (1994). *Beyond Dispute*. Wiley, Chichester.

Belsey, C. (1980). *Critical Practice*. Methuen, London.

Bennett, P., Cropper, S. and Huxham, C. (1989). Modelling interactive decisions: The hypergame focus. In, *Rational Analysis for a Problematic World*. Rosenhead, J. (ed.). Wiley, Chichester.

Bennett, P., Huxham, C. and Cropper, S. (1989). Using the hypergame perspective: A case study. In, *Rational Analysis for a Problematic World*. Rosenhead, J. (ed.). Wiley, Chichester.

Bennett, P., Ackermann, F., Eden, C. and Williams, T. (1997). Analysing litigation and negotiation: Using a combined methodology. In, *Multimethodology: The Theory and Practice of Combining Management Science Methodologies*. Mingers, J. and Gill, A. (eds.). Wiley, Chichester.

Bentham, J. (1997). One size doesn't fit all: Reflections on using systems techniques in an operational setting. In, *Multimethodology: The Theory and Practice of Combining Management Science Methodologies*. Mingers, J. and Gill, A. (eds.). Wiley, Chichester.

Berger, P.L. (1974). *Pyramids of Sacrifice: Political Ethics and Social Change*. Penguin, Harmondsworth.

Bergson, H. (1911). *Creative Evolution*. Macmillan, London.

Berkeley, G. (1710). *The Principles of Human Knowledge*. 1962 ed. Warnock, G.J. (ed.). Fontana Press, London.

Bernstein, R.J. (1983). *Beyond Objectivism and Relativism: Science, Hermeneutics and Praxis*. Basil Blackwell, Oxford.

Bernstein, R.J. (1991). *The New Constellation*. Polity Press, Cambridge.

Berry, B.J.L. and Kim, H. (1994). Leadership generations: A long-wave macrohistory. *Technological Forecasting and Social Change*, **46**, 1-9.

Bertalanffy, L. von (1950). The theory of open systems in physics and biology. *Science*, **111**, 23-29.

Bertalanffy, L. von (1956). General systems theory. *General Systems Year Book*, **1**, 1-10.

Bertalanffy, L. von (1968). *General Systems Theory*. Penguin, London.

Bhaskar, R. (1975). *A Realist Theory of Science*. Leeds Books, Leeds.

Bhaskar, R. (1986). *Scientific Realism and Human Emancipation*. Verso, London.

Bilson, A. (1996). *Bringing Forth Organisational Realities: Guidelines for a Constructivist Approach to the Management of Change in Human Services*. Ph.D. thesis, Lancaster University.

Bilson, A. (1997). Guidelines for a constructivist approach: Steps toward the adaptation of ideas from family therapy for use in organizations. *Systems Practice*, **10**, 153-177.

Bogdanov, A.A. (1913-1917). *Bogdanov's Tektology*. 1996 ed. Dudley, P. (ed.). Centre for Systems Studies Press, Hull.

Bohm, D. (1980). *Wholeness and the Implicate Order*. Ark, London.

Bohr, N. (1963). *Essays 1958/1962 on Atomic Physics and Human Knowledge*. Interscience Wiley and Sons, London.

Bolweg, F.J. (1976). *Job Design and Industrial Democracy*. Martinus Nijhoff, Leiden.

Booth Fowler, R. (1991). *The Dance with Community: The Contemporary Debate in American Political Thought*. University Press of Kansas, Lawrence.

Boyd, A., Brown, M. and Midgley, G. (1999a). *Home and Away: Developing Services with Young People Missing from Home or Care*. Centre for Systems Studies, Hull.

Boyd, A., Brown, M. and Midgley, G. (1999b). *Manchester Times*. Centre for Systems Studies, Hull.

Bradburn, N.M. and Sudman, S. (1980). *Improving Interview Method and Questionnaire Design*. Jossey-Bass, London.

Brauer, T. (1995). Do brahmins dream of electric sheep? In, *Systems Thinking, Government Policy and Decision Making*. Bergvall-Kareborn, B. (ed.). International Society for the Systems Sciences, Louisville.

Brewer, J. and Hunter, A. (1989). *Multimethod Research: A Synthesis of Styles*. Sage, London.

Brier, S. (1999). Foreword. *Cybernetics and Human Knowing*, **6**, 3-4.

Brocklesby, J. (1994). Let the jury decide: Assessing the cultural feasibility of Total Systems Intervention. *Systems Practice*, **7**, 75-86.

Brocklesby, J. (1995). From single to multi-paradigm systems research. In, *Systems for Sustainability: People, Organizations, and Environments*. Stowell, F.A., Ison, R.L., Armson, R., Holloway, J., Jackson, S. and McRobb, S. (eds.). Plenum, New York.

Brocklesby, J. (1997). Becoming multimethodology literate: An assessment of the cognitive difficulties of working across paradigms. In, *Multimethodology: The Theory and Practice of Combining Management Science Methodologies*. Mingers, J. and Gill, A. (eds.). Wiley, Chichester.

Brocklesby, J. and Cummings, S. (1995). Combining hard, soft and critical methodologies in systems research: The cultural constraints. *Systems Research*, **12**, 239-244.

Brodie, D.A., Williams, J.G. and Owens, R.G. (1994). *Research Methods for the Health Sciences*. Harwood Academic, Chur, Switzerland.

Broskowski, A. (1976). Program evaluation as distinct from evaluation research. *Division of Community Psychology Newsletter*, #10, 3.

Brown, H. and Smith, H. (1989). Whose 'ordinary life' is it anyway? *Disability, Handicap and Society*, **4**, 105-119.

Brown, H. and Smith H. (1992). Introduction. In, *Normalisation: A Reader for the Nineties*. Brown, H. and Smith, H. (eds.). Routledge, London.

Brown, M. (1996). A framework for assessing participation. In, *Critical Systems Thinking: Current Research and Practice*. Flood, R.L. and Romm, N.R.A. (eds.). Plenum, New York.

Brown, R.H. (1977). *A Poetic for Sociology: Toward a Logic of Discovery for the Human Sciences*. University of Chicago Press, Chicago.

Burrell, G. and Morgan, G. (1979). *Sociological Paradigms and Organizational Analysis*. Heinemann, London.

Burton, M. (1983). Understanding mental health services: Theory and practice. *Critical Social Policy*, **3**, 54-74.

Bynoe, I. (1992). *Treatment, Care and Security*. MIND, London.

Caldwell, M. (1977). *The Wealth of Some Nations*. Zed Press, London.

Cameron, J. (1989). *JSP & JSD: The Jackson Approach to Software Development*. IEEE Computer Society Press, Washington DC.

Campbell, D., Coldicott, T. and Kinsella, K. (1994). *Systemic Work with Organizations: A New Model for Managers and Change Agents*. Karnac Books, London.

Campbell, D.T. and Cook, T.D. (1976). *Quasi-Experiments and True Experiments in Field Settings*. Rand McNally, Chicago.

Campbell, D.T. and Stanley, J.C. (1966). *Experimental and Quasi-Experimental Designs for Research*. Rand McNally, Chicago.

Campbell, P. (1987). Survivors speak out. *Open Mind*, #26, 8.

Caney, S. (1992). Liberalism and communitarianism: A misconceived debate. *Political Studies*, **40**, 273-289.

Capek, M. (1971). *Bergson and Modern Physics*. Reidel, Dordrecht.

Capra, F. (1996). *The Web of Life: A New Synthesis of Mind and Matter*. HarperCollins, London.

Carr, E.H. (1961). *What is History?* Penguin, Harmondsworth.

Carter, P., Jackson, M.C., Jackson, N. and Keys, P. (1987). Community OR at Hull University. *Dragon*, **2**, special issue.

Cecchin, G. (1987). Hypothesising, circularity and neutrality revisited: An invitation to curiosity. *Family Process*, **26**, 405-413.

Charlton, M. (1993). Unpublished paper circulated to staff and research students in the Centre for Systems Studies, University of Hull.

Checkland, P. (1975). The development of systems thinking by systems practice—A methodology from an action research program. In, *Progress in Cybernetics and Systems Research, Volume II*. Trappl R. and Hanika F. de P. (eds). Hemisphere, Washington.

Checkland, P. (1981). *Systems Thinking, Systems Practice*. Wiley, Chichester.

Checkland, P. (1993). Book review of 'Practical Soft Systems Analysis' by D. Patching. *Systems Practice*, **6**, 435-438.

Checkland, P. (1997). Rhetoric and reality in contracting: Research in and on the NHS. In, *Contracting for Health*. Flynn, R. and Williams, G. (eds). Oxford University Press, Oxford.

Checkland, P. and Holwell, S. (1998). *Information, Systems and Information Systems: Making Sense of the Field*. Wiley, Chichester.

Checkland, P. and Scholes, J. (1990). *Soft Systems Methodology in Action*. Wiley, Chichester.

Church, K. (1992). *Moving Over: A Commentary on Power-Sharing*. Psychiatric Survivor Leadership Facilitation Program, Toronto.

Churchman, C.W. (1968a). *Challenge to Reason*. McGraw-Hill, New York.

Churchman, C.W. (1968b). *The Systems Approach*. Dell, New York.

Churchman, C.W. (1970). Operations research as a profession. *Management Science*, **17**, B37-53.

Churchman, C.W. (1971). *The Design of Inquiring Systems*. Basic Books, New York.

Churchman, C.W. (1979). *The Systems Approach and its Enemies*. Basic Books, New York.

Churchman, C.W. (1987). Systems profile: Discoveries in an exploration into systems thinking. *Systems Research*, **4**, 139-146.

Cohen, C. and Midgley, G. (1994). *The North Humberside Diversion from Custody Project for Mentally Disordered Offenders: Research Report*. Centre for Systems Studies, Hull.

Cohen, J. and Stewart, I. (1994). *The Collapse of Chaos: Discovering Simplicity in a Complex World*. Viking, London.

Cordoba, J., Midgley, G. and Torres, D. (2000). Rethinking stakeholder involvement: An application of the theories of autopoiesis and boundary critique in IS planning. In, *Human-Centred Methods in Information Systems: Current Research and Practice*. Clarke, S. and Lehaney, B. (eds.). Idea Group, Hershey, PA.

Couzens Hoy, D. (1994). *Critical Theory*. Blackwell, Cambridge MA.

Coyle, R.G. (1977). *Management Systems Dynamics*. Wiley, New York.

Cronbach, L.J. (1982). *Designing Evaluations of Educational and Social Programs*. Jossey-Bass, San Francisco.

Crowe, M. (1996). Heraclitus and information systems. *Systemist*, **18**, 157-176.

Cummings, S. (1994). An open letter to total systems intervention (TSI) and friends: A postmodern remedy to make everybody feel better. *Systems Practice*, **7**, 575-587.

Daly, H.E. and Cobb, J.B. (1989). *For the Common Good: Redirecting the Economy towards*

Community, the Environment and a Sustainable Future. Green Print, London.
Dando, M.R. and Bennett, P.G. (1981). A Kuhnian crisis in management science? *Journal of the Operational Research Society*, **32**, 91-104.
Dando, M.R., DeFrenne, A. and Sharp, R.G. (1977). Could OR be a science? *Omega*, **5**, 89-92.
Darier, É. (1999). Foucault and the environment: An introduction. In, *Discourses of the Environment*. Darier, É. (ed.). Blackwell, Oxford.
Davies, C.G. and Layzell, P.J. (1993). *The Jackson Approach to System Development: An Introduction*. Chartwell-Bratt, Bromley.
Dávila, J. (1993). Foucault's interpretive analytics of power. *Systems Practice*, **6**, 383-405.
Dell, P. (1982a). Beyond homeostasis: Towards a concept of coherence. *Family Process*, **21**, 407-414.
Dell, P. (1982b). Family theory and the epistemology of Humberto Maturana. *Family Therapy Networker*, **6**, 26, 39-41.
Derrida, J. (1976). *Of Grammatology*. John Hopkins University Press, Baltimore.
Derrida, J. (1978). *Writing and Difference*. Routledge & Kegan Paul, London.
Descartes, R. (1642). Meditations on first philosophy. In, *Philosophical writings*. 1954 ed. Open University Press, Middlesex.
Descartes, R. (1644). Principles of philosophy. In, *Philosophical writings*. 1954 ed. Open University Press, Middlesex.
Dewdney, A., Grey, C., Minnion, A. and the Residents of the Rufford Street Hostel (1994). *Down but Not Out: Young People, Photography and Images of Homelessness*. Trentham, Stoke-on-Trent.
Dewey, J. (1946). *Problems of Men*. Philosophical Library, New York.
Donaldson, P. (1986). *Worlds Apart*. 2nd ed. Penguin, Harmondsworth.
Douglas, M. (1966). *Purity and Danger: An Analysis of the Concepts of Pollution and Taboo*. Ark, London.
Douglas, M. (1986). *How Institutions Think*. Routledge and Kegan Paul, London.
Douglas, M. (1992). *Risk and Blame: Essays in Cultural Theory*. Routledge, London.
Drever, E. (1995). *Using Semi-Structured Interviews in Small-Scale Research: A Teacher's Guide*. The Scottish Council for Research in Education, Edinburgh.
Dreyfus, H.L. and Rabinow, P. (1982). *Michel Foucault: Beyond Structuralism and Hermeneutics*. Harvester Press, Hemel Hempstead.
Dudley, P. (1996). Back to basics? Tektology and general systems theory (GST). *Systems Practice*, **9**, 273-284.
Dutt, P.K. (1994). Problem contexts—A consultant's perspective. *Systems Practice*, **7**, 539-550.
Dworkin, R. (1989). Liberal community. *California Law Review*, **77**, 479-509.
Eckersley, R. (1992). *Environmentalism and Political Theory: Toward an Ecocentric Approach*. UCL Press, London.
Eden, C. (1988). Cognitive mapping: A review. *European Journal of Operational Research*, **36**, 1-13.
Eden, C. (1989). Using cognitive mapping for strategic options development and analysis. In, *Rational Analysis for a Problematic World*. Rosenhead, J. (ed.). Wiley, Chichester.
Eden, C., Jones, S. and Sims, D. (1983). *Messing About in Problems*. Pergamon, Oxford.
Eden, C. and Simpson, P. (1989). SODA and cognitive mapping in practice. In, *Rational Analysis for a Problematic World*. Rosenhead, J. (ed.). Wiley, Chichester.
Einstein, A. (1934). *The World as I See It*. Covici Friede, New York.
Elden, M. and Levin, M. (1991). Cogenerative learning: Bringing participation into action research. In, *Participatory Action Research*. Whyte, W.F. (ed.). Sage, London.
Elstob, M. (1992). Review of 'Creative Problem Solving: Total Systems Intervention' by R.L. Flood and M.C. Jackson. *Kybernetes*, **21**, 62-63.
Emery, F.E. and Thorsrud, E. (1969). *Form and Content in Industrial Democracy*. Tavistock, London.
Emery, F.E. and Thorsrud, E. (1976). *Democracy at Work*. Nijhoff, Leiden.
Emery, F.E. and Trist, E.L. (1965). The causal texture of organizational environments. *Human Relations*, **18**, 21-32.
Engel, E. (1956). The role of content in binocular resolution. *American Journal of Psychology*, **69**, 87-91.

English, C.J. (1973). Leaving home: A typology of runaways. *Society*, **10**, #5, 22-24

Espejo, R. and Harnden, R. (eds.) (1989). *The Viable System Model: Interpretations and applications of Stafford Beer's VSM*. Wiley, Chichester.

ESRC Global Environmental Change Programme (1999). *The Politics of GM Food: Risk, Science and Public Trust*. Special Briefing #5. University of Sussex, Brighton.

Fals Borda, O. and Rahman, M.A. (eds.) (1991). *Action and Knowledge: Breaking the Monopoly with Participatory Action Research*. Apex Press, New York.

Fargason, C.A., Barnes, D., Schneider, D. and Galloway, B.W. (1994). Enhancing multiagency collaboration in the management of child sexual abuse. *Child Abuse and Neglect*, **18**, 859-869.

Farmer, E. and Eyre, A.B. (1922). An investigation into the packing of chocolates (1). *Journal of the National Institute of Industrial Psychology*, **1**, 12-14.

Fay, B. (1975). *Social Theory and Political Practice*. George Allen and Unwin, London.

Fay, B. (1987). *Critical Social Science*. Polity Press, Cambridge.

Ferguson, G.A. (1981). *Statistical Analysis in Psychology and Education*. 5th ed. McGraw-Hill, New York.

Festinger, L. (1957). *A Theory of Cognitive Dissonance*. Stanford University Press, Stanford.

Festinger, L. and Katz, D. (1954). *Research Methods in the Behavioral Sciences*. Holt-Saunders, Eastbourne.

Fine, G.A. and Sandstrom, K.L. (1988). *Knowing Children: Participant Observation with Minors*. Sage, California.

Fitzgerald, L.A. (1999). Why there's nothing wrong with systems thinking a little *chaos* won't fix? A critique of modern systems theory and the practice of organizational change it informs. *Systemic Practice and Action Research*, **12**, 229-235.

Flood, R.L. (1989a). Six scenarios for the future of systems 'problem solving', part 1. *Systems Practice*, **2**, 75-99.

Flood, R.L. (1989b). Archaeology of (systems) inquiry. *Systems Practice*, **2**, 117-124.

Flood, R.L. (1990). *Liberating Systems Theory*. Plenum Press, New York.

Flood, R.L. (1995a). *Solving Problem Solving*. Wiley, Chichester.

Flood, R.L. (1995b). Total systems intervention (TSI): A reconstitution. *Journal of the Operational Research Society*, **46**, 174-191.

Flood, R.L. (1999a). *Rethinking the Fifth Discipline: Learning within the Unknowable*. Routledge, London.

Flood, R.L. (1999b). Knowing of the unknowable. *Systemic Practice and Action Research*, **12**, 247-256.

Flood, R.L. and Carson, E.R. (1993). *Dealing with Complexity: An Introduction to the Theory and Application of Systems Science*. 2nd ed. Plenum, New York.

Flood, R.L. and Gregory, W.J. (1988). Systems: Past, present and future. In, *Systems Prospects: The Next Ten Years of Systems Research*. Flood, R.L., Jackson, M.C. and Keys, P. (eds.). Plenum, New York.

Flood, R.L. and Jackson, M.C. (eds.) (1991a). *Critical Systems Thinking: Directed Readings*. Wiley, Chichester.

Flood, R.L. and Jackson, M.C. (1991b). *Creative Problem Solving: Total Systems Intervention*. Wiley, Chichester.

Flood, R.L. and Jackson, M.C. (1991c). Critical systems heuristics: Application of an emancipatory approach for police strategy toward the carrying of offensive weapons. *Systems Practice*, **4**, 283-302.

Flood, R.L. and Robinson, S.A. (1990). New domains for analogy: Systemic dialectics and theory development. *International Journal of General Systems*, **18**, 113-123.

Flood, R.L. and Romm, N.R.A. (1995a). Enhancing the process of choice in TSI, and improving chances of tacking coercion. *Systems Practice*, **8**, 377-408.

Flood, R.L. and Romm, N.R.A. (1995b). Diversity management: Theory in action. *Systems Practice*, **8**, 469-482.

Flood, R.L. and Romm, N.R.A. (1996a). *Diversity Management: Triple Loop Learning*. Wiley, Chichester.

Flood, R.L. and Romm, N.R.A. (eds.) (1996b). *Critical Systems Thinking: Current Research and*

Practice. Plenum, New York.

Flood, R.L. and Romm, N.R.A. (1997). From meta-theory to "multimethodology". In, *Multimethodology: The Theory and Practice of Combining Management Science Methodologies*. Mingers, J. and Gill, A. (eds.). Wiley, Chichester.

Flood, R.L. and Ulrich, W. (1990). Testament to conversations on critical systems theory between two systems practitioners. *Systems Practice*, **3**, 7-29.

Flood, R.L. and Zambuni, S. (1990). Viable system diagnosis I: Application with a major tourism services group. *Systems Practice*, **3**, 225-248.

Foddy, W. (1993). *Constructing Questions for Interviews and Questionnaires*. Cambridge University Press, Cambridge.

Fodor, J.A. (1974). Special sciences (or: the disunity of science as a working hypothesis). *Synthese*, **28**, 97-115.

Foerster, H. von (1984). *Observing Systems*. 2nd ed. Intersystems, California.

Forrester, J.W. (1961). *Industrial Dynamics*. MIT Press, Cambridge MA.

Foucault, M. (1961). *Madness and Civilization*. English ed., 1965. Pantheon, New York.

Foucault, M. (1970). *The Order of Things*. Tavistock, London.

Foucault, M. (1972). *The Archaeology of Knowledge*. Tavistock, London.

Foucault, M. (1973). *The Birth of the Clinic*. Tavistock, London.

Foucault, M. (1976). *The History of Sexuality, Volume 1: Introduction*. Pantheon, New York.

Foucault, M. (1977). *Discipline and Punish*, Pantheon, New York.

Foucault, M. (1980). *Power/Knowledge: Selected Interviews and Other Writings, 1972-1977*. Gordon, C. (ed.). Harvester Press, Brighton.

Foucault, M. (1984a). What is enlightenment? In, *The Foucault Reader*. Rabinow, P. (ed.). Penguin, London.

Foucault, M. (1984b). *The Care of the Self*. Pantheon, New York.

Foucault, M. (1984c). *The Use of Pleasure*. Pantheon, New York.

Francescato, D. (1992). A multi-dimensional perspective of organizational change. *Systems Practice*, **5**, 129-146.

Frank, H.D. (1997). Redrawing the boundaries: Can action research and action learning be seen as parts of a single system? Some initial thinking. In, *Forum Two: Action Research and Critical Systems Thinking*. Wilby, J. (ed.). Centre for Systems Studies, Hull.

Freire, P. and Faundez, A. (1989). *Learning to Question: A Pedagogy of Liberation*. World Council of Churches, Geneva.

Freud, A. (ed.) (1986). *Sigmund Freud: The Essentials of Psycho-Analysis*. Penguin, London.

Freud, S. (1915). *On Meta-Psychology: The Theory of Psychoanalysis*. The Pelican Freud Library, Vol. 11 (translated into English by Angela Richards, 1957). Penguin, Harmondsworth.

Friend, J.K. (1989). The strategic choice approach. In, *Rational Analysis for a Problematic World*. Rosenhead, J. (ed.). Wiley, Chichester.

Friend, J.K. and Hickling, A. (1987). *Planning under Pressure: The Strategic Choice Approach*. Pergamon, Oxford.

Fuenmayor, R.L. (1991a). The roots of reductionism: A counter-ontoepistemology for a systems approach. *Systems Practice*, **4**, 419-448.

Fuenmayor, R.L. (1991b). The self-referential structure of an everyday living situation: A phenomenological ontology for interpretive systemology. *Systems Practice*, **4**, 449-472.

Fuenmayor, R.L. (1991c). Truth and openness: An epistemology for interpretive systemology. *Systems Practice*, **4**, 473-490.

Fuenmayor, R.L. (1993). The trap of evolutionary organicism. *Systems Practice*, **6**, 469-499.

Fuenmayor, R.L. (1994). "Systems science: Addressing global issues"—the death rattle of a dying era? *Systemist*, **16**, 110-157.

Gare, A. (1996). *Nihilism Inc.: Environmental Destruction and the Metaphysics of Sustainability*. Eco-Logical Press, Como NSW.

George, S. (1976). *How the Other Half Dies: The Real Reasons for World Hunger*. Penguin, Harmondsworth.

Gergen, K.J. (1991). *The Saturated Self: Dilemmas of Identity in Contemporary Life*. Basic Books, New York.

Gergen, K.J. (1994). The limits of pure critique. In, *After Postmodernism: Reconstructing Ideology Critique*. Simons, H.W. and Billig, M. (eds.). Sage, London.

Geus, A.P. de (1994). Modeling to predict or to learn? In, *Modeling for Learning Organizations*. Morecroft, J.D.W. and Sternman, J.D. (eds.). Productivity Press, Portland, Oregon.

Ghosal, A. (1992). Review of 'Creative Problem Solving: Total Systems Intervention' by R.L. Flood and M.C. Jackson. *SCIMA*, **21**, 39-40.

Giddens, A. (1985). Reason without revolution? Habermas's theorie des kommunikativen handelns. In, *Habermas and Modernity*. Bernstein, R.J. (ed.). Polity Press, Cambridge.

Giddens, A. (1991). *Modernity and Self-Identity: Self and Society in the Late Modern Age*. Polity Press, Cambridge.

Giddens, A. (1998). *The Third Way*. Polity Press, Cambridge.

Gill, A. (1997). Managing a virtual organization. In, *Multimethodology: The Theory and Practice of Combining Management Science Methodologies*. Mingers, J. and Gill, A. (eds.). Wiley, Chichester.

Glansdorff, P. and Prigogine, I. (1971). *Stability and Fluctuations*. Wiley Interscience, London.

Glasersfeld, E. von (1984). An introduction to radical constructivism. In, *The Invented Reality*. Watzlawick, P. (ed.). Norton, New York.

Glasersfeld, E. von (1999). How do we mean? A constructivist sketch of semantics. *Cybernetics and Human Knowing*, **6**, 9-16.

Gleick, J. (1987). *Chaos*. Penguin, Harmondsworth.

Gödel, K (1931). *On Formally Undecidable Propositions*. 1962 trans. Basic Books, New York

Goldner, V. (1985). Feminism and family therapy. *Family Process*, **24**, 31-47.

Goldner, V. (1988). Generation and gender: Normative and covert hierarchies. *Family Process*, **27**, 17-31.

Goldner, V. (1991). Feminism and systemic practice: Two critical traditions in transition. *Journal of Family Therapy*, **13**, 95-104.

Goldner, V., Penn, P., Sheinberg, M. and Walker, G. (1990). Love and violence: Gender paradoxes in volatile attachments. *Family Process*, **29**, 343-364.

Goodwin, B. (1992). *Green Political Theory*. Polity Press, Cambridge.

Gorelik, G. (1987). Bogdanov's tektologia, general systems theory, and cybernetics. *Cybernetics and Systems*, **18**, 157-175.

Graue, M.E. and Walsh, D.J. (1998). *Studying Children in Context: Theories, Methods, and Ethics*. Sage, California.

Gray, J. (1991a). On the morality of speciesism. *The Psychologist*, **4**, 196-198.

Gray, J. (1991b). On speciesism and racism: Reply to Singer and Ryder. *The Psychologist*, **4**, 202-203.

Green, S.M. (1993a). Total systems intervention: A practitioner's critique. *Systems Practice*, **6**, 71-79.

Green, S.M. (1993b). Total systems intervention: A trial by jury. *Systems Practice*, **6**, 295-299.

Gregory, A.J. (1997). Evaluation practice and the tricky issue of coercive contexts. *Systems Practice*, **10**, 589-609.

Gregory, A.J. and Jackson, M.C. (1992a). Evaluating organizations: A systems and contingency approach. *Systems Practice*, **5**, 37-60.

Gregory, A.J. and Jackson, M.C. (1992b). Evaluation methodologies: A system for use. *Journal of the Operational Research Society*, **43**, 19-28.

Gregory, W.J. (1989). Critical theory and critical systems heuristics: The history and development of an emancipatory systems approach to social change. *Proceedings of the 33rd Annual Meeting of the International Society for General Systems Research (Volume II)*, held in Edinburgh, Scotland, on 2-7 July 1989.

Gregory, W.J. (1990). Critical systems thinking and LST: How "liberating" are contemporary critical and liberating systems approaches? In, *Toward a Just Society for Future Generations. Volume I: Systems Design*. Banathy, B.H. and Banathy, B.A. (eds.). International Society for the Systems Sciences, Pomona, CA.

Gregory, W.J. (1992). *Critical Systems Thinking and Pluralism: A New Constellation*. Ph.D. thesis, City University, London.

Gregory, W.J. (1996a). Discordant pluralism: A new strategy for critical systems thinking? *Systems Practice*, **9**, 605-625.

Gregory, W.J. (1996b). Dealing with diversity. In, *Critical Systems Thinking: Current Research*

and Practice. Flood, R.L. and Romm, N.R.A. (eds.). Plenum, New York.
Gregory, W.J. (2000). (Personal communication).
Gregory, W.J. and Midgley, G. (1994). Planning for disaster: Developing a multi-agency, post-disaster counselling service. In, *Community Works: 26 Case Studies showing Community Operational Research in Action*. Ritchie, C., Taket, A. and Bryant, J. (eds.). Pavic Press, Sheffield.
Gregory, W.J. and Midgley, G. (1999). *Planning for Disaster: Developing a Counselling Service*. Centre for Systems Studies Research Memorandum #23. Centre for Systems Studies, Business School, University of Hull, Hull.
Gregory, W.J. and Midgley, G. (2000). Planning for disaster: Developing a counselling service. *Journal of the Operational Research Society*, **51**, 278-290.
Gregory, W.J., Romm, N.R.A. and Walsh, M.P. (1994). *The Trent Quality Initiative: A Multi-Agency Evaluation of Quality Standards in the National Health Service*. Centre for Systems Studies, Hull.
Gu, J. and Zhu, Z. (1995). The *Wu-li Shi-li Ren-li* approach (WSR): An Oriental systems methodology. In, *Systems Methodology: Possibilities for Cross-Cultural Learning and Integration*. Midgley, G. and Wilby, J. (eds.). Centre for Systems Studies, Hull.
Gu, J. and Zhu, Z. (2000). Knowing *Wuli*, sensing *Shili*, caring for *Renli*: Methodology of the WSR approach. *Systemic Practice and Action Research*, **13**, 11-20.
Gustavsen, B. (1992). *Dialogue and Development: Social Science for Social Action*. Van Gorcum, Assen.
Gustavsen, B. and Engelstad, P. (1986). The design of conferences and the evolving role of democratic dialogue in changing working life. *Human Relations*, **39**, 101-116.
Habermas, J. (1971). *Toward a Rational Society*. Beacon Press, Boston.
Habermas, J. (1972). *Knowledge and Human Interests*. Heinemann, London.
Habermas, J. (1976). *Communication and the Evolution of Society*. English ed., 1979. Heinemann, London.
Habermas, J. (1984a). *The Theory of Communicative Action, Volume One: Reason and the Rationalisation of Society*. Polity Press, Cambridge.
Habermas, J. (1984b). *The Theory of Communicative Action, Volume Two: The Critique of Functionalist Reason*. Polity Press, Cambridge.
Habermas, J. (1985). *The Philosophical Discourse of Modernity: Twelve Lectures*. English ed., 1987. Polity Press, Cambridge.
Haley, J. (1962). Family experiments: A new type of experimentation. *Family Process*, **1**, 265-293.
Haley, J. (1963). *Strategies of Psychotherapy*. Grune & Stratton, New York.
Hall, A.D. (1962). *A Methodology for Systems Engineering*. Van Nostrand, Princeton.
Han, J. (2000). A framework of value-focused systems thinking. *Systemic Practice and Action Research*, **13**, 97-109.
Harré, R. and Gillett, G. (1994). *The Discursive Mind*. Sage, London.
Hastorf, A.H. and Myro, G. (1959). The effect of meaning on binocular rivalry. *American Journal of Psychology*, **72**, 393-400.
Hawking, S.W. (1988). *A Brief History of Time: From the Big Bang to Black Holes*. Bantam, London.
Hays, W.L. (1974). *Statistics for the Social Sciences*. 2nd ed. Holt, Rinehart and Winston, London.
Heermann, E.F. and Braskamp, L.A. (eds.) (1970). *Reading in Statistics for the Behavioral Sciences*. Prentice Hall, Hemel Hempstead.
Hegel, G.W.F. (1807). *The Phenomenology of Mind*. 2nd ed. English ed., 1931. George Allen and Unwin, London.
Heidegger, M. (1927). *Being and Time*. 7th ed. 1962 trans. Basil Blackwell, Oxford.
Heidegger, M. (1954). Logos. In, *Vorträge und Ausfsätze*. 3rd ed. (1967), Günther Neske, Pfullingen.
Heraclitus (600-500 B.C.). *The Cosmic Fragments*. 1954 ed. Kirk, G.S. (ed.). Cambridge University Press, Cambridge.
Heron, J. (1996). *Co-operative Inquiry: Research into the Human Condition*. Sage, London.
Hickling, A. (1989). Gambling with frozen fire? In, *Rational Analysis for a Problematic World*. Rosenhead, J. (ed.). Wiley, Chichester.

Hillman, J. (1995). A psyche the size of the Earth: A psychological foreword. In, *Ecopsychology: Restoring the Earth, Healing the Mind.* Roszak, T., Gomes, M.E. and Kanner, A.D. (eds.). Sierra Club Books, San Francisco.

Ho, C.H. (1997). *A Critical Process for the Evaluation of Methodology.* Ph.D. thesis, University of Hull.

Ho, J.K.K. (1994). Is total systems intervention (TSI) no better than common sense and not necessarily related to critical systems thinking (CST)? *Systems Practice,* **7**, 569-573.

Ho, M.W. (1989). Reanimating nature: The integration of science with human experience. *Beshara,* #8, 16-25.

Hoffman, L. (1988). A constructivist position for family therapy. *Irish Journal of Psychology,* **9**, 110-129.

Hofstadter, D.R. (1979). *Gödel, Escher, Bach: An Eternal Golden Braid.* Penguin, London.

Holland, R. (1970). George Kelly: Constructive innocent and reluctant existentialist. In, *Perspectives in Personal Construct Theory.* Bannister, D. (ed.). Academic Press, London.

Hollway, W. (1989). *Subjectivity and Method in Psychology: Gender, Meaning and Science.* Sage, London.

Hollway, W. (1991). *Work Psychology and Organizational Behaviour: Managing the Individual at Work.* Sage, London.

Holstein, J.A. and Gubrium, J.F. (1995). *The Active Interview.* Sage, London.

Horkheimer, M. (1937). Traditionelle und Kritische Theorie. *Zeitschrift fur Sozialforschung,* **6**, 245-294.

Howard, N. (1989a). The manager as politician and general: The metagame approach to analysing cooperation and conflict. In, *Rational Analysis for a Problematic World.* Rosenhead, J. (ed.). Wiley, Chichester.

Howard, N. (1989b). The CONAN play: A case study illustrating the process of metagame analysis. In, *Rational Analysis for a Problematic World.* Rosenhead, J. (ed.). Wiley, Chichester.

Hutchinson, W.E. (1996). Making systems thinking relevant. *Systemist,* **18**, 196-201.

Ivanov, K. (1991). Critical systems thinking and information technology. *Journal of Applied Systems Analysis,* **18**, 39-55.

Jackson, D.D. (ed.) (1960). *The Aetiology of Schizophrenia.* Basic Books, New York.

Jackson, M.A. (1983). *System Development.* Prentice Hall, New York.

Jackson, M.C. (1982). The nature of soft systems thinking: The work of Churchman, Ackoff and Checkland. *Journal of Applied Systems Analysis,* **9**, 17-29.

Jackson, M.C. (1985a). Systems inquiring competence and organisational analysis. *Proceedings of the 1985 Meeting of the Society for General Systems Research,* pp. 522-530.

Jackson, M.C. (1985b). Social systems theory and practice: The need for a critical approach. *International Journal of General Systems,* **10**, 135-151.

Jackson, M.C. (1985c). The itinerary of a critical approach: Review of Ulrich's 'Critical Heuristics of Social Planning'. *Journal of the Operational Research Society,* **36**, 878-881.

Jackson, M.C. (1987a). Present positions and future prospects in management science. *Omega,* **15**, 455-466.

Jackson, M.C. (1987b). New directions in management science, In, *New Directions in Management Science.* Jackson, M.C. and Keys, P. (eds). Gower, Aldershot.

Jackson, M.C. (1988). Some methodologies for community operational research. *Journal of the Operational Research Society,* **39**, 715-724.

Jackson, M.C. (1990). Beyond a system of systems methodologies. *Journal of the Operational Research Society,* **41**, 657-668.

Jackson, M.C. (1991). *Systems Methodology for the Management Sciences.* Plenum, New York.

Jackson, M.C. (1992). With friends like this.... A comment on Mingers' 'Recent developments in critical management science'. *Journal of the Operational Research Society,* **43**, 729-731.

Jackson, M.C. (1993a). The system of systems methodologies: A guide to researchers. *Journal of the Operational Research Society,* **44**, 208-209.

Jackson, M.C. (1993b). How to cause anguish without even trying: A reply to Graham Jones. *Journal of the Operational Research Society,* **44**, 848-849.

Jackson, M.C. (1993c). Don't bite my finger: Haridimos Tsoukas' critical evaluation of total systems intervention. *Systems Practice*, **6**, 289-294.
Jackson, M.C. (1995). Beyond the fads: Systems thinking for managers. *Systems Research*, **12**, 25-42.
Jackson, M.C. (1997). Pluralism in systems thinking and practice. In, *Multimethodology: The Theory and Practice of Combining Management Science Methodologies*. Mingers, J. and Gill, A. (eds.). Wiley, Chichester.
Jackson, M.C. (1999). Towards coherent pluralism in management science. *Journal of the Operational Research Society*, **50**, 12-22.
Jackson, M.C. and Keys, P. (1984). Towards a system of systems methodologies. *Journal of the Operational Research Society*, **35**, 473-486.
Jackson, N. and Carter, P. (1991). In defence of paradigm incommensurability. *Organization Studies*, **12**, 109-127.
Jahoda, A., Cattermole, M. and Markova, I. (1989). Day services for people with mental handicaps: A purpose in life? *Mental Handicap*, **17**, 136-139.
James, K. and McIntyre, D. (1983). The reproduction of families: The social role of family therapy? *Journal of Marital and Family Therapy*, **9**, 119-129.
James, W. (1904). The pragmatic method. *Journal of Philosophy*, **1**, 673-687.
Jenkins, G. (1969). The systems approach. *Journal of Systems Engineering*, **1**, 3-49.
Jennings, D. and Wattan, S. (1994). *Decision Making*. Pitman, London.
Jones, G.C. (1993). OR practice, systems methodologies, and the need to do better. *Journal of the Operational Research Society*, **44**, 845-848.
Jones, H.G. (1992). Early OR in the steel company of Wales. *Journal of the Operational Research Society*, **43**, 563-567.
Julienne, L. (1998). Homelessness and young single people from black and minority ethnic communities. *Youth & Policy*, **59**, Spring, 23-27.
Jung, C.G. (1946). On the nature of the psyche. In, *On the Nature of the Psyche*. 1960 ed. Ark, London.
Kant, I. (1787). *The Critique of Pure Reason*. 2nd ed, 1933. Smith N.K. (trans.). Macmillan, Basingstoke.
Kant, I. (1788). *Critique of Practical Reason and Other Writings in Moral Philosophy*. Beck, L. (trans. and ed.). 1949 ed. University of Chicago Press, Chicago.
Kauffman, L.H. (1999). Virtual logic—The Flagg resolution. *Cybernetics and Human Knowing*, **6**, 87-96.
Keeney, B. (1982). What is an epistemology of family therapy? *Family Process*, **21**, 153-168.
Keeney, B. (1983). *Aesthetics of Change*. Guilford, New York.
Kelly, G.A. (1955). *The Psychology of Personal Constructs. Volume One: A Theory of Personality*. W.W. Norton, New York.
Kelly, G.A. (1970). A brief introduction to personal construct theory. In, *Perspectives in Personal Construct Theory*. Bannister, D. (ed.). Academic Press, London.
Kelly, M. (1994). *Critique and Power: Recasting the Foucault/Habermas Debate*. MIT Press, Cambridge, MA.
Keys, P. (1987). Management and management support in community service agencies. *Dragon*, **2**, 19-45.
Keys, P. (1988). A methodology for methodology choice. *Systems Research*, **5**, 65-76.
Keys, P. (1989). OR as technology: Some issues and implications. *Journal of the Operational Research Society*, **40**, 753-759.
Keys, P. (1991). *Operational Research and Systems: The Systemic Nature of Operational Research*. Plenum, New York.
Keys, P. (1998). OR as technology revisited. *Journal of the Operational Research Society*, **49**, 99-108.
Kintrea, K. (1996). Whose partnership? Community interests in the regeneration of a Scottish housing scheme. *Housing Studies*, **11**, 287-306.
Koch, S. and Leary, D.E. (eds.) (1985). *A Century of Psychology as Science*. McGraw-Hill, New York.
Koehler, W. (1938). *The Place of Values in the World of Fact*. Liveright, New York.
Kolb, D. (1984). *Experimental Learning*. Prentice Hall, Hemel Hempstead.

Krell, D.F., and Capuzzi, F.A. (1975). English translation from the original German of Martin Heidegger's *Early Greek Thinking: The Dawn of Western Philosophy*. Harper and Row, San Francisco.

Kremyanskiy, V.I. (1958). Certain peculiarities of organisms as a "system" from the point of view of physics, cybernetics and biology. English trans., 1960. *General Systems*, **5**, 221-230.

Krishnamurti, J. (1991). *Meeting Life*. Arkana, London.

Krull, M., Luhmann, N. and Maturana, H. (1989). Basic concepts of the theory of autopoietic systems. *Systems Studies*, **1**, 79-104.

Kuhn, T. (1962). *The Structure of Scientific Revolutions*. University of Chicago Press, Chicago.

Kvale, S. (1996). *InterViews: An Introduction to Qualitative Research Interviewing*. Sage, California.

Laing, R.D. and Esterson, A. (1964). *Sanity, Madness and the Family*. Penguin, Harmondsworth.

Lander, R., McRobb, S. and Stowell, F.A. (1997). Bridging the gap between IS definition and IS specification. In, *Systems for Sustainability: People, Organizations, and Environments*. Stowell, F.A., Ison, R.L., Armson, R., Holloway, J., Jackson, S. and McRobb, S. (eds.). Plenum, New York.

Lane, D.C. (2000). Should system dynamics be described as a 'hard' or 'deterministic' systems approach? *Systems Research and Behavioral Science*, **17**, 3-22.

Larner, G. (1995). The real as illusion: Deconstructing power in family therapy. *Journal of Family Therapy*, **17**, 191-217.

Lathrop, J.B. (1959). Operations research looks to science. *Operations Research*, **7**, 423-429.

Leach, E. (1976). *Culture and Communication: The Logic by which Symbols are Connected*. Cambridge University Press, Cambridge.

Leary, D.E. (1980). One hundred years of experimental psychology. *Psychological Research*, **42**, 175-189.

Leclerc, I. (1972). *The Nature of Physical Existence*. George Allen & Unwin, London.

Leclerc, I. (1986). *The Philosophy of Nature*. Catholic University of America Press, Washington DC.

Leonard, A. (1997). Using models in sequence: A case study of a post-acquisition intervention. In, *Multimethodology: The Theory and Practice of Combining Management Science Methodologies*. Mingers, J. and Gill, A. (eds.). Wiley, Chichester.

Levin, M. (1994). Action research and critical systems thinking: Two icons carved out of the same log? *Systems Practice*, **7**, 25-41.

Levy, N. (1999). Foucault's unnatural ecology. In, *Discourses of the Environment*. Darier, É. (ed.). Blackwell, Oxford.

Lewin, K. (1947). Frontiers in group dynamics. *Human Relations*, **1**, 2-38.

Lewin, K. (1948). *Resolving Social Conflicts*. Harper and Brothers, New York.

Lewin, K. (1952). *Field Theory in the Social Sciences*. Tavistock Publications, London.

Lewis, P.J. (1993). Linking soft systems methodology with data-focused information system development. *Journal of Information Systems*, **3**, 169-186.

Lewis, P.J. (1994). *Information-Systems Development*. Pitman, London.

Lilienfeld, R. (1978). *The Rise of Systems Theory: An Ideological Analysis*. Wiley, New York.

Linstone, H.A. (1999). *Decision Making for Technology Executives: Using Multiple Perspectives to Improve Performance*. Artech House, Boston.

Lleras, E. (1995). Towards a methodology for organisational intervention in Colombian enterprises. *Systems Practice*, **8**, 169-182.

Locke, J. (1689). *Two Treatises of Government*. 1967 ed. Laslett, P. (ed.). Cambridge University Press, Cambridge.

Lovelock, J. (1979). *Gaia: A New Look at Life on Earth*. Oxford University Press, Oxford.

Lovelock, J. (1988). *The Ages of Gaia: A Biography of our Living Earth*. Oxford University Press, Oxford.

Lovelock, J. (1990). The greening of science. *Resurgence*, #138, 12-19.

Luhmann, N. (1986). *Ecological Communication*. 1989 trans. University of Chicago Press, Chicago.

Lyotard, J.F. (1979). *The Post-Modern Condition: A Report on Knowledge*. 1984 trans. Manchester University Press, Manchester.

MacIntyre, A. (1985). *After Virtue: A Study in Moral Theory*. Duckworth, London.

MacKinnon, L. and Miller, D. (1987). The new epistemology and the Milan approach: Feminist and sociopolitical considerations. *Journal of Marital and Family Therapy*, **13**, 139-155.

Malin, H. (1981). Of kings and men, especially OR men. *Journal of the Operational Research Society*, **32**, 953-965.

Mandel, E. (1975). *Late Capitalism*. New Left Books, London.

Mansell, G. (1991). Methodology choice in a coercive context. *Systems Practice*, **4**, 37-46.

Marcuse, H. (1964). *One Dimensional Man: Studies in the Ideology of Advanced Industrial Societies*. Routledge and Kegan Paul, London.

Margulis, L. and Sagan, D. (1987). *Microcosmos: Four Billion Years of Microbial Evolution*. Allen & Unwin, London.

Mar Molinero, C. (1992). Operational research: From war to community. *Socio-Economic Planning Sciences*, **26**, 203-212.

Mar Molinero, C. (1993). Aldermoor School: The operational researcher on the side of the community. *Journal of the Operational Research Society*, **44**, 237-245.

Marx, K. (1887). *Capital: A Critique of Political Economy, Volume I*. 4th ed. Engels, F. (ed.). Moore, S. and Aveling, E. (trans.). Lawrence & Wishart, London.

Marx, K. and Engels, F. (1888). *The Communist Manifesto*. Revised ed. Lawrence & Wishart, London.

Mason, R.O. (1969). A dialectical approach to strategic planning. *Management Science*, **15**, B403-414.

Mason, R.O. and Mitroff, I.I. (1981). *Challenging Strategic Planning Assumptions*. Wiley, New York.

Masterman, M. (1970). The nature of a paradigm. In, *Criticism and the Growth of Knowledge*. Lakatos, I. and Musgrave, A. (eds.). Cambridge University Press, London.

Mathews, F. (1991). *The Ecological Self*. Routledge, London.

Maturana, H. (1988a). *Ontology of Observing: The Biological Foundations of Self Consciousness and the Physical Domain of Existence*. http://www.inteco.cl/biology/

Maturana, H. (1988b). Reality: The search for objectivity or the quest for a compelling argument. *Irish Journal of Psychology*, **9**, 25-82.

Maturana, H.R. and Varela, F.J. (1992). *The Tree of Knowledge: The Biological Roots of Human Understanding*. Revised ed. Shambhala, Boston.

Mayo, E. (1949). *The Social Problems of an Industrial Civilization*. Routledge & Kegan Paul, London.

Maze, J.R. (1983). *The Meaning of Behaviour*. George Allen and Unwin, London.

McBurney, J. (1990). *Ecology into Economics Won't Go: Or, Life is not a Concept*. Green Books, Hartland.

McCormick, B.J., Kitchin, P.D., Marshall, G.P., Sampson, A.A. and Sedgwick, R. (1974). *Introducing Economics*. Penguin, Harmondsworth.

McGill, I. and Beaty, L. (1992). *Action Learning: A Practitioner's Guide*. Kogan Page, London.

McGuigan, F.J. (1968). *Experimental Psychology: A Methodological Approach*. Prentice Hall, Hemel Hempstead.

McKnight, C. (1976). *Purposive Preferences for Multi-Attributed Alternatives: A Study of Choice Behaviour using Personal Construct Theory in Conjunction with Decision Theory*. Ph.D. Thesis, Brunel University.

McLoughlin, B. (1995). *Developing Psychodynamic Counselling*. Sage, London.

McNamee, S. and Gergen, K.J. (eds.) (1992). *Therapy as Social Construction*. Sage, London.

Meadows, D.H., Meadows, D.L., Randers, J. and Behrens, W.W. (1972). *The Limits to Growth*. Earth Island, London.

Meadows, D.H., Meadows, D.L. and Randers, J. (1992). *Beyond the Limits to Growth*. Chelsea Green, Post Mills.

Mendez, C., Coddou, F. and Maturana, H. (1988). The bringing forth of pathology. *Irish Journal of Psychology*, **9**, 14-172.

Merleau-Ponty, M. (1962). *The Phenomenology of Perception*. Routledge and Kegan Paul, London.

Middleton, D. and Edwards, D. (eds.) (1990). *Collective Remembering*. Sage, London.

Midgley, G. (1988). *A Systems Analysis and Evaluation of Microjob—A Vocational Rehabilitation and Information Technology Training Centre for People with Disabilities*. M. Phil. Thesis, City University, London.

Midgley, G. (1989a). Critical systems: The theory and practice of partitioning methodologies. *Proceedings of the 33rd Annual Meeting of the International Society for General Systems Research (Volume II)*, held in Edinburgh, Scotland, on 2-7 July 1989.

Midgley, G. (1989b). Critical systems and the problem of pluralism. *Cybernetics and Systems*, **20**, 219-231.

Midgley, G. (1990a). Creative methodology design. *Systemist*, **12**, 108-113.

Midgley, G. (1990b). Critical systems and methodological pluralism. In, *Toward a Just Society for Future Generations. Volume I: Systems Design*. Banathy, B.H. and Banathy, B.A. (eds.). International Society for the Systems Sciences, Pomona, CA.

Midgley, G. (1991a). The legitimation of systems science. In, *Systems Science in the 21st Century: Integrating the New Sciences of Complexity in Service of Humans and their Environment*. Proceedings of the 35th Annual Conference of the International Society for Systems Science, Östersund, Sweden, June 13-21, 1991.

Midgley, G. (1991b). The sacred and profane in critical systems thinking. In, *Systems Thinking in Europe*. Jackson, M.C., Mansell, G.J., Flood, R.L., Blackham, R.B. and Probert, S.V.E. (eds.). Plenum, New York.

Midgley, G. (1992a). Pluralism and the legitimation of systems science. *Systems Practice*, **5**, 147-172.

Midgley, G. (1992b). The sacred and profane in critical systems thinking. *Systems Practice*, **5**, 5-16.

Midgley, G. (1992c). Breaking the bonds of language. Unpublished paper (available on request from the author).

Midgley, G. (1992d). *Unity and Pluralism*. Ph.D. thesis, City University, London.

Midgley, G. (1993). A contextual theory of ethics. *The Psychologist*, **6**, 175-178.

Midgley, G. (1994). Ecology and the poverty of humanism: A critical systems perspective. *Systems Research*, **11**, 67-76.

Midgley, G. (1995a). Evaluation and change in service systems for people with disabilities. In, *Evaluating Quality in Services for Disabled and Older People*. Pilling, D. and Watson, G. (eds.). Jessica Kingsley Publishers, London.

Midgley, G. (1995b). Systemic intervention: A critical systems perspective. In, *Systems Methodology: Possibilities for Cross-Cultural Learning and Integration*. Midgley, G. and Wilby, J. (eds.). Centre for Systems Studies, Hull.

Midgley, G. (1996a). The ideal of unity and the practice of pluralism in systems science. In, *Critical Systems Thinking: Current Research and Practice*. Flood, R.L. and Romm, N.R.A. (eds.). Plenum, New York.

Midgley, G. (1996b). Presentation: Gerald Midgley. In, *Forum One: Transcripts and Reflections*. Wilby, J. (ed.). Centre for Systems Studies, Hull.

Midgley, G. (1996c). What is this thing called critical systems thinking? In, *Critical Systems Thinking: Current Research and Practice*. Flood, R.L. and Romm, N.R.A. (eds.). Plenum, New York.

Midgley, G. (1996d). Evaluation and change in service systems for people with disabilities: A critical systems perspective. *Evaluation*, **2**, 67-84.

Midgley, G. (1997a). Mixing methods: Developing systemic intervention. In, *Multimethodology: The Theory and Practice of Combining Management Science Methodologies*. Mingers, J. and Gill, A. (eds.). Wiley, Chichester.

Midgley, G. (1997b). Developing the methodology of TSI: From the oblique use of methods to creative design. *Systems Practice*, **10**, 305-319.

Midgley G. (1997c). Dealing with coercion: Critical systems heuristics and beyond. *Systems Practice*, **10**, 37-57.

Midgley, G. (1998). *Rethinking the Unity of Science*. Centre for Systems Studies Research Memorandum #19. Centre for Systems Studies, University of Hull, Hull.

Midgley, G. and Floyd, M. (1988). *Microjob: A Computer Training Service for People with*

Disabilities. Rehabilitation Resource Centre, London.

Midgley, G. and Floyd, M. (1990). Vocational training in the use of new technologies for people with disabilities. *Behaviour and Information Technology*, **9**, 409-424.

Midgley, G., Gu, J. and Campbell, D. (2000). Dealing with human relations in Chinese systems practice. *Systemic Practice and Action Research*, **13**, 71-96.

Midgley, G., Kadiri, Y., Vahl, M. (1996). Managing stories about quality. *International Journal of Technology Management*, **11**, 140-150.

Midgley, G. and Milne. A, (1995). Creating employment opportunities for people with mental health problems: A feasibility study for new initiatives. *Journal of the Operational Research Society*, **46**, 35-42.

Midgley, G. and Munlo, I. (1996). The theory and practice of boundary critique. In, *Systems Methodology II: Possibilities for Learning and Cross-Cultural Integration*. Wilby, J. (ed.). Centre for Systems Studies, Hull.

Midgley, G., Munlo, I. and Brown, M. (1997). *Sharing Power: Integrating User Involvement and Multi-Agency Working to Improve Housing for Older People*. Policy Press, Bristol.

Midgley, G., Munlo, I. and Brown, M. (1998). The theory and practice of boundary critique: Developing housing services for older people. *Journal of the Operational Research Society*, **49**, 467-478.

Midgley, G. and Ochoa-Arias, A.E. (1999). Visions of community for community OR. *Omega*, **27**, 259-274.

Midgley, G., Ritchie, C. and White, L. (1994). Community OR in the UK: A variety of experiences and learning. Unpublished paper presented to the Joint National Meeting of the Institute of Management Science and the Operations Research Society of America, in Boston (USA), April 24-27, 1994.

Mill, J.S. (1859). *On Liberty*. 1974 ed. Penguin, Harmondsworth.

Miller, G.A. (1956). The magic number seven, plus or minus two. *Psychological Review*, **63**, 81-97.

Miller, G.A. (1968). *The Psychology of Communication*. Penguin, London.

Miller, J.G. (1978). *Living Systems*. McGraw-Hill, New York.

Mingers, J.C. (1980). Towards an appropriate social theory for applied systems thinking: Critical theory and soft systems methodology. *Journal of Applied Systems Analysis*, **7**, 41-50.

Mingers, J.C. (1984). Subjectivism and soft systems methodology—A critique. *Journal of Applied Systems Analysis*, **11**, 85-103.

Mingers, J.C. (1988). Comparing conceptual models and data flow diagrams. *The Computer Journal*, **31**, 376-379.

Mingers, J.C. (1992a). Criticising the phenomenological critique: Autopoiesis and critical realism. *Systems Practice*, **5**, 173-180.

Mingers, J.C. (1992b). Recent developments in critical management science. *Journal of the Operational Research Society*, **43**, 1-10.

Mingers, J.C. (1992c). Technical, practical and critical OR—Past, present and future? In, *Critical Management Studies*. Alvesson, M. and Willmott, H. (eds.). Sage, London.

Mingers, J.C. (1992d). What are real friends for? A reply to Mike Jackson. *Journal of the Operational Research Society*, **43**, 732-735.

Mingers, J.C. (1993). The system of systems methodologies—A reply to Schecter. *Journal of the Operational Research Society*, **44**, 206-208.

Mingers, J.C. (1995). *Self-Producing Systems: Implications and Applications of Autopoiesis*. Plenum, New York.

Mingers, J.C. (1997a). Towards critical pluralism. In, *Multimethodology: The Theory and Practice of Combining Management Science Methodologies*. Mingers, J. and Gill, A. (eds.). Wiley, Chichester.

Mingers, J.C. (1997b). A critical evaluation of Maturana's constructivist family therapy. *Systems Practice*, **10**, 137-151.

Mingers, J.C. (1997c). Multi-paradigm multimethodology. In, *Multimethodology: The Theory and Practice of Combining Management Science Methodologies*. Mingers, J. and Gill, A. (eds.). Wiley, Chichester.

Mingers, J. and Brocklesby, J. (1996). Multimethodology: Towards a framework for critical pluralism. *Systemist*, **18**, 101-131.

Mingers, J. and Gill, A. (eds.) (1997). *Multimethodology: The Theory and Practice of Combining Management Science Methodologies*. Wiley, Chichester.

Minuchin, S. (1974). *Families and Family Therapy*. Tavistock, London.

Miser, H.J. and Quade, E.S. (eds.) (1985). *Handbook of Systems Analysis: Overview of Uses, Procedures, Applications and Practice*. North Holland, New York.

Miser, H.J. and Quade, E.S. (eds.) (1988). *Handbook of Systems Analysis: Craft Issues and Procedural Choices*. Wiley, New York.

Mitchell, J. (ed.) (1986). *The Selected Melanie Klein*. Penguin, London.

Mitroff, I.I. and Emshoff, J.R. (1979). On strategic assumption-making: A dialectical approach to policy and planning. *Academy of Management Review*, **6**, 649-651.

Mitroff, I.I., Emshoff, J.R. and Kilmann, R.H. (1979). Assumptional analysis: A methodology for strategic problem-solving. *Management Science*, **25**, 583-593.

Mitroff, I.I. and Linstone, H.A. (1993). *The Unbounded Mind: Breaking the Chains of Traditional Business Thinking*. Oxford University Press, Oxford.

Morecroft, J.D.W. and Sternman, J.D. (eds.) (1994). *Modeling for Learning Organizations*. Productivity Press, Portland, Oregon.

Morgan, G. (ed.) (1983). *Beyond Method: Strategies for Social Research*. Sage, London.

Morgan, G. (1986). *Images of Organization*. Sage, London.

Morgan-Klein, B. (1990). *Young Runaways: A Scottish Review*. HMSO, London.

Morrissette, P. (1992). Engagement strategies with reluctant homeless young people. *Psychotherapy*, **29**, 447-451.

Müller-Merbach, H. (1994). A system of systems approaches. *Interfaces*, **24**, 16-25.

Mumford, A. (ed.) (1997). *Action Learning at Work*. Gower, Aldershot.

Munlo, I.G. (1997). *Critical Systems Thinking, Theory and Practice: A Case Study of an Intervention in Two British Local Authorities*. Ph.D. Thesis, University of Hull.

Munro, I. (1999). Man-machine systems: People and technology in OR. *Systemic Practice and Action Research*, **12**, 513-532.

Mvula, A. (1999). *Critical Systems Thinking and Social Systemic Transformation: The Case of Malawi*. Ph.D. Thesis, University of Hull.

Myers, C.S. (1920). *Mind and Work: The Psychological Factors in Industry and Commerce*. University of London Press, London.

Myers, C.S. (1923). The efficiency engineer and the industrial psychologist. *Journal of the National Institute of Industrial Psychology*, **1**, 168-172.

Myers, C.S. (1926). *Industrial Psychology in Great Britain*. Jonathan Cape, London.

Nicolis, G. and Prigogine, I. (1977). *Self-Organization in Non-Equilibrium Systems*. Wiley, New York.

Nicolis, G. and Prigogine, I. (1989). *Exploring Complexity: An Introduction*. W.H. Freeman and Co., New York.

Nirje, B. (1960): The normalization principle and its human management implications. In, *Changing Patterns in Residential Services for the Mentally Retarded*. Kugel, R. and Wolfensberger, W. (eds.). President's Committee on Mental Retardation, Washington, DC.

Noad, N. and King, L. (1977). Area co-ordination: Some examples compared. *Linkage*, **2**, 10-13.

Norman, M. and Stoker, B. (1991). *Data Envelopment Analysis: The Assessment of Performance*. Wiley, Chichester.

Northrop, F.S.C. (1967). The method and theories of physical science and their bearing upon biological organization. In, *Great Ideas in Modern Science*. Marks, R.W. (ed.). Bantam Books, New York.

Nozick, R. (1974). *Anarchy, State and Utopia*. Basic Books, New York.

Ochoa-Arias, A.E. (1999). (Personal communication).

Oliga, J.C. (1988). Methodological foundations of systems methodologies. *Systems Practice*, **1**, 87-112.

Optner, S.L. (ed.) (1973). *Systems Analysis*. Penguin, Harmondsworth.

Ormerod, R. (1996). New methods for old. *Journal of the Operational Research Society*, **47**, 1317-1318.

Ormerod, R. (1997). Mixing methods in practice: A transformation-competence perspective. In, *Multimethodology: The Theory and Practice of Combining Management Science Methodologies*. Mingers, J. and Gill, A. (eds.). Wiley, Chichester.

Ormerod, R. (1999). Ethical dilemmas. *Journal of the Operational Research Society*, **50**, 546-548.

Pace, L.A. and Argona, D.R. (1991). Participatory action research: A view from Xerox. In, *Participatory Action Research*. Whyte, W.F. (ed.). Sage, London.

Parry, R. and Mingers, J. (1991). Community operational research: Its context and its future. *Omega*, **19**, 577-586.

Patton, M.Q. (1978). *Utilization-Focused Evaluation*. Sage, Beverly Hills, CA.

Patton, M.Q. (1980). *Qualitative Evaluation Methods*. Sage, London.

Patton, M.Q. (1987). *How to Use Qualitative Methods in Evaluation*. Sage, London.

Pepper, S.C. (1942). *World Hypotheses: A Study in Evidence*. University of California Press, Berkeley, CA.

Perelberg, R. and Miller, A. (1990). *Gender and Power in Families*. Routledge, London.

Pierce, C.S. (1934). *The Collected Papers of C.S. Pierce*. Harvard University Press, Cambridge, MA.

Pilling, D. and Watson, G. (eds.) (1995). *Evaluating Quality in Services for Disabled and Older People*. Jessica Kingsley, London.

Pindar, S. (1994). Planning a network response to racial harassment. In, *Community Works: 26 Case Studies showing Community Operational Research in Action*. Ritchie, C., Taket, A. and Bryant, J. (eds.). Pavic Press, Sheffield.

Pinzón, L. and Midgley, G. (2000). Developing a systemic model for the evaluation of conflicts. *Systems Research and Behavioral Science*, **17**(6), in press.

Ploeg, J. van der and Scholte, E. (1997). *Homeless Youth*. Sage, London.

Plutchick, R. (1968). *Foundations of Experimental Research*. Harper and Row, San Francisco.

Pols, E. (1967). *Whitehead's Metaphysics*. Southern Illinois University Press, Carbondale.

Popper, K.R. (1959). *The Logic of Scientific Discovery*. Originally published as *Logik de Forschung*, 1935. Harper, New York.

Popper, K.R. (1966). *The Open Society and Its Enemies. Volume II. The High Tide of Prophesy: Hegel, Marx, and the Aftermath*. 5th ed. Routledge, London.

Popper, K.R. (1972). *Objective Knowledge*. Oxford University Press, Oxford.

Power, S., Whitty, G. and Youdell, D. (1995). *No Place to Learn: Homelessness and Education*. Shelter, London.

Powney, J. and Watts, M. (1987). *Interviewing in Educational Research*. Routledge and Kegan Paul, London.

Prigogine, I. (1947). *Étude Thermodynamique des Phénomènes Irréversibles*. Dunod, Paris.

Prigogine, I. (1989). The rediscovery of time: Science in a world of limited predictability. *Beshara*, #9, 28-32.

Prigogine, I. and Stengers, I. (1984). *Order out of Chaos: Man's New Dialogue with Nature*. Fontana, London.

Prior, R. (1990). Deriving data flow diagrams from a 'soft systems' conceptual model. *Systemist*, **12**, 65-75.

Prosser, J. (ed.) (1998). *Image-Based Research: A Sourcebook for Qualitative Researchers*. Falmer Press, London.

Quade, E.S. and Boucher, W.I. (1968). *Systems Analysis and Policy Planning: Applications in Defence*. Elsevier, New York.

Quade, E.S., Brown, K., Levien, R., Majone, G. and Rakhmankulov, V. (1978). Systems analysis: An outline for the IIASA international series of monographs. *Journal of Applied Systems Analysis*, **5**, 91-98.

Raadt de, J.D.R. (1997). Faith and the normative foundation of systems science. *Systems Practice*, **10**, 13-35.

Rahman, M.A. (1991). The theoretical standpoint of PAR. In, *Action and Knowledge: Breaking the Monopoly with Participatory Action Research*. Fals Borda, O. and Rahman, M.A. (eds.). Apex Press, New York.

Raitt, R.A. (1979). OR and science. *Journal of the Operational Research Society*, **30**, 835-836.

Rawls, J. (1971). *A Theory of Justice*. Harvard University Press, Cambridge, MA.

Reason, P. (1988a). The co-operative inquiry group. In, *Human Inquiry in Action: Developments in New Paradigm Research*. Reason, P. (ed.). Sage, London.

Reason, P. (ed.) (1988b). *Human Inquiry in Action: Developments in New Paradigm Research.* Sage, London.

Reason, P. (ed.) (1994). *Participation in Human Inquiry.* Sage, London.

Reason, P. (1996). Comments on Midgley's paper, 'The Theory and Practice of Boundary Critique'. In, *Forum One: Transcripts and Reflections.* Wilby, J. (ed.). Centre for Systems Studies, Hull.

Reason, P. and Heron, J. (1995). Co-operative inquiry. In, *Rethinking Methods in Psychology.* Smith, J.A., Harré, R. and Van Langenhove, L. (eds.). Sage, London.

Reed, J. (1992). *Review of Health and Social Services for Mentally Disordered Offenders and Others Requiring Similar Services: A Note on the Visit Programme.* Department of Health/Home Office, London.

Reed, M. (1985). *Redirections in Organizational Analysis.* Tavistock, London.

Resnik, D.B. (1998). *The Ethics of Science: An Introduction.* Routledge, London.

Revens, R.W. (1982). *The Origins and Growth of Action Learning.* Chartwell-Bratt, Bromley.

Revens, R.W. (1983). *The ABC of Action Learning.* Chartwell-Bratt, Bromley.

Reynolds, M. (1998). "Unfolding" natural resource-use information systems: Fieldwork in Botswana. *Systemic Practice and Action Research,* **11**, 127-152.

Reynolds, M. (1999). Unpublished seminar delivered in the Business School, University of Hull, November 1999.

Rich, J. (1968). *Interviewing Children and Adolescents.* Macmillan, London.

Ritchie, C. (1994). Community OR—Five years of organised activities and beyond. *International Transactions in Operational Research,* **1**, 41-49.

Ritchie, C., Taket, A. and Bryant, J. (eds.) (1994). *Community Works: 26 Case Studies showing Community Operational Research in Action.* Pavic Press, Sheffield.

Roberts, N., Andersen, D., Deal, R., Garet, M. and Shaffer, W. (1983). *Introduction to Computer Simulation: A System Dynamics Modeling Approach.* Addison-Wesley, Reading, MA.

Robson, C. (1973). *Experiment, Design and Statistics in Psychology.* Penguin, Harmondsworth.

Rolston, H. (1983). Are values in nature subjective or objective? In, *Environmental Philosophy.* Elliot, R. and Gare, A. (eds.). Open University Press, Milton Keynes.

Romm, N.R.A. (1994). *Continuing Tensions between Soft Systems Methodology and Critical Systems Heuristics.* Centre for Systems Studies Working Paper #5. Centre for Systems Studies, University of Hull, Hull.

Romm, N.R.A. (1995a). Knowing as intervention: Reflections on the application of systems ideas. *Systems Practice,* **8**, 137-167.

Romm, N.R.A. (1995b). Some anomalies in Ulrich's critical inquiry and problem-solving approach. In, *Critical Issues in Systems Theory and Practice.* K. Ellis, A. Gregory, B. Mears-Young and G. Ragsdell (eds.). Plenum, New York.

Romm, N.R.A. (1996). Inquiry-and-intervention in systems planning: Probing methodological rationalities. *World Futures,* **47**, 25-36.

Rorty, R. (1989). *Contingency, Irony and Solidarity.* Cambridge University Press, Cambridge.

Rose, M. (1988). After the revolution. *New Internationalist,* #182, 13-14.

Rose, N. (1990). *Governing the Soul: The Shaping of the Private Self.* Routledge, London.

Rosenblatt, P.C. (1994). *Metaphors of Family Systems Theory: Toward New Constructions.* Guilford, New York.

Rosenhead, J. (1986). Custom and practice. *Journal of the Operational Research Society,* **37**, 335-343.

Rosenhead, J. (1987). From management science to workers' science. In, *New Directions in Management Science.* Jackson, M.C. and Keys, P. (eds.). Gower, Aldershot.

Rosenhead, J. (ed.) (1989a). *Rational Analysis for a Problematic World.* Wiley, Chichester.

Rosenhead, J. (1989b). Robustness analysis: Keeping your options open. In, *Rational Analysis for a Problematic World.* Rosenhead, J. (ed.). Wiley, Chichester.

Rosenhead, J. (1989c). Robustness to the first degree. In, *Rational Analysis for a Problematic World.* Rosenhead, J. (ed.). Wiley, Chichester.

Rosenhead, J. and Thunhurst, C. (1982). A materialist analysis of operational research. *Journal of the Operational Research Society,* **33**, 111-122.

Roszak, T. (1993). *The Voice of the Earth: An Exploration of Eco-Psychology*. Touchstone, New York.
Rothschuh, K.E. (1973). The mind-body problem. In, *Unity through Diversity: A Festschrift for Ludwig von Bertalanffy, Part II*. Gray, W. and Rizzo, N.D. (eds.). Gordon and Breach, New York.
Roy-Chowdhury, S. (1997). Systemic therapy: An overview. *Systems Practice*, **10**, 127-135.
Ryder, R. (1991). Sentientism: A comment on Gray and Singer. *The Psychologist*, **4**, 201.
Sachs, W. (1999). Rich in things, poor in time. *Resurgence*, #196, 14-16.
Samuels, A. (1993). *The Political Psyche*. Routledge, London.
Sandel, M. (1982). *Liberalism and the Limits of Justice*. Cambridge University Press, Cambridge.
Savage, A. and Mingers, J. (1996). A framework for linking soft systems methodology and Jackson system development. *Information Systems Journal*, **6**, 109-129.
Schecter, D. (1991). Critical systems thinking in the 1980s: A connective summary. In, *Critical Systems Thinking: Directed Readings*. Flood, R.L. and Jackson, M.C. (eds.). Wiley, Chichester.
Schecter, D. (1993). In defence of the system of systems methodologies: Some comments on the Mingers/Jackson debate. *Journal of the Operational Research Society*, **44**, 205-206.
Schein, E.H. (1969). *Process Consultation: Its Role in Organizational Development*. Addison-Wesley, Reading, MA.
Schön, D.A. (1983). *The Reflective Practitioner: How Professionals Think in Action*. Maurice Temple Smith, London.
Schwaninger, M. (1997). Status and tendencies of management research: A systems oriented perspective. In, *Multimethodology: The Theory and Practice of Combining Management Science Methodologies*. Mingers, J. and Gill, A. (eds.). Wiley, Chichester.
Schwarz, E. (1994). A metamodel to interpret the emergence, evolution and functioning of viable natural systems. In, *Cybernetics and Systems '94*. Trappl, R. (ed.). World Scientific, Singapore.
Schwarz, E. (1995). Where is the paradigm? In the people's mind or in the social system? *Separata de Revista International de Systemas*, **7**, 5-54.
Scriven, M. (1991). *Evaluation Thesaurus*. 4th ed. Sage, London.
Seidman, E. (1988). Back to the future, community psychology: Unfolding a theory of social intervention. *American Journal of Community Psychology*, **16**, 3-24.
Selvini-Palazzoli, M., Cecchin, G., Prata, G. and Boscolo, L. (1974). The treatment of children through brief therapy of their parents. *Family Process*, **13**, 429-442.
Selvini-Palazzoli, M., Cecchin, G., Prata, G. and Boscolo, L. (1978). *Paradox and Counterparadox*. Jason Aronson, New York.
Selvini-Palazzoli, M., Cecchin, G., Prata, G. and Boscolo, L. (1980). Hypothesising-circularity-neutrality: Three guidelines for the conductor of the session. *Family Process*, **19**, 3-12.
Shiva, V. (1990). Cry foul, cry freedom. *New Internationalist*, #206, 20-21.
Shotter, J. (1993). *Conversational Realities: Constructing Life through Language*. Sage, London.
Sieber, J.E. (1992). *Planning Ethically Responsible Research: A Guide for Students and Internal Review Boards*. Sage, London.
Siegel, S. (1956). *Non-Parametric Statistics for the Behavioral Sciences*. McGraw-Hill, New York.
Simons, H.W. and Billig, M. (1994). *After Postmodernism: Reconstructing Ideology Critique*. Sage, London.
Singer, A.E. (1959). *Experience and Reflection*. Churchman, C.W. (ed.). University of Pennsylvania Press, Philadelphia.
Singer, P. (1990). *Animal Liberation*. 2nd ed. Jonathan Cape, London.
Singer, P. (1991). Speciesism, morality and biology: A response to Jeffrey Gray. *The Psychologist*, **4**, 199-200.
Skinner, B.F. (1971). *Beyond Freedom and Dignity*. Penguin, Harmondsworth.
Smart, B. (1983). *Foucault, Marxism and Critique*. Routledge and Kegan Paul, London.
Sommerhoff, G. (1969). The abstract characteristics of living systems. In, *Systems Thinking, Volume One*. Emery, F.E. (ed.). Penguin, Harmondsworth.
Soper, K. (1995). *What is Nature? Culture, Politics and the Non-Human*. Blackwell, Oxford.
Spash, C.L. (1997). Ethics and environmental attitudes with implications for economic valuation. *Journal of Environmental Management*, **50**, 403-416.

Spaul, M. (1997). Multimethodology and critical theory: An intersection of interests? In, *Multimethodology: The Theory and Practice of Combining Management Science Methodologies*. Mingers, J. and Gill, A. (eds.). Wiley, Chichester.

Spencer Brown, G. (1972). *Laws of Form*. Julian Press, New York.

Stacey, R. (1992). *Managing the Unknowable: Strategic Boundaries between Order and Chaos in Organisations*. Jossey-Bass, San Francisco.

Stein, M., Rees, G. and Frost, N. (1994). *Running—The Risk: Young People on the Streets of Britain Today*. The Children's Society, London.

Stein, M., Rees, G., Wallis, L., Wade, J., Franks, M., Stevenson, J., Mumtaz, S., Lynes, D., Johnson, L. and Raws, P. (1999). *Still Running: Children on the Streets in the UK*. The Children's Society, London.

Stewart, I. (1989). *Does God Play Dice?* Blackwell, Cambridge, MA.

Storr, A. (ed.) (1983). *Jung: Selected Writings*. Fontana, London.

Stowell, F.A. (1985). Experience with soft systems methodology and data analysis. *Information Technology Training*, May 1985, 48-50.

Strauss, W. and Howe, N. (1991). *Generations: The History of America's Future, 1584-2069*. William Morrow and Co., New York.

Straussfogel, D. and Becker, M.L. (1996). An evolutionary systems approach to policy intervention for achieving ecologically sustainable societies. *Systems Practice*, **9**, 441-468.

Stumpf, S. and Dunbar, R. (1991). The effects of personality type on choices made in strategic decision situations. *Decision Sciences*, **22**, 1047-1069.

Sutcliffe, A. (1988). *Jackson System Development*. Prentice Hall, New York.

Sutton, D.C. (1995). The enterprise design framework meets the system of systems methodologies. *Systems Practice*, **8**, 409-439.

Swenson, R. (1991). Order, evolution, and natural law: Fundamental relations in complex systems theory. In, *Handbook of Systems and Cybernetics*. Negoita, C. (ed.). Marcel Dekker, New York.

Swenson, R. (1992). Autocatakinetics, yes—Autopoiesis, no: Steps towards a unified theory of evolutionary ordering. *International Journal of General Systems*, **21**, 207-228.

Taggert, M. (1985). The feminist critique in epistemological perspective: Questions of context in family therapy. *Journal of Marital and Family Therapy*, **11**, 113-126.

Taket, A. (1992). Review of 'Creative Problem Solving: Total Systems Intervention' by R.L. Flood and M.C. Jackson. *Journal of the Operational Research Society*, **43**, 1013-1016.

Taket, A. (1994a). Evaluation in the Health Service. In, *Community Works: 26 Case Studies showing Community Operational Research in Action*. Ritchie, C., Taket, A. and Bryant, J. (eds.). Pavic Press, Sheffield.

Taket, A. (1994b). Starting from where I was: Working in a feminist collective. In, *Community Works: 26 Case Studies showing Community Operational Research in Action*. Ritchie, C., Taket, A. and Bryant, J. (eds.). Pavic Press, Sheffield.

Taket, A. and White, L. (1993). After OR: An agenda for postmodernism and poststructuralism in OR. *Journal of the Operational Research Society*, **44**, 867-881.

Taket, A. and White, L. (2000). *Partnership and Participation: Decision-Making in the Multi-Agency Setting*. Wiley, Chichester.

Targett, D. (1996). *Analytical Decision Making*. Pitman, London.

Taylor, F.W. (1947). *Scientific Management*. Harper and Row, London.

Teubner, G. (1993). *Law as an Autopoietic System*. Blackwell, Oxford.

Thomas, A.R. and Lockett, M. (1979). Marxism and systems research: Values in practical action. In, *Improving the Human Condition*. Ericson, R.F. (ed.). Society for General Systems Research, Louisville, Kentucky.

Thomas, G.V. and Blackman, D. (1991). Are animal experiments on the way out? *The Psychologist*, **4**, 208-212.

Thompson, J. (1995). *User Involvement in Mental Health Services: The Limits of Consumerism, the Risks of Marginalisation and the Need for a Critical Approach*. Centre for Systems Studies Research Memorandum #8. Centre for Systems Studies, University of Hull, Hull.

Thunhurst, C., Ritchie, C., Friend, J. and Bocher, P. (1992a). Housing in Dearne Valley:

Doing community OR with the Thurnscoe Tenants Housing Cooperative, part 1. *Journal of the Operational Research Society*, **43**, 81-94.
Thunhurst, C., Ritchie, C., Friend, J. and Bocher, P. (1992b). Housing in Dearne Valley: Doing community OR with the Thurnscoe Tenants Housing Cooperative, part 2. *Journal of the Operational Research Society*, **43**, 677-690.
Trefethen, F.N. (1954). A history of operational research. In, *Operations Research for Management, Volume 1*. McCloskey, J.F. and Trefethen, F.N. (eds.). John Hopkins University Press, Baltimore.
Trist, E.L. and Bamforth, K.W. (1951). Some social and psychological consequences of the longwall method of coal-getting. *Human Relations*, **4**, 3-38.
Trist, E.L. and Burgess, S. (1978). Multiple deprivation: A human and economic approach. *Linkage*, **3**, 8-9.
Trist, E.L., Higgin, G.W., Murray, H. and Pollock, A.B. (1963). *Organizational Choice*. Tavistock, London.
Tsoukas, H. (1992). Panoptic reason and the search for totality: A critical assessment of the critical systems perspective. *Human Relations*, **45**, 637-657.
Tsoukas, H. (1993a). By their fruits shall ye know them: A reply to Jackson, Green, and Midgley. *Systems Practice*, **6**, 311-317.
Tsoukas, H. (1993b). The road to emancipation is through organizational development: A critical evaluation of total systems intervention. *Systems Practice*, **6**, 53-70.
Tsouvalis, C. and Checkland, P. (1996). Reflecting on SSM—The dividing line between real world and systems thinking world. *Systems Research*, **13**, 35-45.
Ulrich, W. (1983). *Critical Heuristics of Social Planning: A New Approach to Practical Philosophy*. Haupt, Berne.
Ulrich, W. (1986). *Critical Heuristics of Social Systems Design*. Working Paper #10, Department of Management Systems and Sciences, University of Hull, Hull.
Ulrich, W. (1988). Systems thinking, systems practice and practical philosophy: A program of research. *Systems Practice*, **1**, 137-163.
Ulrich, W. (1990). Critical systems thinking and ethics: The role of contemporary practical philosophy for developing an "ethics of whole systems". In, *Toward a Just Society for Future Generations. Volume I: Systems Design*. Banathy, B.H. and Banathy, B.A. (eds.). International Society for the Systems Sciences, Pomona, CA.
Ulrich, W. (1993). Some difficulties of ecological thinking, considered from a critical systems perspective: A plea for critical holism. *Systems Practice*, **6**, 583-611.
Ulrich, W. (1994). Can we secure future-responsive management through systems thinking and design? *Interfaces*, **24**, 26-37.
Ulrich, W. (1996a). *A Primer to Critical Systems Heuristics for Action Researchers*. Centre for Systems Studies, Hull.
Ulrich, W. (1996b). *Critical Systems Thinking for Citizens: A Research Proposal*. Centre for Systems Studies Research Memorandum #10. Centre for Systems Studies, University of Hull, Hull.
Ussher, J.M. (1994). Women's conundrum: Feminism or therapy? *Clinical Psychology Forum*, #54, 2-5.
Vahl, M. (1994). *Improving Mental Health Service in Calderdale: An Evaluation of 5 Schemes Funded through the Mental Illness Specific Grant*. Centre for Systems Studies, Hull.
Valero-Silva, N. (1996). Towards a critique of critical systems thinking within a Foucauldian framework: A "demystification process" or an "instrumental use" of critical theory. *Systems Practice*, **9**, 539-546.
Valero-Silva, N. (1998). *A Critical History of Critical Systems Thinking*. Ph.D. Thesis, University of Hull.
Vega, R.R. (1999). *Health Care and Social Justice Evaluation: A Critical and Pluralist Approach*. Ph.D. Thesis, University of Hull.
Vélez, J.I. (1999). *Autopoiesis and Power*. MA Dissertation, University of Hull.
Vickers, G. (1970). *Freedom in a Rocking Boat: Changing Values in an Unstable Society*. Penguin, Harmondsworth.
Wang, H. (1995). The spiral propulsion principle. In, *Systems Methodology: Possibilities for Cross-Cultural Learning and Integration*. Midgley, G. and Wilby, J. (eds.). Centre for Systems Studies, Hull.

Wanner, E. (1975). Do we understand sentences from the outside-in or from the inside-out? In, *Language as a Human Problem*. Bloomfield, M. and Haugen, E. (eds.). Lutterworth Press, Guildford.

Watson, L. (1997). *High Hopes: Making Housing and Community Care Work*. Joseph Rowntree Foundation, York.

Watzlawick, P. (ed.) (1984). *The Invented Reality*. Norton, New York.

Watzlawick, P., Beavin, J. and Jackson, D.D. (1968). *Pragmatics of Human Communication: A Study of Interactional Patterns, Pathologies and Paradoxes*. Faber and Faber, London.

Weakland, J.H. and Jackson, D.D. (1958). Patient and therapist observations on the circumstances of a schizophrenic episode. *Archives of Neurology and Psychiatry*, **79**, 554-574.

Weber, L.R., Miracle, A. and Skehan, T. (1994). Interviewing early adolescents: Some methodological considerations. *Human Organization*, **53**, 42-47.

Weil, S. (1998a). Rhetorics and realities in public service organizations: Systemic practice and organizational learning *as* critically reflexive action research (CRAR). *Systemic Practice and Action Research*, **11**, 37-62.

Weil, S. (1998b). Our concerns as researchers. Workshop with research students, held at the School of Management, University of Hull, May 15th, 1998.

Weimer, W.B. (1979). *Notes on the Methodology of Scientific Research*. Lawrence Erlbaum Associates, NJ.

Weiss, C. (1972). *Evaluation Research: Methods of Assessing Program Effectiveness*. Prentice-Hall, Englewood Cliffs, NJ.

Weiss, C. (1973). Where politics and evaluation meet. *Evaluation*, **1**, 37-45.

Weiss, C. (1977). Between cup and lip. In, *Resource Materials for Community Mental Health Program Evaluation*. DHEW Publication ADM77-328. Hargreaves, W.A. *et al* (eds.). Government Printing Office, Washington DC.

Wellmer, A. (1970). *The Critical Theory of Society*. Herder and Herder, New York.

White, D.J. (1970). A critique of 'Research Methodology in the Management Sciences' by A.G. Beged-Dov and T.A. Klein. *Operational Research Quarterly*, **21**, 327-334.

White, L. and Taket, A. (1993). The death of the expert. *Journal of the Operational Research Society*, **45**, 733-748.

White, L. and Taket, A. (1994). Facilitating an organisational review. In, *Community Works: 26 Case Studies showing Community Operational Research in Action*. Ritchie, C., Taket, A. and Bryant, J. (eds.). Pavic Press, Sheffield.

White, L. and Taket, A. (1997). Critiquing multimethodology as metamethodology: Working towards pragmatic pluralism. In, *Multimethodology: The Theory and Practice of Combining Management Science Methodologies*. Mingers, J. and Gill, A. (eds.). Wiley, Chichester.

Whitehead, A.N. (1929). *Process and Reality*. 1978 corrected ed. Free Press, New York.

Whittàker, A., Gardner, S. and Kershaw, J. (1991). *Service Evaluation by People with Learning Difficulties*. King's Fund Centre, London.

Whyte, W.F. (ed.) (1991a). *Participatory Action Research*. Sage, London.

Whyte, W.F. (1991b). Introduction. In, *Participatory Action Research*. Whyte, W.F. (ed.). Sage, London.

Whyte, W.F., Greenwood, D.J. and Lazes, P. (1991). Participatory action research: Through practice to science in social research. In, *Participatory Action Research*. Foote Whyte, W. (ed.). Sage, London.

Wiener, N. (1948). *Cybernetics*. MIT Press, Cambridge, MA.

Wilby, J. (ed.) (1996a). *Forum One: Action Research and Critical Systems Thinking*. Centre for Systems Studies, Hull.

Wilby, J. (ed.) (1996b). *Forum One: Transcripts and Reflections*. Centre for Systems Studies, Hull.

Wilby, J. (1996c). Developing total systems intervention (TSI): The critical review mode. *Systems Practice*, **9**, 231-261.

Wilby, J. (ed.) (1997). *Forum Two: Action Research and Critical Systems Thinking*. Centre for Systems Studies, Hull.

Williams, I. (2000). (Personal communication).

Williams, P. (1995). The results from PASS and PASSING evaluations. In, *Evaluating Quality in Services for Disabled and Older People*. Pilling, D. and Watson, G. (eds.). Jessica Kingsley Publishers, London.

Willmott, H. (1989). OR as a problem situation: From soft systems methodology to critical science. In, *OR and the Social Sciences*. Jackson, M.C., Keys, P. and Cropper, S.A. (eds.). Plenum, New York.

Willmott, H. (1993). Breaking the paradigm mentality. *Organization Studies*, **14**, 681-719.

Wilsdon, C. (1994). Hassocks village appraisal: A community making its voice heard. In, *Community Works: 26 Case Studies showing Community Operational Research in Action*. Ritchie, C., Taket, A. and Bryant, J. (eds.). Pavic Press, Sheffield.

Wittgenstein, L. (1953). *Philosophical Investigations*. Basil Blackwell, Oxford.

Wolfensberger, W. (1972). *The Principle of Normalization in Human Services*. National Institute on Mental Retardation, Toronto.

Wolfensberger, W. (1983). Social role valorization: A proposed new term for the principle of normalization. *Mental Retardation*, **21**, 234-239.

Wolfensberger, W. and Glenn, I. (1975). *Program Analysis of Service Systems, Volumes I and II*. 3rd ed. National Institute on Mental Retardation, Toronto.

Wolfensberger, W. and Thomas, S. (1983). *PASSING: A Method of Evaluating the Quality of Human Services according to the Principle of Normalization. Normalization Criteria and Ratings Manual*. 2nd ed. National Institute on Mental Retardation, Toronto.

Wong, N. and Mingers, J. (1994). The nature of community OR. *Journal of the Operational Research Society*, **45**, 245-254.

Wooliston, G. (1992). *The Architecture and Acuity of Critical Systems Thinking*. Ph.D. Thesis, University of Hull.

Wright, D.S., Taylor, A., Davies, D.R., Sluckin, W., Lee, S.G.M. and Reason, J.T. (1970). *Introducing Psychology: An Experimental Approach*. Penguin, Harmondsworth.

Yerbury, M. (1997). Issues in multidisciplinary teamwork for children with disabilities. *Childcare Health & Development*, **23**, 77-86.

Yolles, M. (1996). Critical systems thinking, paradigms, and the modelling space. *Systems Practice*, **9**, 549-570.

Yolles, M. (1999a). Management systems, conflict, and the changing roles of the military. *Journal of Conflict Processes*, **4**, 13-28.

Yolles, M. (1999b). *Management Systems: A Viable Approach*. Financial Times, London.

Yolles, M. (2000). Extending the theory and practice of boundary critique: Exploring the Liverpool dock strike. *Journal of the Operational Research Society*, **51**, in press.

Zeeuw de, G. (1992). Soft knowledge accumulation, or the rise of competence. *Systems Practice*, **5**, 193-214.

Zhu, Z. (2000). Dealing with a differentiated whole: The philosophy of the WSR approach. *Systemic Practice and Action Research*, **13**, 21-57.

Zich, A. (1997). China's Three Gorges: Before the flood. *National Geographic*, **192**(3), 2-33.

Author Index

Subject Index

Printed in the United Kingdom
by Lightning Source UK Ltd.
113888UKS00001B/13

9 780306 464881